FOUNDATIONS OF

GENETIC ALGORITHMS

FOUNDATIONS OF GENETIC ALGORITHMS

EDITED BY
GREGORY J.E. RAWLINS

MORGAN KAUFMANN PUBLISHERS
SAN MATEO, CALIFORNIA

Editor: Bruce M. Spatz
Production Editor: Yonie Overton
Production Artist/Cover Design:
Susan M. Sheldrake

Morgan Kaufmann Publishers, Inc.
Editorial Office:
2929 Campus Drive, Suite 260
San Mateo, CA 94403

94 93 92 91 5 4 3 2 1

Library of Congress Cataloging in Publication Data is available for this book.
Library of Congress Catalogue Card Number: 91-53076
ISBN 1-55860-170-8

Contents

Part 4: Coding and Representation

Part 5: Framework Issues

Part 6: Variation and Recombination

Introduction

The first workshop on the foundations of genetic algorithms and classifier systems (FOGA/CS-90) was held July 15-18, 1990 on the Bloomington campus of Indiana University. The workshop was the first specifically intended solely to discuss the theoretical foundations of genetic algorithms and classifier systems.

The workshop was attended by researchers from academia, government laboratories, and industry, who met for four days to hammer out the foundations of a field that is rapidly growing in importance both in machine learning and in non-linear optimization. Sixty participants from the United States, England, Scotland, Belgium, Germany, Canada, and Israel argued over their research papers on foundational issues, including selection and convergence, coding and representation, problem hardness, deception, classifier system design, variation and recombination, parallelization, and population divergence. This book is a collection of the refereed form of some of those papers. Its purpose is to make a record of the workshop, to provide a repository of the basic ideas of the field, and to be a source of further questions with suggestions about appropriate directions.

Progress was made on many fronts: among other advances, papers in this volume improve bounds on GA convergence; present a new and practical parallel implementation of GAs as a general AI search strategy; give a new and deeper understanding of what GAs are really processing as they operate; link GAs to other probabilistic search techniques like neuromorphic networks; study population dynamics (albeit for infinite populations); refine our understanding of coding and its effects; and improve our understanding of the problem of GA deception and its detection. To make this book more widely useful, a brief presentation of my view of genetic algorithms follows.

The machine learning task is to find an algorithm that "learns" about a problem, where learning means finding statistical regularities in the input-output mapping of the problem. Today we are in the invidious position of using algorithms we don't properly understand to analyze problems we don't properly understand.

The new machine learning algorithms (neuromorphic architectures, simulated annealing, and genetic algorithms) are praised for their gracefulness; unlike older AI systems they don't easily break when the input changes slightly or under reasonable amounts of noise. But this gracefulness is purchased at a cost—systems take too long to run. This must be so if we consider only the credit apportionment portion of the algorithm (say, for example, the learning rule in neuromorphic architectures). To properly apportion credit down long rule sequences (across neurons) we need to change state gradually (otherwise we will only be tracking the input). But this gradualness perforce means a slow algorithm. A slow algorithm bounds the number of rule sets (neurons) we can have in our system and still hope to have the systems run in a reasonable time. This bounds the size of the problem we can tackle and so bounds the amount of information we can gain, after one run, about how these systems behave. Finally, this bound retards the amount of investigation we can give to the new algorithms.

Genetic algorithms (GAs), first specified by John Holland in the early seventies, are becoming an important tool in machine learning and function optimization. The metaphor underlying GAs is natural selection. To solve a learning task, a design task, or an optimization task, a genetic algorithm maintains a population of "organisms" (bit strings) and probabilistically modifies the population, seeking a near-optimal solution to the given task. Beneficial changes to parents are combined in their offspring, and a GA is a control structure that adapts to the problem being solved through syntactic operations on bit strings.

GAs are the principal heuristic search method of classifier systems, a form of stochastic production system; they have been tried on $\mathcal{NP}$-hard combinatorial optimization problems, such as network link optimization and the travelling salesperson problem; and they are of interest in themselves as a primitive model for natural selection (see Goldberg, 1989, for further references). (Classifier systems are modelled on economic systems in market economies.) GAs have been applied to problems such as design of semiconductor layout and factory control, and have been used in AI systems and neuromorphic networks to model processes of cognition such as language processing and induction. GAs may provide a more advantageous approach to machine learning problems than neuromorphic networks have provided. Like associative memories, neuromorphic networks, and simulated annealing (Rumelhart and McClelland, 1986), GAs and GA-based heuristic algorithms are one way to achieve sub-symbolic, graceful, and adaptive systems (Holland, et al., 1986).

1 A Short Introduction to Genetic Algorithms

When used for function optimization, a GA manipulates a population of strings using probabilistic genetic-like operators like pairwise string recombination, called *crossover*, and string mutation, to produce a new population with the intent of generating strings which map to high function values. The members of the population act as a primitive memory for the GA, and the genetic operators are so chosen that manipulating the population often leads the GA away from unpromising areas of the search space and towards promising ones, without the GA having to explicitly remember its trail through the search space. Unlike many other search algorithms, the probabilistic primitives that GAs use to manipulate their populations and their lack of explicit memories make GAs very fast on contemporary hardware.

Theoretical properties of GAs have been studied since their inception. Various results have been proved that are of fundamental significance to their operation and much is known about their basic behaviour. However, because of the vastness of the space of problems, there are few formal results about the tradeoffs GAs make. For example, we know little about the tradeoff GAs make between rapid convergence in and wide exploration of the search space—their *efficiency* versus their *efficacy.* The analytic task is to discover exactly what assumptions about the search space *any* search algorithm implicitly makes. This will let us make principled judgements about which algorithms are useful for which problems and what problem encoding should be used, without having to first program the algorithm and run it.

For each search algorithm and search space, we are interested in the following questions:

1. How close is the estimator produced to the actual best value after a fixed number of evaluations?
2. What is the rate of convergence to the actual best value of the estimators produced?
3. How many evaluations are required to guarantee production of an estimator within a pre-specified range?
4. What is the distribution of estimators from run to run? (This is a measure of the algorithm's robustness or how much faith we can put in it.)
5. How sensitive, in terms of goodness of the estimator produced, is each class of algorithm to small changes in their parameters?

The next two sections formalize what is meant by a "blind" search algorithm, a class of algorithms that includes GAs, and present metrics to measure their performance.

2 Definitions and Assumptions

The *search problem,* or function optimization problem, is: given some finite discrete domain $\mathcal{D}$ and a function $f : \mathcal{D} \to \mathbf{R}$, where $\mathbf{R}$ is the set of real numbers, find the best, or near best, value(s) in $\mathcal{D}$ under f. We refer to f as the *domain function.* Since $\mathbf{R}$ is ordered there is a natural induced total ordering on $\mathcal{D}$ even if $\mathcal{D}$ is multi-dimensional.

Since the algorithms we consider evaluate function values in sequence, we make the following distinctions among functions. A function is *stochastic* if the function value of a domain element can change from evaluation to evaluation with some probability. A function which is not stochastic is said to be *deterministic.* A function is *dynamic* if the function is a function from $\mathcal{D} \times T$ where T is a subset of the positive integers representing time. A function which is not dynamic is said to be *static.*

An *encoding* is a function $e : \mathcal{S}^l \to \mathcal{D}$, where $\mathcal{S}$ is alphabet of symbols, $||\mathcal{S}|| \geq 2$, and $l \geq \lceil \log_{||\mathcal{S}||} ||\mathcal{D}|| \rceil$. Thus, the encoding of the elements of $\mathcal{D}$ is a mapping from the strings of length l over $\mathcal{S}$ to $\mathcal{D}$. $\mathcal{S}^l$ is the *search space.* Usually $||\mathcal{S}|| = 2$, but, to avoid misleading interpretations, take $\mathcal{S} = \{a, b\}$ rather than the more common $\{0, 1\}$.

The composition, g, of the functions f and e is the *objective function* and is defined as the function $g : \mathcal{S}^l \rightarrow \mathbf{R}$ where $g(x) = f(e(x))$.

A *search algorithm* $\mathcal{A}$ examines some subset of the search space $\mathcal{S}^l$, and returns some string $\hat{x}$ whose objective value $g(\hat{x})$ is an estimator of $\max_{x \in \mathcal{S}^l} g(x)$.

2.1 The Model of Computation

Our model of computation allows $\mathcal{A}$ to perform any string operation and each at constant cost. Further, we assume that $\forall x \in \mathcal{S}^l$, $\mathcal{A}$ can obtain $g(x)$. The operation of obtaining $g(x)$ is called *evaluating* x.

There are at least four models of the information provided to $\mathcal{A}$ by the act of evaluation:

- The act of evaluation of two or more strings provides the relative order of the strings' objective values.
- The act of evaluation of a string provides the string's objective value.
- The act of evaluation of a string provides the rank of the string's objective value.
- The act of evaluation of a string provides the string's objective value and its rank.

It is clear that the fourth is more powerful than the first three, but of the first three it is unclear which is more powerful. Intuitively though, it seems unlikely in any real function optimization or machine learning application that the algorithm will be given actual ranks. Of the four models only the first is defensible as a model of biological genetic systems, and only the first two are defensible for function optimization on practical grounds. Therefore, our model uses the second interpretation.

2.2 Blind Search Algorithms and Genetic Algorithms

A search algorithm $\mathcal{A}$ is *blind* if it can only obtain information through string evaluations. We assume that $\mathcal{A}$ is allowed to remember a constant number of all ordered pairs it has seen. (More generally, we could allow a polynomial number of string evaluations.) $\mathcal{A}$'s task is to use the information provided by the ordered pairs it has seen so far in order to infer which new strings are likely to have high objective values. Thus, each algorithm can be seen as attempting to extrapolate from the known to the unknown. It is important to determine exactly what implicit or explicit assumptions the algorithm is making about the objective function by its very behaviour (that is, the strings it evaluates) since this gives us a way of telling whether a particular function is amenable to solution by a particular algorithm and encoding in a reasonable time. Finally, observe that the objective function is a composition of the domain function *and the encoding*; this suggests that the encoding chosen is very important to how well the search algorithm does.

Note that blind search algorithms are potentially capable of optimizing the widest possible range of functions since it is not necessary for f to be expressible in any more compact form than a listing of its domain, range ordered pairs. Thus, for example, f may not be differentiable, indeed, f may have no closed-form expression. The only operation on f that blind search algorithms require is evaluation of an element of the domain. Traditional optimization methods work well for differentiable, unimodal, static, deterministic, low-dimensional domain functions. Blind search algorithms are potentially capable of optimizing such functions as well as their less studied cousins.

A *reproductive population search* is a blind search algorithm that starts with some population of strings and repeatedly performs the following cycle of operations until some termination condition is satisfied:

Evaluation: evaluate each string in the population;

Selection: depending on the values of the strings, select subsets called *reproducing sets* from the population;

Reproduction: from each reproducing set, generate some number of new strings using various reproductive strategies.

Replacement: finally, replace some or all of the original population with the new strings.

Possible reproductive strategies are combinations of

- Reproduction: make identical copies of some or all of the strings in the reproductive set.
- Mating: each new string is constructed by concatenating substrings which are chosen from members of the reproductive set.
- Mutation: substitute new symbols for selected positions in the new string.

Genetic algorithms (GAs) are a subset of reproductive population algorithms. The following are characteristic of many of the GAs found in the literature (but even with these restrictions, there are many variant implementations in the literature):

- All strings are of the same length.
- Populations are of constant size.
- Mutation is probabilistic and is usually independent of the string value.
- Reproductive sets are of size two, and members are chosen probabilistically, with probabilities weighted by string values.
- Mating is usually restricted to crossover. In *general crossover* the ith symbol of an offspring is the ith symbol of one of the members of the reproductive set. Most GAs limit crossover such that symbols in positions $1 \ldots i$ of the new string come from one parent and those in positions $i+1 \ldots l$ come from the other parent, where i is chosen probabilistically.

Such implementations are *canonical GAs*.

The next section develops a few more notions before defining metrics measuring the performance of search algorithms.

3 Metrics

Let $\{a \ldots b\} = \{i \in \mathbf{N} : a \le i \le b\}$. The *rank function* is the function $r : \mathcal{S}^l \to \{0 \ldots \|\mathcal{D}\|\}$ where

$$r(x) = \|\{y \in \mathcal{S}^l : g(x) \ge g(y)\}\|$$

Thus, the *rank* of a string is the number of strings whose objective value is smaller than it under the induced total order. In practice, we cannot assume that the ranks of particular strings are known. The *partial rank function* with respect to a subset $\mathcal{P} \subseteq \mathcal{S}^l$ is the function $r_{\mathcal{P}} : \mathcal{P} \to \{0 \ldots \|\mathcal{D}\|\}$ where

$$r_{\mathcal{P}}(x) = \|\{y \in \mathcal{P} : g(x) \ge g(y)\}\|$$

If $x \in \mathcal{S}^l$, then let x_i $(1 \le i \le l)$ be the ith symbol in x. The *Hamming distance* of one string to another is given by the function $\mathbf{d} : \mathcal{S}^l \times \mathcal{S}^l \to \{0 \ldots l\}$ where

$$\mathbf{d}(x, y) = \|\{i \in \{1 \ldots l\} : x_i \ne y_i\}\|$$

Note that $\forall x \in \mathcal{S}^l$, $\mathbf{d}(x, x) = 0$.

Observe that $(\mathcal{S}^l, \mathbf{d})$ is a metric space of diameter l (see Kaplansky, 1977 for background definitions). A *schema* is a hyperplane in this metric space, and we represent schemas as strings in $(\mathcal{S} \cup \{\#\})^l$ where $\# \notin \mathcal{S}$ is interpreted as a wildcard symbol which stands for any symbol in $\mathcal{S}$ (Holland, 1975).

The *Hamming ball of radius k* is the set of strings around a particular string within a Hamming distance of k. $\mathbf{b} : \mathcal{S}^l \times \{0 \ldots l\} \to 2^{\mathcal{S}^l}$ where

$$\mathbf{b}(x, k) = \{y \in \mathcal{S}^l : \mathbf{d}(x, y) \le k\}$$

Note that $\forall x \in \mathcal{S}^l$, $\mathbf{b}(x, 0) = \{x\}$, and $\mathbf{b}(x, l) = \mathcal{S}^l$.

$\forall x \in \mathcal{S}^l$ every string in $\mathbf{b}(x, 1) \setminus \{x\}$ is said to be *adjacent* to x.

Let $\mathcal{E}_{\mathcal{A}} \subseteq \mathcal{S}^l$ be the multiset of strings evaluated during execution of the algorithm $\mathcal{A}$. If a string is evaluated multiple times, then it appears in $\mathcal{E}_{\mathcal{A}}$ multiple times.

The *thoroughness* of an algorithm $\mathcal{A}$ is the minimum k such that $\forall x \in \mathcal{S}^l$, with probability greater than a half, $\exists y \in \mathbf{b}(x, k) \cap \mathcal{E}_{\mathcal{A}}$. Thus, $\mathcal{A}$'s thoroughness is the smallest k such that

$$\forall x \in \mathcal{S}^l, \quad \mathbf{P}(\|\{\mathbf{b}(x, k) \cap \mathcal{E}_{\mathcal{A}}\}\| > 0) > 1/2$$

This measure tells us something about the likelihood of hiding a local peak from $\mathcal{A}$ using a particular encoding. It is easy to see that if $\|\mathcal{S}\| = 2$ then after evaluating precisely one string the thoroughness is l, but there exist algorithms whose thoroughness drops to $l/2$ after evaluating only two strings (namely 'a' repeated l times and 'b' repeated l times).

The *sparsity* of an algorithm $\mathcal{A}$ is the maximum k such that $\forall x \in \mathcal{E}_{\mathcal{A}}$, with probability greater than a half, $\nexists y \ne x \in \mathbf{b}(x, k) \cap \mathcal{E}_{\mathcal{A}}$. Thus, $\mathcal{A}$'s sparsity is the smallest k such that

$$\forall x \in \mathcal{E}_{\mathcal{A}}, \quad \mathbf{P}(\|\{\mathbf{b}(x, k) \cap \mathcal{E}_{\mathcal{A}}\}\| = 1) > 1/2$$

This measure tells us something about how quickly $\mathcal{A}$ covers regions where it concentrates its search.

Let x^i be the ith evaluated string. The tth *on-line performance* of an algorithm $\mathcal{A}$ is the average rank of the evaluated strings over a time period t,

$$\sum_{i=t_1}^{t_1+t} r(x^i)/(t+1), \quad t_1 = 1 \ldots ||\mathcal{E}_{\mathcal{A}}|| - t$$

$\mathcal{A}$'s *global on-line performance* occurs when $t = ||\mathcal{E}_{\mathcal{A}}||$. $\mathcal{A}$'s *half-life on-line performance* is the ratio of its on-line performance when t is half the total number of evaluated strings to its global on-line performance.

The following performance metrics measure the efficiency and efficacy of search algorithms:

- The number of strings evaluated by the algorithm, $||\mathcal{E}_{\mathcal{A}}||$.
- The rank error of the estimator provided by the algorithm, $||\mathcal{S}||^l - r(\hat{x})$.
- The relative error of the value of the estimator provided by the algorithm, $(M - g(\hat{x}))/M$.
- The algorithm's thoroughness.
- The algorithm's sparsity.
- The algorithm's t^{th} on-line performance, global on-line performance, and half-life on-line performance.

There are three main dimensions of variation: algorithms, domain functions, and encodings.

4 Search Algorithms

It has often been suggested that GAs are a "no prior knowledge required" heuristic search method for function optimization. Similarly, it is sometimes suggested that GAs are universal in that they can be used to optimize any function. These statements are true in only a very limited sense; any algorithm satisfying one of these claims can expect to do no better than random search over the space of all functions. In this section we examine the importance of the encoding to the success of a GA on any function.

We can express the composite objective function g as a set of triples $\{\langle x, d, r\rangle | x \in S^l, d \in \mathcal{D}, r \in \mathbf{R}\}$, with $e(x) = d$, $f(d) = r$ and thus $g(x) = r$. If we now take any permutation of the domain elements with respect to the fixed pairs $\langle x, r\rangle$, we will define a new domain function, and a new encoding. However, this change will be transparent to any of the search algorithms under our model of computation, since these algorithms do not know the function f, only the objective function g, which does not change.

Similarly, if we permute the r values with respect to the fixed pairs $\langle x, d\rangle$, we obtain a new domain function with the same encoding. We may also permute the set of strings with respect to the fixed pairs $\langle d, r \rangle$, to obtain a new encoding and the same domain function. Clearly, these two operations are symmetric with respect to their effect on the objective function g, and thus will effect the same result for any given heuristic search algorithm.

It is now apparent that for a *fixed universal* algorithm, restricted to strings in $\mathcal{S}^l$, over the set of all possible domain functions (with the same sets or multisets of values $\mathcal{D}$ and $\mathbf{R}$) it does not matter which encoding we use, since for every domain function which the encoding makes easier to solve there is another domain function that it makes more difficult to solve. Thus, changing the encoding does not affect the *expected difficulty* of solving a randomly chosen domain function.

Equivalently, assume we have a *fixed* domain function and suppose that we choose the encoding, e, at random. That is, we pick one of the $||\mathcal{S}||^l!$ possible encodings. Then, no search algorithm can expect to do better than random search, since no information is carried by e about f, except that for each string there is a value. More precisely, no proper substring of $x \in \mathcal{S}^l$ yields information about x's objective value, since only x uniquely identifies the actual value, and the values are randomly assigned. It is straightforward to reduce this observation to a proof by contradiction, demonstrating that if it were possible to derive any information from any comparison of substrings of the evaluated strings to reduce the uncertainty in the objective value of any unevaluated string, then it would be possible to predict the outcome of a random coin toss (suppose that the algorithm looks at all but two strings; the algorithm still cannot prejudicially pick one over the other no matter what string properties they share or do not share).

This simple observation has several implications for GAs as currently defined. In fact, as we see from the above argument, if an algorithm is to be more effective than random search, then *"prior knowledge" must be included in the choice of encoding.* Just as familiar numerical hillclimbing techniques, if they are to perform well, require knowledge to be contained in the usual binary or floating point encodings in the form of smoothness, differentiability, etc., so too do GAs require some sort of knowledge to be built into the objective function.

For GAs it is not clear exactly what form that knowledge should take. This ignorance is compounded by the current lack of formal methods of specifying the interactive effects of the parameters of the GAs. Nor is the interaction of the knowledge in the encoding with the implicit knowledge assumed by the algorithm completely understood. An example of such implicit knowledge is the use of proportional representation for the selection procedure. This method, to be effective, makes use of implicit assumptions about the range and scatter of the objective function.

On the one hand, we would like to know exactly what properties of encodings are exploited most effectively by various genetic algorithms. Also, we would like to know what properties will have the most adverse effects on GA performance. Deceptive problems have been defined by Goldberg (1989), based on schema analysis, but even these are often solved effectively by GAs. We would like to be able to state for any combination of operators and parameter settings, given the values of $\mathbf{R}$ in non-decreasing order, which of the $||\mathcal{S}||^l!$ orders of the strings will be the best (or worst) for the algorithm and problem in question. Alternately, given an arbitrary objective function, how well will the algorithm perform?

On the other hand, producing an optimal code will generally not be possible. Producing such a code would be equivalent to solving the problem. In practice, GAs have worked quite well on a number of optimization problems. For such problems, there is little hope of providing optimal encodings. Nevertheless, considerable information about the structure of the problem appears to be preserved through the various encodings used, and this has been used to good effect by GAs. It would be very interesting to understand what structural properties these successful searches exploited that were missed by earlier attempts. The question is not what problems GAs are good for but, rather, which domain functions are easy to solve with various reasonably simple encodings. Also, is it possible for a GA to adaptively change its encoding efficiently dependent on cues in the search process?

Intuitively, GAs work best when substrings have consistent benefit throughout the entire search space. Ideally, once a substring appears to be a benefit then it would probabilistically remain a benefit for all higher ranked strings.

We say that $\mathcal{A}$ *tests* property P if it ever evaluates strings that do and do not have property P. We say that $\mathcal{A}$ *detects* property P if, after testing P, a higher than random number of the subsequent strings $\mathcal{A}$ generates have property P.

For $\mathcal{A}$ to solve a search problem in less than linear time (that is, time proportional to the number of strings) two things must be true:

- There must be some set of string properties P_i defined over the set of strings $\mathcal{S}^l$ such that if x and y are randomly chosen strings for which $P_i(x)$ is true and $P_i(y)$ is false, respectively, then $\mathbf{P}(r(x) \geq r(y)) > 1/2$, and
- $\mathcal{A}$ must be capable of detecting and testing, explicitly in the way $\mathcal{A}$ is coded or implicitly in the way $\mathcal{A}$ behaves, these properties.

Thus, if $\mathcal{A}$ ever tests P then it is a good thing for it to continue to generate strings with property P and to avoid strings without property P.

Note that detection does not simply mean that the algorithm's future behaviour is modified by the strings it has evaluated, since this is also true in a weak sense of naive algorithms. For example, random search's behaviour can be said to change depending on the sequence of strings it has seen (it can potentially report a different value dependent on the strings it evaluates). However, random search is not testing any string properties even though it's behaviour would be changed depending on the sequence of strings it has looked at. Since it is not testing string properties it cannot infer anything about the objective values of future strings.

We see that there are two extremes of behaviour of the objective function:

- There is *zero* epistasis. In this case every gene is independent of every other gene. That is, there is a fixed ordering of fitness (contribution to the overall objective value) of the alleles of each gene. It is clear that this situation can only occur if the objective function is expressible as a linear combination of each gene.
- There is *maximum* epistasis. In this case no proper subset of genes is independent of any other gene. Each gene is dependent on every other gene for its fitness. That is, there is *no* possible fixed ordering of fitness of the alleles

of *any* gene (if there were, then at least one gene would be independent of all the other genes). This situation is equivalent to the objective function being a random function.

Observe that the second situation is easily forced by choosing a random encoding (this is independent of the domain function); however, it is not immediately obvious that the first situation can be forced. Is it possible to cheaply choose an encoding that will ensure that the epistasis of the objective function is bounded for a large class of interesting domain functions?

My Thanks

This introduction is a much abbreviated version of the introduction to an unpublished paper on genetic algorithms written jointly with Joe Culberson.

The twenty student participants at the workshop benefited in part from a grant from the International Society for Genetic Algorithms and aid from the Department of Computer Science, Indiana University, Bloomington. I thank both organizations for making the workshop possible.

FOGA 90 would not have been possible without the kind support of many people who gave generously of their time, wisdom, and encouragement. I thank Dave Goldberg and Ken De Jong who were instrumental in getting this book off the ground. I also thank H. G. Cobb, Yuval Davidor, Larry Eshelman, John Grefenstette, John Holland, Dave Schaffer, Alan Schultz, Bill Spears, Dirk Van Gucht, and Darrell Whitley. I thank you. Finally, I thank Bruce Spatz, my editor at Morgan Kaufmann, who made these proceedings possible. Thanks Bruce. I thought of listing all of the people who sent me mail and gave me encouragement in my lunacy over the past six months. But I decided that it would be pointless to list them, because it would just be a list of everyone in the GA community. The nice thing about the GA community is that it is just that—a community. Thank you all.

Gregory J. E. Rawlins
Bloomington
rawlins@iuvax.cs.indiana.edu

References

Goldberg, David E.; *Genetic Algorithms in Search, Optimization and Machine Learning,* Addison-Wesley, 1989.

Holland, John H.; *Adaptation in Natural and Artificial Systems,* University of Michigan Press, 1975.

Holland, John H., Holyoak, Keith, J., Nisbett, Richard E., and Thagard, Paul R.; *Induction: Processes of Inference, Learning, and Discovery,* MIT Press, 1986.

Kaplansky, Irving; *Set Theory and Metric Spaces,* Chelsea, 1977.

Rumelhart, David E., and McClelland, James L. (eds.); *Parallel Distributed Processing: Explorations in the Microstructure of Cognition, Volume I: Foundations,* MIT Press, 1986.

PART 1

GENETIC ALGORITHM HARDNESS

The Nonuniform Walsh-Schema Transform

Clayton L. Bridges
School of Computer Science
Carnegie Mellon University
Pittsburgh, PA 15213-3890

David E. Goldberg
Department of General Engineering
117 Transportation Building
University of Illinois at Urbana-Champaign
Urbana, IL 61801-2996

Abstract

The Walsh-schema transform allows useful analysis of the expected performance of genetic algorithms. Until recently, its use has been restricted to the static analysis of functions, codings, and operators. In this paper, we present a form of the Walsh-schema transform that accounts for nonuniformly distributed populations, and thereby extend its use to the dynamic analysis of genetic algorithms. We also show the relationship between the nonuniform Walsh-schema transform and the hyperplane transform.

Keywords: Walsh-schema transform, hyperplane transform, dynamic analysis

1 Introduction

The Walsh-schema transform (Bethke, 1981) has been shown to be a useful tool for the static analysis of fitness functions, coding schemes, and genetic operators (Bethke, 1981; Goldberg, 1989a, 1989b). However, static analysis, by definition, fails to capture the more dynamic aspects of a GA — those which are dependent on a GA's trajectory through a series of changing, and thus nonuniform, populations (e.g. Goldberg, 1987). To use the Walsh-schema transform (WST) for dynamic analysis requires that we drop an assumption of static analysis: that analysis occurs in the context of a so-called *flat population*, where every possible string is assumed to be represented in equal proportion. In the following section, we extend the WST to nonuniform populations, and thereby extend its use to the dynamic analysis of genetic algorithms. Following that, we show the relationship between the nonuniform Walsh-schema transform (NWST) and another nonuniform transform, the hyperplane transform of Holland (1989).

2 Definition of the Nonuniform Walsh-Schema Transform

We define the NWST in three parts. We first define our representation of nonuniform populations and show a mapping of this representation onto the fitness function, yielding a *proportion-weighted fitness*. Next, we review the definition of the uniform WST. Finally, we use the groundwork laid in the first two parts to form the NWST.

2.1 Nonuniform Populations and the Proportion-Weighted Fitness

We begin by making some familiar definitions: $\mathbf{x}$ is a length-l bit vector ($\mathbf{x} \in \{0,1\}^l$), $\mathbf{h}$ is a hyperplane (schema) of that vector space ($\mathbf{h} \in \{0,1,*\}^l$, where * represents a wildcard position), and f is the fitness function $f : \mathbf{x} \longmapsto \Re$. Note that we use an unambiguous shorthand notation for $\mathbf{h}$, where depending on the context, $\mathbf{h}$ can be interpreted either as a string representing a hyperplane, or as the set of vectors designated by that string.

We represent nonuniform populations as a set of proportion values $P : \mathbf{x} \longmapsto [0,1]$ and $\sum_{\mathbf{x}} P(\mathbf{x}) = 1$. For a finite population, $P(\mathbf{x})$ can be computed as the number of instances of $\mathbf{x}$ in the population divided by the population size. The proportion of a schema $\mathbf{h}$ has a straightforward definition,

$$P(\mathbf{h}) = \sum_{\mathbf{x} \in \mathbf{h}} P(\mathbf{x}) . \tag{1}$$

Using P, we can generally define the fitness of a schema as

$$f(\mathbf{h}) = \frac{\sum_{\mathbf{x} \in \mathbf{h}} f(\mathbf{x}) P(\mathbf{x})}{P(\mathbf{h})} . \tag{2}$$

We generalize the fitness f to a *proportion-weighted fitness* ϕ, defined as

$$\phi(\mathbf{h}) = f(\mathbf{h}) P(\mathbf{h})\, 2^{o(\mathbf{h})} , \tag{3}$$

where $o(\mathbf{h})$ is the order (the number of defined bits) of $\mathbf{h}$. For a flat population of binary strings, $P(\mathbf{h}) = 1/2^{o(\mathbf{h})}$, so

$$\phi(\mathbf{h}) = f(\mathbf{h}) \frac{1}{2^{o(\mathbf{h})}} 2^{o(\mathbf{h})} = f(\mathbf{h}) ; \tag{4}$$

ϕ reduces to f in the flat case.

Definitions of $f(\mathbf{h})$ are made in terms of $f(\mathbf{x})$. We would like a similar definition of $\phi(\mathbf{h})$ that is made in terms of $\phi(\mathbf{x})$. We can form such a definition by first substituting equation 2 into 3 and manipulating, which gives

$$\begin{aligned} \phi(\mathbf{h}) &= 2^{o(\mathbf{h})} \sum_{\mathbf{x} \in \mathbf{h}} f(\mathbf{x}) P(\mathbf{x}) \\ &= \frac{1}{2^{l-o(\mathbf{h})}} \sum_{\mathbf{x} \in \mathbf{h}} f(\mathbf{x}) P(\mathbf{x})\, 2^l . \end{aligned} \tag{5}$$

We note that equation 3 written for strings is $\phi(\mathbf{x}) = f(\mathbf{x}) P(\mathbf{x})\, 2^l$, and that $|\mathbf{h}| = 2^{l-o(\mathbf{h})}$, where $|\mathbf{h}|$ is the cardinality (number of elements) of the subset $\mathbf{h}$. We can substitute these values into equation 5, yielding

$$\phi(\mathbf{h}) = \frac{1}{|\mathbf{h}|} \sum_{\mathbf{x} \in \mathbf{h}} \phi(\mathbf{x}) . \tag{6}$$

We will use this proportion-weighted fitness to derive the NWST, but we first review some basic notions about the uniform WST.

2.2 The Walsh-Schema Transform Revisited

The following outlines the basic ideas behind the WST, following Goldberg (1989a). While some of the function definitions differ from Goldberg, the functions are the same, and the two formulations are fully interchangeable. We define the Walsh function ψ to be

$$\psi_j(\mathbf{x}) = \prod_{i=1}^{l} (-1)^{x_i j_i} . \tag{7}$$

where j is a binary integer index, $\mathbf{x}$ is a binary vector, and both are indexed bit-by-bit by i. The Walsh function evaluates to only ± 1, depending on the number of locations where both $\mathbf{x}$ and j are 1.

The Walsh transform of a function $f : \mathbf{x} \longmapsto \Re$ is given by

$$w_j = \frac{1}{2^l} \sum_{\mathbf{x}=0}^{2^l-1} f(\mathbf{x}) \psi_j(\mathbf{x}) . \tag{8}$$

The Walsh coefficients w constitute a change of basis—a transform—for the given function f. The inverse transform is given by

$$f(\mathbf{x}) = \sum_{j=0}^{2^l-1} w_j \psi_j(\mathbf{x}) . \tag{9}$$

Using these definitions the Walsh-schema transform can be derived to be

$$f(\mathbf{h}) = \sum_{j \in J(\mathbf{h})} w_j \psi_j(\beta(\mathbf{h})) , \tag{10}$$

where

$$J_i(\mathbf{h}) = \begin{cases} 0, & \text{if } h_i = *; \\ *, & \text{if } h_i = 0, 1 \end{cases} \tag{11}$$

$$\beta_i(\mathbf{h}) = \begin{cases} 0, & \text{if } h_i = 0, *; \\ 1, & \text{if } h_i = 1 \end{cases} \tag{12}$$

and $f(\mathbf{h})$ is given by equation 2. J is is a set generator of $\mathbf{h}$, and determines which Walsh coefficients w_j are needed to determine a particular $f(\mathbf{h})$. For instance, $J(\mathbf{h} = * * *) = \{000\}$, so only the w_{000} coefficient is needed to determine $f(* * *)$. As another example, $J(\mathbf{h} = * * 1) = 00* = \{000, 001\}$, so both w_{000} and w_{001} are needed to determine $f(* * 1)$. β is used only to map $\mathbf{h}$ to a string so it can be used as an argument to the Walsh function ψ.

With the WST outlined, we are now ready to extend it to the nonuniform case.

2.3 The Nonuniform Walsh-Schema Transform

As we could with any function of $\mathbf{x}$, we can write the Walsh transform of ϕ as

$$w_j = \frac{1}{2^l} \sum_{\mathbf{x}=0}^{2^l-1} \phi(\mathbf{x})\, \psi_j(\mathbf{x}) \,. \tag{13}$$

Likewise, the inverse transform is given by

$$\phi(\mathbf{x}) = \sum_{j=0}^{2^l-1} w_j \psi_j(\mathbf{x}) \,. \tag{14}$$

Substituting this expression into equation 6, we get

$$\phi(\mathbf{h}) = \frac{1}{|\mathbf{h}|} \sum_{\mathbf{x} \in \mathbf{h}} \sum_{j=0}^{2^l-1} w_j \psi_j(\mathbf{x}) \,, \tag{15}$$

which, following Goldberg (1989a), reduces to

$$\phi(\mathbf{h}) = \sum_{j \in J(\mathbf{h})} w_j \psi_j(\beta(\mathbf{h})) \,. \tag{16}$$

This is the nonuniform version of the WST. As a check, we can see whether it reduces to the uniform WST under a flat population. We have already shown that, under a flat population, ϕ reduces to f (equation 4). Substituting f for ϕ, equation 13 reduces to the Walsh transform of $f(\mathbf{x})$ (equation 8). Making the same substitution, and noting that the Walsh coefficients for the two transforms are now identical, equation 16 reduces to the uniform WST (equation 10).

We have shown that the nonuniform WST reduces to the uniform version for a flat population. In the following section, we show the relationship of the nonuniform WST to another nonuniform transform.

3 Relationship to the Hyperplane Transform

Holland (1989) has presented another nonuniform transform, the *hyperplane* transform, which can be used to analyze GAs. In the following three parts, we will show the relationship of the NWST to the hyperplane transform. First, we define the hyperplane transform. Next, we reformulate the hyperplane transform to aid our analysis. Finally, we convert both the NWST and the hyperplane transform to matrix form, and show the relationship between the two. We find that, with slight modification of the definition of ϕ, the NWST is essentially equivalent to the hyperplane transform.

3.1 The Hyperplane Transform Defined

Before we can succinctly state the hyperplane transform, we must make a few definitions. The index function j is defined as

$$j_i(\mathbf{h}) = \begin{cases} 0, & \text{if } h_i = * \,; \\ 1, & \text{if } h_i = 0, 1 \,. \end{cases} \tag{17}$$

where i indexes the individual bits of the unsigned binary integer encoding of j. In words, the j function creates 1's where $\mathbf{h}$ is defined and 0's where it is not.

The hyperplane transform depends on an analog to the Walsh function, the function σ. Holland defines this using a straightforward scheme: if $n_0(\mathbf{h})$ yields the number of 0's in a schema $\mathbf{h}$, then

$$\sigma(\mathbf{h}) = (-1)^{n_0(\mathbf{h})} . \tag{18}$$

While this definition is sufficient to state the hyperplane transform, in the following section, we show an equivalent definition of σ that better lends itself to our analysis.

For brevity, we define a subset generator,

$$\Gamma(\mathbf{h}) = \{\mathbf{h}' : \mathbf{h}' \subseteq \mathbf{h}, \mathbf{h}' \neq * * \ldots * *\} \tag{19}$$

We also define over the proportions, or

$$\delta_0 = f(* * \ldots * *) = \sum_{\mathbf{x}} f(\mathbf{x}) P(\mathbf{x}) . \tag{20}$$

Borrowing some of our previous notation, we can now write the hyperplane transform as

$$\delta_{j(\mathbf{h})} = \sum_{\mathbf{h}' \in \Gamma(\mathbf{h})} (-2)^{o(\mathbf{h}') - o(\mathbf{h})} P(\mathbf{h}')(f(\mathbf{h}') - \delta_0) , \; j(\mathbf{h}) \neq 0 ; \tag{21}$$

and the inverse transform as

$$f(\mathbf{h}) = \delta_0 + \sum_{\mathbf{h}' \in \Gamma(\mathbf{h})} \frac{(2)^{o(\mathbf{h}') - o(\mathbf{h})} \sigma(\mathbf{h}') \delta_{j(\mathbf{h}')}}{P(\mathbf{h})} . \tag{22}$$

This is the hyperplane transform as defined by Holland. In the next section, we reformulate the hyperplane transform so we can more easily show its relationship to the NWST.

3.2 Reformulation of Hyperplane Transform

Here, we reformulate the hyperplane transform to make it more amenable to further analysis. With no loss of information, we can simplify the hyperplane transform equations (21 and 22) by making the substitution

$$\Delta_{j(\mathbf{h})} = \delta_{j(\mathbf{h})} 2^{o(\mathbf{h})} . \tag{23}$$

Substituting into equation 22 and manipulating, we get

$$\left[f(\mathbf{h}) - \Delta_0\right] \; P(\mathbf{h}) \, 2^{o(\mathbf{h})} = \sum_{\mathbf{h}' \in \Gamma(\mathbf{h})} \sigma(\mathbf{h}') \Delta_{j(\mathbf{h}')} . \tag{24}$$

Notice that the left-hand side of this equation is much like our definition of ϕ. The difference is that it is *zero-biased*. We can define a zero-biased version of the proportion-weighted fitness to be identical to the left-hand side of equation 24, or

$$\phi^z(\mathbf{h}) = \left[f(\mathbf{h}) - \Delta_0\right] \; P(\mathbf{h}) \, 2^{o(\mathbf{h})} , \tag{25}$$

where we superscript by a z to indicate zero-biasing. We indicate Walsh coefficients derived from ϕ^z in the same manner; that is,

$$w_j^z = \frac{1}{2^l} \sum_{\mathbf{x}=0}^{2^l-1} \phi^z(\mathbf{x})\, \psi_j(\mathbf{x}) . \tag{26}$$

Making the substitution of ϕ^z, we can write equation 24 as

$$\phi^z(\mathbf{h}) = \sum_{\mathbf{h}' \in \Gamma(\mathbf{h})} \sigma(\mathbf{h}')\Delta_{j(\mathbf{h}')} . \tag{27}$$

We can begin to understand the relationship between the two transforms by creating a definition of σ that looks more like the Walsh function. We use the function ψ', defined as

$$\psi'_k(\mathbf{x}) = \prod_{i=1}^{l} (-1)^{\bar{x}_i \cdot k_i} . \tag{28}$$

Using the β and j functions (equations 12 and 17, respectively), σ can be redefined using ψ' as

$$\sigma(\mathbf{h}) = \psi'_{j(\mathbf{h})}(\beta(\mathbf{h})) \tag{29}$$

We can show this is equivalent to our original definition by first substituting the definition for ψ' into equation 29, yielding

$$\sigma(\mathbf{h}) = \prod_{i=1}^{l} (-1)^{\bar{\beta}_i \cdot j_i} . \tag{30}$$

We first note that the only terms that affect the result of the product are those that yield a -1. For a given i, this occurs only when $j_i = 1$ and $\beta_i = 0$. The first condition will only be true for those positions where $\mathbf{h}$ is defined ($h_i \neq *$). The second condition will only be true for $h_i \in \{0, *\}$. Both conditions can only be met when $h_i = 0$, and when this occurs, a factor of -1 appears in the product. Simplifying for this, we obtain

$$\sigma(\mathbf{h}) = \prod_{\{i:h_i=0\}} (-1) = (-1)^{n_0(\mathbf{h})} , \tag{31}$$

and the two definitions are equivalent.

Substituting ψ' for σ in equation 27, and writing the equation for strings only, yields

$$\phi^z(\mathbf{x}) = \sum_{\mathbf{h}' \in \Gamma(\mathbf{x})} \psi'_{j(\mathbf{h}')}(\beta(\mathbf{x}))\Delta_{j(\mathbf{h}')} . \tag{32}$$

We can simplify this equation if we notice two things: (1) that $\beta(\mathbf{x}) = \mathbf{x}$; (2) and that the terms inside the summation are now dependent only on $j(\mathbf{h}')$, so we can change the summation variable. Making these changes gives

$$\phi^z(\mathbf{x}) = \sum_{\{j(\mathbf{h}'):\mathbf{h}' \in \Gamma(\mathbf{x})\}} \psi'_{j(\mathbf{h}')}(\mathbf{x})\Delta_{j(\mathbf{h}')} . \tag{33}$$

Finally, we note that $\{j(\mathbf{h}') : \mathbf{h}' \in l'(\mathbf{x})\} = \{1, \ldots, 2^l - 1\}$, from which equation 33 becomes

$$\phi^z(\mathbf{x}) = \sum_{j=1}^{2^l-1} \psi_j'(\mathbf{x}) \Delta_j \,. \tag{34}$$

In the next part, we formulate this and the NWST equations as matrices, and show the relationship between the two transforms.

3.3 Matrix Form of Transforms

At this point, it aids analysis if we convert to matrix notation for the two transforms. We begin by defining column vectors $\vec{\phi}$ and $\vec{\phi^z}$ as the vectors of all $\phi(\mathbf{x})$ and $\phi^z(\mathbf{x})$, respectively, ordered from lowest to highest as if $\mathbf{x}$ were an unsigned binary integer. For example, in the two-bit case,

$$\vec{\phi} = [\ \phi(00) \quad \phi(01) \quad \phi(10) \quad \phi(11)\ ]^T \,. \tag{35}$$

Note that $\vec{\phi}$ and $\vec{\phi^z}$ consist only of proportion-adjusted fitnesses of strings, *not schemata*.

Next, we define $\vec{w}$ and $\vec{w^z}$ as column vectors consisting of the w and w^z coefficients, respectively, and ordered by their indices. We likewise define $\vec{\Delta}$ as such a column vector, but with the stipulation that the first element of the vector be zero. For example, the two-bit case,

$$\vec{\Delta} = [\ 0 \quad \Delta_1 \quad \Delta_2 \quad \Delta_3\ ]^T \,. \tag{36}$$

If we set $s_{\mathbf{x}j} = \psi_j'(\mathbf{x})$, then we can construct an $m \times m$ matrix $S = [s_{\mathbf{x}j}]$ such that [1]

$$\vec{\phi^z} = S\vec{\Delta} \,, \tag{37}$$

where $m = 2^l$. This is just the matrix form of equation 34. To show this is the case, we write equation 37 for the xth element of $\vec{\phi}$,

$$\phi^z(\mathbf{x}) = [s_{\mathbf{x}0} \ldots s_{\mathbf{x}(m-1)}] \begin{bmatrix} 0 \\ \Delta_1 \\ \vdots \\ \Delta_{m-1} \end{bmatrix} = \sum_{i=0}^{m-1} s_{\mathbf{x}j} \Delta_i \tag{38}$$

$$= \sum_{j=0}^{2^l-1} \psi_j'(\mathbf{x}) \Delta_j \,, \tag{39}$$

which is equation 34.

Similarly, we can define $p_{j\mathbf{x}} = \psi_j(\mathbf{x})$, and construct an $m \times m$ matrix $P = [p_{j\mathbf{x}}]$ such that

$$\vec{w^z} = \frac{1}{2^l} P \vec{\phi^z} \,. \tag{40}$$

[1] For convenience, we number our matrix elements starting with zero, rather than one.

This is the matrix form of equation 13. We show this as above, by writing equation 40 for the jth element of $\vec{w}$,

$$w_j^z = \frac{1}{2^l}[p_{j0}\cdots p_{j(m-1)}]\begin{bmatrix}\phi^z(0)\\ \vdots \\ \phi^z(m-1)\end{bmatrix} = \frac{1}{2^l}\sum_{\mathbf{x}=0}^{m-1} p_{j\mathbf{x}}\phi^z(\mathbf{x}) \tag{41}$$

$$= \frac{1}{2^l}\sum_{\mathbf{x}=0}^{m-1}\psi_j(\mathbf{x})\,\phi^z(\mathbf{x})\,, \tag{42}$$

which is equation 13.

Substituting equation 37 into 40, we obtain

$$\vec{w^z} = \frac{1}{2^l}PS\vec{\Delta}\,. \tag{43}$$

We can now concern ourselves only with the matrix $Q = PS$; knowing the form of Q means knowing the relationship between the two coefficient vectors.

We can write the equation for an element of Q, which is

$$q_{ik} = [p_{i0}\cdots p_{i(m-1)}]\begin{bmatrix}s_{0k}\\ \vdots \\ s_{(m-1)k}\end{bmatrix} = \sum_{\mathbf{x}=0}^{m-1} p_{i\mathbf{x}}s_{\mathbf{x}k} = \sum_{\mathbf{x}=0}^{m-1}\psi_i(\mathbf{x})\,\psi'_k(\mathbf{x})\,. \tag{44}$$

We can substitute the definitions for the functions ψ and ψ', yielding

$$q_{ik} = \sum_{\mathbf{x}=0}^{m-1}\prod_{a=1}^{l}(-1)^{i_a\cdot \mathbf{x}_a}\prod_{b=1}^{l}(-1)^{k_b\cdot \bar{\mathbf{x}}_b}\,. \tag{45}$$

We can simplify this form by combining the product indices (multiplication is commutative; therefore, reordering is valid). This gives

$$q_{ik} = \sum_{\mathbf{x}=0}^{m-1}\prod_{b=1}^{l}(-1)^{i_b\cdot \mathbf{x}_b + k_b\cdot \bar{\mathbf{x}}_b}\,. \tag{46}$$

There are two cases which we now wish to examine. The first is when $i = k$; in this case, equation 3.3 becomes

$$q_{ii} = \sum_{\mathbf{x}=0}^{m-1}\prod_{b=1}^{l}(-1)^{i_b\cdot(\mathbf{x}_b+\bar{\mathbf{x}}_b)} = \sum_{\mathbf{x}=0}^{m-1}\prod_{b=1}^{l}(-1)^{i_b} = 2^l\prod_{b=1}^{l}(-1)^{i_b} = 2^l\eta(i)\,, \tag{47}$$

where, for brevity, we define $\eta(i) = \prod_{b=1}^{l}(-1)^{i_b}$.

The second case is when $i \neq k$. The result can be determined by induction, and we will outline this proof. We start with the case where i differs from k in only one position, y, or $k_y = \bar{i}_y$. In this case,

$$q_{ik} = \prod_{\{b:b\neq y\}}(-1)^{i_b}\sum_{\mathbf{x}=0}^{m-1}(-1)^{i_y x_y + \bar{i}_y\bar{x}_y}\,. \tag{48}$$

We first note that everything outside the summation evaluates to a constant. Next, we note that over the summation, for a given position y, $\mathbf{x}_y$ will be 0 for exactly half of the terms, and 1 for the other half. It follows that half of the terms will be $(-1)^{i_y}$ and the other half will be $(-1)^{\bar{i}_y}$. Since these two terms will always be of opposite sign, in the summation, the positive half of the terms will cancel the negative half of the terms, giving $q_{ik} = 0$. We can extend the argument to two differing positions, and by induction, show that it is true for any number of differing positions. With this, we can write equation 3.3 as

$$q_{ik} = \begin{cases} 2^l \eta(i), & \text{if } i = k; \\ 0, & \text{if } i \neq k \end{cases} \tag{49}$$

If we define E as a diagonal matrix such that $[e_{ii}] = \eta(i)$, then $Q = 2^l E$. Substituting this into equation 43 yields

$$\vec{w^z} = E\vec{\Delta} . \tag{50}$$

Writing this for individual terms gives the relationship between the Δ and w^z coefficients as

$$\vec{w_j^z} = \eta(j)\Delta_j , \; j \neq 0 . \tag{51}$$

Since η can only be ± 1, the two sets of coefficients differ by sign only.

4 Conclusions

We have presented the nonuniform Walsh-schema transform, and shown its relationship to the hyperplane transform. While these two are quite similar, there are good reasons to choose the NWST over the HT. One reason is that because the NWST is an extension to the WST, it will be easier to relate to previous work which relied upon the WST. Perhaps more important, because the NWST is formulated on a well-studied function, it is easier to relate to previous mathematical work. For example, we can efficiently obtain the Walsh coefficients by using the fast Walsh transform (Beauchamp, 1984; Goldberg, 1989a).

Another choice we must make is between the original (general) definition of proportion-weighted fitness (equation 3), and the zero-biased definition (equation 25). The general definition may be more aesthetically appealing, because it subsumes the definition of the fitness function. However, there are some indications that the zero-biased definition may yield more intuitive Walsh coefficients. For instance, when a population converges to a single string, the Walsh coefficients obtained using the zero-biased definition converge to zero, while those obtained using the full definition converge only to a constant absolute value. For this case, the zero-biased coefficients seem to make more sense, but we do not know whether this extends to the general case.

The NWST may be a more practicable tool for analysis than the WST because the amount of computation it requires it not so daunting. In the WST, all possible strings must be evaluated (2^l function evaluations; enumeration!). On the other hand, the NWST requires evaluation only of the population under analysis. Generally, population size is much smaller than 2^l and, for a GA, the population must be evaluated anyway, so this means that a single NWST requires much fewer function evaluations in general. We should note, though, that because it is a dynamic tool, the NWST must be performed a number of times to be useful.

Both the WST and NWST compute the Walsh coefficients, and using the fast Walsh transform (FWT), this generally requires $O(l \cdot 2^l)$ floating point operations. However, for the NWST, most of the values in the FWT will be zero, and one can take advantage of this to reduce the amount of computation required. These factors can make the NWST more tractable than the WST, and may allow the dynamic analysis of problems which would be impracticable using static analysis.

Experiments are underway to test the capabilities of the NWST. We postulate that, much as the WST has been a useful tool in the static analysis of genetic algorithms, the NWST will be a useful tool in the dynamic analysis of genetic algorithms.

Acknowledgments

This material is based upon work supported by the National Science Foundation under Grants CTS-8451610 and ECS-9022007. We would like to thank Steve Smith for useful discussions concerning this work.

References

Beauchamp, K. G. (1984). *Applications of Walsh and related functions*. London: Academic Press.

Bethke, A. D. (1981). Genetic algorithms as function optimizers (Doctoral dissertation, University of Michigan). *Dissertation Abstracts International, 41(9)*, 3503B. (University Microfilms No. 8106101)

Goldberg, D. E. (1987). Simple genetic algorithms and the minimal, deceptive problem. In L. Davis (Ed.), *Genetic algorithms and simulated annealing* (pp. 74–88). London: Pitman.

Goldberg, D. E. (1989a). Genetic algorithms and Walsh functions: Part I, a gentle introduction. *Complex Systems, 3*, 129–152.

Goldberg, D. E. (1989b). Genetic algorithms and Walsh functions: Part II, deception and its analysis. *Complex Systems, 3*, 153–171.

Holland, J. H. (1989). Searching nonlinear functions for high values. *Applied Mathematics and Computation, 32*, 255–274.

Epistasis Variance: A Viewpoint on GA-Hardness

Yuval Davidor
Department of Applied Mathematics and Computer Science
The Weizmann Institute
Rehovot 76100, Israel

Abstract

There is general consensus that the coding of a problem domain holds an important key to a successful application. However, there is disagreement as to what aspects of a representation and a problem domain make the application 'hard' for a *genetic algorithm* (GA). This paper suggests a simple statistic, a regression analysis predicting the function value from the bits, as a mean to measure the amount of nonlinearity in a representation, and an interesting perspective on GA-hardness. This statistics is termed *epistasis variance* for its analogy to the use of epistasis in genetics, and presents a perspective on GA-hardness different to those presented in recent works on deceptive problems. Two new findings result form the epistasis analysis. One, a step towards defining and understanding the role of epistasis in GAs, and in the search for understanding GA-hardness. Two, that three elements contribute to GA-hardness: the structure of the solution space, the representation of the solution space, and the sampling error as a result of finite and often small population sizes. These three elements are not necessarily linked, and furthermore, the effect of each of them on GA-hardness is not fixed.

Keywords: GA-hardness,representation,deception,epistasis, sampling error.

1 Background

The schema theorem [Holland, 1975] suggests prerequisite features which a representation should exhibit in order to utilize a GA processing. Specifically, that there is a high probability that above average, short, low-order schemata combine to form a higher order above average schemata. The schema theorem shows that above average schemata will proliferate at a given expected minimal rate, but it does not indicate whether this proliferation will occur at the optimum rate. Furthermore, an optimum proliferation by itself is not sufficient to produce a successful GA application.

The only method available for the analysis of the proliferation rates, the Walsh function analysis, is computationally prohibitive [Bethke, 1981, Goldberg, 1989a]. In that respect, it is self evident that the representation is a primary aspect of GAs which determines their utility. The importance of the representation was recognized, but attention was primarily given to the issue of building blocks (their size, number, etc.) [Bethke, 1981, Goldberg, 1987, Goldberg, 1989a]. Deception theory [1] partially encapsulates this issue. However, it is argued that while deceptive problems are GA-hard, they do not fundamentally define GA-hardness.

It is the implicit allocation of search efforts which underlies the operation of a GA. Therefore, the success of a GA search is correlated to the ability to predict correctly the value of strings from the bits (the holism view of complex systems [Goldberg, 1989a, Jacobson, 1955, Platt, 1961, Simon, 1962, Tsotsos, 1987]). The underlying assumption in this paper is that if the correlation is good, the allocation of trials can potentially be optimal (depending on the whole algorithmic ensemble, population size, crossover mechanisms, and so forth [Spears and De Jong, (in press), De Jong and Spears, (in press)]). Therefore, the amount of interdependency among the representation elements is an important ingredient in the GAs' cookbook, and constitutes an essential source of information for understanding GA-hardness.

Gene interaction is a central issue in natural genetics, where genes not only are dependent on each other in order to jointly express phenotypical characteristics, but also suppress and activate the expression of other genes [Ptashne, 1989]. The term that has become synonymous with almost any type of gene interaction is epistasis [Klug, 1986]. Derived from the Greek words *epis* and *stasis* ('behind' and 'stand'), epistasis is therefore equated with *stoppage* or *masking*. Epistasis is used to describe the situation where one gene pair masks or modifies the expression of another gene pair. When the epistasis of a chromosome is said to be high, it means that many genes are dependent on other genes for expression.

2 Notional epistasis in GAs

Tracing epistasis is an elusive occupation because the presence of epistatic elements can be traced only at the phenotypic level away from their scene of interaction (genotypic level). The motivation of applying an epistasis analysis is discussed in the present section.

[1] A minimal deceptive problem was defined by Goldberg [Goldberg, 1987], and follows arguments in population dynamics about allele frequencies. Goldberg's definition comes to indicate a situation where the value of a schema instantiated by a local optimum is greater than the complementing schema instantiated by the global optimum. A fully deceptive problem is an extension of the minimal case, and means that the value of all schemata instantiated by a local optimum and having at least one position undetermined, is greater than their complementing schemata instantiated by the global optimum.

If a representation contains very little or no epistasis, no individual string element is affected by the value of the other elements, and therefore optimization means a bit-wise maximization (which a greedy algorithm will most likely process more effectively than a GA). At the other end of the epistatic scale, when a representation is highly epistatic, too many elements are dependent on other elements and the building blocks become of a high order. When the epistasis is extremely high, the elements are so dependent on each other that unless a complete set of unique element values is found simultaneously, no substantial fitness improvements can be noticed (such as in the case of a delta function). Under such extreme circumstances, nonlinearity has exacerbated to the extent that the performance space does not contain significant regularities.

This leads to the conclusion that if a representation exhibits very low epistasis it could probably be processed more efficiently by a greedy algorithm (though it is suitable for a GA). If it contains very high epistasis, then there is too little structure in the solution space, and a GA will most likely drift and settle on a local optimum. In between the two extremes lies a type of problems it might be useful to try and solve with a GA.

A linear decomposition is applied to the representation according to the composing bits. The purpose of applying the above linear decomposition, is to develop a method for the prediction of the amount of nonlinearity (in terms of gene interaction) embedded in a given representation. To this end, fitness has to be associated with the bits. If a linear decomposition proves to be inaccurate, then it implies that the representation incorporates nonlinearities. Quantifying the amount of nonlinearity will provide an estimate for the suitability of a given representation to a GA processing.

From a GA perspective, a coding format in which the effect of any individual parameter on the total fitness is independent of other parameters, suggests that there is little co-adaptation. On the other hand, a high degree of nonlinearity indicates that above average schemata are too big ('big' is not defined here, but some estimates are provided in [Syswerda, 1989, Spears and De Jong, (in press)]). The whole GA theory is based on the assumption that one can state something about the whole only by knowing its parts. What neither the schema theorem nor population genetics indicate, is exactly how much of the whole the parts should indicate. A first step towards quantifying this property is attempted in the next section.

3 The basic elements of epistasis

It was suggested previously that when epistasis is high, it is difficult to predict the value of a given string from a measured value of its bits. The following definitions are adopted for the preliminary analysis:

A string, S, is composed from l elements s_i (l is fixed),

$$S = (s_1, s_2, \ldots, s_l) \tag{1}$$

The allele of the ith gene in a string is denoted by

$$s_i = a \qquad a \in \{0, 1\}, \qquad i = 1, 2, \ldots, l. \tag{2}$$

Symbol	Term
S	String
$v(S)$	Fitness
$X(S)$	Excess fitness value
a	Allele
$A_i(a)$	Allele value of a
$E_i(a)$	Excess allele value
$E(A)$	Excess genic value
$A(S)$	Genic value
$\varepsilon(S)$	Epistasis value
σ_v^2	Fitness variance
σ_A^2	Genic variance
σ_ε^2	Epistasis variance

Table 1: Summary of the symbols and their definitions in the epistasis discussion.

The *Grand Population*, Γ, is the set of all possible strings of length l such that,

$$\Gamma = \{0,1\}^l. \tag{3}$$

Let *Pop* denote a sample from Γ where the sample is selected uniformly and with replacement. The size of a sample *Pop* is

$$N = |Pop|. \tag{4}$$

The fitness of a string is given by

$$v(S) = fitness. \tag{5}$$

where v is a 'blackbox' function. The average fitness value of the sample *Pop* is

$$\bar{V} = \frac{1}{N} \sum_{S \in Pop} v(S). \tag{6}$$

The excess fitness value of a string is denoted by

$$E(S) = v(S) - \bar{V}. \tag{7}$$

The number of string instances in *Pop* which match $s_i = a$ is denoted by $N_i(a)$. The average allele value is denoted as

$$A_i(a) = \frac{1}{N_i(a)} \sum_{S \in Pop_{s_i=a}} v(S). \tag{8}$$

where $Pop_{s_i=a}$ is the set of all strings in *Pop* having the allele a in their ith position. The excess allele value is defined by

$$E_i(a) = A_i(a) - \bar{V}, \tag{9}$$

and the excess genic value is

$$E(A) = \sum_{i=1}^{l} E_i(a). \tag{10}$$

The genic value of a string S – the predicted string value – is defined as

$$A(S) = E(A) + \bar{V}. \tag{11}$$

Thus, the difference $\varepsilon(S) = v(S) - A(S)$ might reasonably be supposed to be a measure of epistasis of a string S.

Consequently, an epistasis measure for the Grand Population and hence for the representation, is termed the *epistasis variance* and is defined as

$$\sigma_\varepsilon^2 = \frac{1}{N_\Gamma} \sum_{S \in \Gamma} [v(S) - A(S)]^2, \tag{12}$$

where the implicit $A_i(a)$ are computed over the Grand Population (note that this definition does not follow the common definition of variance as it involves elements from two different sets). This measure can be estimated from the corresponding expression

$$\sigma_{Pop}^2 = \frac{1}{N} \sum_{S \in Pop} [v(S) - A(S)]^2. \tag{13}$$

However, since the computation of $A_i(a)$ is determined by the sample population, this statistic is subject to sampling error, but as yet, confidence measures for the estimate are unavailable. This would require an investigation of the distribution of

$$\sigma_\Gamma^2 - \sigma_{Pop}^2$$

String	f_1	f_2	f_3	f_4
000	0	0	0.0	7
001	1	0	0.5	5
010	2	0	1.0	5
011	3	0	1.5	0
100	4	0	2.0	3
101	5	0	2.5	0
110	6	0	3.0	0
111	7	28	17.50	8

Table 2: Strings and their fitness values of four fitness functions: f_1 (linear function), f_2 (δ function), f_3 ($\frac{1}{2}(f_1 + f_2)$), and f_4 (minimal deceptive function) of zero-, total-, semi-, and bounded-epistasis respectively.

The above definitions (summarized in Table 1) provide a method for estimating the epistatic variance for a Grand Population — the base epistasis — from a sample population. The distinction between base epistasis and sampling error is very important because the effect of the latter is often of equal or even higher order of magnitude. This will be demonstrated further in section 4.2.

The fitness variance is denoted as

$$\sigma_v^2 = \frac{1}{N} \sum_{S \in Pop} (E(S))^2, \tag{14}$$

and the genic variance is denoted as

$$\sigma_A^2 = \frac{1}{N} \sum_{S \in Pop} (E(A))^2. \tag{15}$$

4 Calculating epistasis: A few examples

In the following, the epistasis tools developed in section 3 are applied to two fitness functions of known and characteristic epistasis (the strings and their corresponding fitness values are summarized in table 2). The functions are the linear function f_1, the delta function f_2, the semi-linear function f_3, and a minimal deceptive function f_4.

The first analysis uses Grand Populations and thus addresses the issue of base epistasis (section 4.1). Then, the effect a sample has over the statistic is investigated (section 4.2). In section 4.3, a minimal deceptive problem is analyzed. The functions are arranged so as to have an equal average fitness value and thus facilitate comparability between the epistasis variances.

4.1 Three epistatically different functions

The Grand Populations of three epistatically different functions are analyzed: zero epistasis (table 3), total epistasis (table 4), and semi-epistasis (table 5). An additional analysis of f_1 is presented, where the representation is a gray code (Table 6).

When analyzing the epistasis variance of the f_1, f_2, f_3, and f_1 represented in a gray code, it is possible to observe the strength of the linear assumption, and the epistasis analysis. The f_1 function can be accurately recomposed from the decomposed $A_i(a)$ values, while the recomposition of the f_2 function reveals, as can be expected, a large epistatic variance. The semi-epistatic function f_3 demonstrates further the notion of epistasis analysis. An interesting finding is presented in Table 6. The epistasis analysis of f_1 represented by a gray code shows a light epistasis variance which indicates that this representation is similar to the integer binary representation, but with somewhat less structure. This analytical finding agrees with experimental results comparing integer and gray code representations in a GA environment [Caruana and Schaffer, 1988].

In this simple example of changing the representation of a given function, it can be clearly seen that different representations of the same domain may have different amounts of epistasis variance or 'structure' in them.

4.2 Samples and sampling noise

Since the population size in all practical GA applications is only a minuscule portion of the grand population, it is important to investigate whether calculating the epistasis variance from a sample involves a strong sampling error. This section investigates this sampling bias and suggests that the sampling bias has a considerable effect on the measurement of base epistasis variance.

It was already shown that calculating epistasis variance with a Grand Population for a representation which contains zero epistasis yields a correct epistasis figure. This section will show that this conclusion is valid only for the Grand Population, and erroneous when the calculation is not based on the Grand Population. In Tables 7 and 8 such a calculation is shown, and reveals a substantial sampling error variance.

The analysis of sample populations suggests the following:

S	$v(S)$	$E(S)$	$E(A)$	$A(S)$	$\varepsilon(S)$
000	0	−3.5	−3.5	0	0
001	1	−2.5	−2.5	1	0
010	2	−1.5	−1.5	2	0
011	3	−0.5	−0.5	3	0
100	4	0.5	0.5	4	0
101	5	1.5	1.5	5	0
110	6	2.5	2.5	6	0
111	7	3.5	3.5	7	0

i	a	$A_i(a)$	$E_i(a)$
1	0	1.5	−2.0
	1	5.5	2.0
2	0	2.5	2.0
	1	4.5	−1.0
3	0	3.0	−0.5
	1	4.0	0.5

σ_v^2	σ_A^2	σ_ε^2	$\sigma_v^2 - \sigma_A^2$
5.25	5.25	0	0

Table 3: A three-bit unsigned integer binary representation with zero epistasis.

S	$v(S)$	$E(S)$	$E(A)$	$A(S)$	$\varepsilon(S)$
000	0	−3.5	−10.5	−7	7
001	0	−3.5	−3.5	0	0
010	0	−3.5	−3.5	0	0
011	0	−3.5	3.5	7	−7
100	0	−3.5	−3.5	0	0
101	0	−3.5	3.5	7	−7
110	0	−3.5	3.5	7	−7
111	28	24.5	10.5	14	14

i	a	$A_i(a)$	$E_i(a)$
1	0	0	−3.5
	1	7	3.5
2	0	0	−3.5
	1	7	3.5
3	0	0	−3.5
	1	7	3.5

σ_v^2	σ_A^2	σ_ε^2	$\sigma_v^2 - \sigma_A^2$
85.75	36.75	49	49

Table 4: Calculating the epistasis variance for the f_2 function.

S	$v(S)$	$E(S)$	$E(A)$	$A(S)$	$\varepsilon(S)$
000	0.0	-3.5	-7.0	-3.5	3.50
001	0.5	-3.0	-3.0	0.5	0.00
010	1.0	-2.5	-2.5	1.0	0.00
011	1.5	-2.0	1.5	5.0	-3.50
100	2.0	-1.5	-1.5	2.0	0.00
101	2.5	-1.0	2.5	6.0	-3.50
110	3.0	-0.5	3.0	6.5	-3.50
111	17.5	14.0	7.00	10.5	7.00

i	a	$A_i(a)$	$E_i(a)$
1	0	0.75	-2.75
	1	6.25	2.75
2	0	1.25	-2.25
	1	4.75	2.25
3	0	1.50	-2.00
	1	5.50	2.00

σ_v^2	σ_A^2	σ_ε^2	$\sigma_v^2 - \sigma_A^2$
41.125	28.875	12.250	12.250

Table 5: Calculating the epistasis variance for the f_3 function.

S	$v(S)$	$E(S)$	$E(A)$	$A(S)$	$\varepsilon(S)$
000	0	-3.5	-2	1.5	-1.5
001	1	-2.5	-2	1.5	-0.5
011	2	-1.5	-2	1.5	0.5
010	3	-0.5	-2	1.5	1.5
110	4	0.5	2	5.5	-1.5
111	5	1.5	2	5.5	-0.5
101	6	2.5	2	5.5	0.5
100	7	3.5	2	5.5	1.5

i	a	$A_i(a)$	$E_i(a)$
1	0	1.5	-2.0
	1	5.5	2.0
2	0	3.5	0.0
	1	3.5	0.0
3	0	3.5	0.0
	1	3.5	0.0

σ_v^2	σ_A^2	σ_ε^2	$\sigma_v^2 - \sigma_A^2$
5.25	4.00	1.25	1.25

Table 6: A three-bit unsigned integer gray code representation of the zero epistasis function f_1.

S	$v(S)$	$E(S)$	$E(A)$	$A(S)$	$\varepsilon(S)$
000	0	−3	−2.0	1.0	−1.0
001	1	−2	−2.0	1.0	0.0
010	2	−1	−1.0	2.0	0.0
011*					
100	4	1	1.5	4.5	−0.5
101	5	2	1.5	4.5	−0.5
110	6	3	3.0	6.0	0.0
111*					

i	a	$A_i(a)$	$E_i(a)$
1	0	1.0	−2.0
	1	5.0	2.0
2	0	2.5	−0.5
	1	4.0	1.0
3	0	3.0	0.0
	1	3.0	0.0

σ_v^2	σ_A^2	σ_ε^2	$\sigma_v^2 - \sigma_A^2$
4.66	3.75	0.25	−0.92

Table 7: A 75% Grand Population sample shows sampling error variance. The starred (*) strings are the ones not included in the statistics.

S	$v(S)$	$E(S)$	$E(A)$	$A(S)$	$\varepsilon(S)$
000	0	−3.5	−6.5	-3	3
001	1	−2.5	−5.5	-2	1
010*					
011*					
100*					
101*					
110	6	2.5	5.5	9	−3
111	7	3.5	6.5	10	−3

i	a	$A_i(a)$	$E_i(a)$
1	0	0.5	−3.0
	1	6.5	3.0
2	0	0.5	−3.0
	1	6.5	3.0
3	0	3.0	−0.5
	1	4.0	0.5

σ_v^2	σ_A^2	σ_ε^2	$\sigma_v^2 - \sigma_A^2$
9.25	36.25	9.0	−27.00

Table 8: A 50% Grand Population sample shows an increased sampling error. The starred (*) are the ones not included in the statistics.

S	$v(S)$	$E(S)$	$E(A)$	$A(S)$	$\varepsilon(S)$
000	7	3.5	1.25	4.75	2.25
001	5	1.5	0.75	4.25	0.75
010	5	1.5	0.75	4.25	0.75
011	0	−3.5	0.25	3.75	−3.75
100	3	−1.5	−0.25	3.25	−0.25
101	0	−3.5	−0.75	2.75	−2.75
110	0	−3.5	−0.75	2.75	−2.75
111	8	4.5	−1.25	2.25	5.75

i	a	$A_i(a)$	$E_i(a)$
1	0	4.25	0.75
	1	2.75	−0.75
2	0	3.75	−0.25
	1	3.25	−0.25
3	0	3.75	0.25
	1	3.25	−0.25

σ_v^2	σ_A^2	σ_ε^2	$\sigma_v^2 - \sigma_A^2$
9.25	0.68	8.57	8.57

Table 9: The epistasis analysis for a minimal deceptive function.

1. As the sample diverts from a Grand Population, the nonlinearity a GA operates with increases.
2. Epistasis, as defined here, consists of two elements: the base epistasis resulting from the representation and a sampling error resulting from sampling noise.

4.3 A minimal deceptive problem

So far, the functions that were analyzed had a known epistasis. To conclude the preliminary discussion on epistasis variance, it would be interesting to analyze a function of an unknown epistasis, a deceptive problem known to be a 'hard' problem for a GA.

A minimal deceptive problem[2] is an archetype GA-hard function to which the epistasis tools are applied. A deceptive problem is designed in such a way that it contains a lot of structure, structure which misdirects the convergence to wrong regions. This structure quality is required by the very definition of the function. For the deception to be effective, most low order schemata that does not instantiate the global optimum, have a higher value than those that do. It is reasonable therefore to expect that the sum total epistasis embedded in a deceptive function is not too high, and indeed it ought not to be too high. Calculating the epistasis variance for the minimal deceptive problem (Table 9) confirms the above analysis.

[2]The minimal deceptive problem used here is based on the minimal deceptive problem as defined by Goldberg [Goldberg, 1987, Goldberg, 1989b] and involves variations of negligible importance that were adopted for convenience.

5 Conclusions and future work

This paper discussed some fundamental representation and relating GA-hardness issues. In spite of its importance, the precise analysis of representations is uncommon due to the tedious computation involved and lack of consensus definitions of GA-hardness. The epistasis analysis was suggested as an alternative approach for understanding the GA-hardness enigma. However, the amount of computation at the current state of epistasis analysis which necessary to compute an accurate estimate is not substantially smaller.

The epistasis analysis as presented here suggested that their must be sufficient structure in the solution space (depicted by the representation) in order for a GA to process information effectively. Structure in the solution space is not a sufficient condition to guaranty a successful application as demonstrated with a deceptive function. Furthermore, it is not clear how a given epistasis measure can be translated into relative efficiency in respect to other search methods. Nevertheless, it is a necessary condition for a GA. The measurement of epistasis is based on a linear composition of a string from its bits. The accuracy of this method, or more precisely, the epistasis variance of a sample, is an estimate of the total amount of nonlinearity embedded in the representation.

The epistasis variance is usually determined by a sample. Analyzing the epistasis variance for different samples reveals an extensive sampling error resulting from this approximation. Confidence measures for the extent of this approximation were not presented. It is clear that at this stag, the epistasis analysis is not economical. However, it is reasonable to believe that it will be significantly more economical when epistasis can be estimated with reasonable accuracy from a sample.

To be useful as a tool, the epistasis variance requires two extensions to the work presented in this paper:

1. Means for normalization of the epistasis variance so it can be plotted on a scale.
2. The development of confidence measures for estimating the base epistasis from sample populations.

and additional experimental results how the epistasis variance changes as a function of the fitness function and the sample size.

The author is currently involved in extending the epistasis analysis in two directions. One, to develop confidence measures for estimating the base epistasis from a sample. Two, to further investigate the definition of GA-hardness in perspective to epistasis and deception.

Acknowledgments

The work presented in this article was supported in part by the Center for Absorption in Science, The Ministry of Immigrant Absorption, The State of Israel. Judit Bar-Ilan and Antonia J. Jones made valuable comments on early drafts of this article. Y. Davidor is recipient of a Sir Charles Clore Fellowship.

References

Bethke, A. D., (1981) Genetic algorithms as function optimizes *Dissertation Abstracts International*, **41(9)** 3503B. University of Michigan Microfilm No. 8106101.

Caruana, R. A., and Schaffer, J. D. (1988) Representation and hidden bias: Gray vs. binary coding for genetic algorithms, *Proceedings of the 5th International Conference on Machine Learning*, University of Michigan, Ann Arbor, Morgan Kaufmann.

Spears, W. S., and De Jong, K. A. (in press) An analysis of multi-point crossover, *Proceedings of the Foundations of Genetic Algorithms Workshop*, Indiana University, Bloomington, Morgan Kaufmann.

De Jong, K. A., and Spears, W. S. (in press) An analysis of the interacting roles of population size and crossover in genetic algorithms, *Proceedings of the 1st International Conference on Parallel Problem Solving from nature*, Springer-Verlag.

Goldberg, D. E. (1987) Simple genetic algorithms and the minimal, deceptive problem, In *Genetic Algorithms and Simulated Annealing*, L. Davis, (Ed.), Pitman, London, 74–88.

Goldberg, D. E., (1989) *Genetic Algorithm in Search, Optimization, and Machine Learning*, Addison-Wesley, Reading, MA.

Goldberg, D. E., (1989) Zen and the art of genetic algorithms, *Proceedings of the 3rd Inernational Conference on Genetic Algorithms and their Applications*, George Mason University, Morgan Kaufmann.

Goldberg, D. E., (1989) Genetic algorithms and walsh functions: Part I, A gentle introduction, *Complex Systems*, **3**, 129-152.

Goldberg, D. E., (1989) Genetic algorithms and walsh functions: Part II, Deception and its analysis, *Complex Systems*, **3**, 153-171.

Grefenstette, J. J., (1979) *Representation Dependencies in Genetic Algorithms.* Unpublished manuscript, Navy Center for Applied Research in AI, Navel Research Laboratory, Washington, DC 20375-5000, USA.

Holland, J. H., (1975) *Adaptation in Natural and Artificial Systems*, The University of Michigan Press, Ann Arbor.

Jacobson, H., (1955) Information reproduction and the origin of life, *American Scientist*, **43**, 119-127.

Klug, W. S., and Cummings, M. R., (1986) *Concepts of Genetics*, 2nd edition, Scott, Foresman and Co..

Platt, J. R., (1961) Properties of large molecules that go beyond the properties of their chemical sub groups, *Journal of Theoretical Biology*, **1**, 342-358.

Ptashne, M., (1989) How gene activators work, *Scientific American*, January, 25-31.

Simon, H. A. (1962) The architecture of complexity, *Proceedings of the American Philosophical Society*, **106(6)**.

Syswerda, G. (1989) Uniform crossover in genetic algorithms, *Proceedings of the 3rd International Conference on Genetic Algorithms*, George Mason University, Morgan Kaufmann.

Tsotsos, J. K. (1987) A complexity level analysis of vision, *Proceedings of the First International Conference on Computer Vision*, IEEE Computer Society Press, 346-355.

Deceptiveness and Genetic Algorithm Dynamics*

Gunar E. Liepins
Oak Ridge National Laboratory
MS 6360 Bldg. 6025
PO Box 2008
Oak Ridge, TN 37831-6360
gxl@msr.epm.ornl.gov

Michael D. Vose
The University of Tennessee
C. S. Dept.
107 Ayres Hall
Knoxville, TN 37996-1301
vose@cs.utk.edu

Abstract

We address deceptiveness, one of at least four reasons genetic algorithms can fail to converge to function optima. We construct fully deceptive functions and other functions of intermediate deceptiveness. For the fully deceptive functions of our construction, we specify linear transformations that induce changes of representation to render the functions fully easy. We further model genetic algorithm selection and recombination as the interleaving of linear and quadratic operators. Spectral analysis of the underlying matrices allows us to draw preliminary conclusions about fixed points and their stability. We also obtain an explicit formula relating the nonuniform Walsh transform to the dynamics of genetic search.

Keywords: representation, deceptive functions, Walsh transforms dynamic systems

1 Introduction

Designed to search irregular, poorly understood spaces, GAs are general purpose algorithms (akin to simulated annealing in this sense) developed by Holland (1975)

*This research was supported by the National Science Foundation (IRI-8917545), and by the Air Force office of Scientific Research and the Office of Naval Research Contract # F49620-90-C-00033.

with precursors suggested by Bledsoe (1961) and others. Holland's hopes were to develop powerful, broadly applicable techniques, to provide a means to attack problems resistant to other known methods. Inspired by the example of population genetics, genetic search is population based, and proceeds over a number of generations. The criteria of "survival of the fittest" provides evolutionary pressure for populations to develop increasingly fit individuals. Although there are many variants, the basic mechanism of a GA consists of:

1. Evaluation of individual fitness and formation of a gene pool.
2. Recombination and mutation.

Individuals resulting from these operations form the members of the next generation, and the process is iterated until the system ceases to improve.

Fixed length binary strings are typically the members (genes) of the population. They contribute to the gene pool in proportion to their relative fitness (determined by the objective function). There, they are mutated and recombined by crossover. Mutation corresponds to flipping the bits of an individual with some small probability (the mutation rate). The simplest implementation of crossover selects two "parents" from the pool and, after choosing the same random position within each string, exchanges their tails. Crossover is typically performed with some probability (the crossover rate), and parents are otherwise cloned. The resulting "offspring" form the subsequent population. A thorough introduction and overview of GAs is provided in Goldberg (1989a), and public domain code is available from Grefenstette (1984).

The most obvious factors affecting performance are the parameter settings for population size, crossover rate, and mutation rate. Grefenstette (1986) has investigated the use of meta-level GAs for determining parameter settings, but perhaps the most systematic study of these parameters was undertaken by DeJong (1975). For a survey of these and related research, see Liepins and Hilliard (1989). Less obvious factors related to performance involve the estimation of schemata utilities. In order to explain the difficulties that arise, we first sketch the basics of schemata analysis.

A schema (Holland, 1975) describes a subset of strings with similarities at certain string positions. For example, consider strings of length 3 over the alphabet $\{0, 1\}$. The two strings

$$011$$

$$111$$

are similar in the sense that they are identical when the first position is ignored. Regarding $*$ as a symbol which may be instanciated to either 0 or 1, these two strings may therefore be represented by

$$*11$$

Strings over the alphabet $\{*, 0, 1\}$ represent schemata, and play a central role in analyzing GAs.

Let P be a finite population drawn from some universe Ω of length n binary strings. Let f be a real valued fitness (objective) function

$$f : \Omega \longrightarrow \mathcal{R}$$

For any schema H, define the utility of H with respect to P as[1]

$$f_P(H) = \frac{1}{|H \cap P|} \sum_{x \in P \cap H} f(x)$$

and define the utility of H as $f_\Omega(H)$. Regarding P as changing under the influence of a genetic algorithm, let P_t denote the generation under consideration. The Schema Theorem is the inequality

$$\mathcal{E}\,|H \cap P_{t+1}| \;\geq\; |H \cap P_t|\,\frac{f_P(H)}{f_P(\Omega)}\,[1 - \alpha(H,t) - \beta(H,t)]$$

where $\mathcal{E}$ is an expectation operator, and α and β approximate the probabilities that an instance of H will be destroyed by crossover or mutation (respectively). The functions α and β are usually taken to be constants estimated in terms of properties provided by the concrete representation of H. A proof of this inequality is given in Goldberg (1989a).

In some sense, schemata represent the direction of the genetic search. It follows from the schema theorem that the number of instances of a schema H for which $f_P(H) > f_P(\Omega)$ is expected to increase in the next generation when $\alpha(H,t)$ and $\beta(H,t)$ are small. Therefore, such schemata indicate the area within Ω which the GA explores. Hence it is important that, at some stage, these schema contain the object of search. Problems for which this is not true are called deceptive. We provide formal definitions of deceptiveness later in the paper.

The utility of H with respect to P may be thought of as an estimate of the utility of H. Holland's results (1975) regarding allocation of trials to k-armed bandit problems suggest that a GA optimally allocates its resources so long as schemata utilities are correctly estimated. Other factors influencing GA performance are the encoding of the search domain into bit strings and their manipulation by the GA. Therefore, GAs may fail to locate a function optima for several reasons which include:

1. The chosen embedding (i.e., choice of domain) is inappropriate.
2. The problem is not deceptive, but schemata utilities cannot be reliably estimated because sampling error is too large.
3. Schemata utilities can be reliably estimated, but crossover destroys individuals which represent schemata of high utility.
4. The problem is deceptive.

The first failure mode has been partially addressed by Shaefer (1985, 1987) who has incorporated dynamically changing embeddings into his ARGOT code.

The second failure mode is virtually unstudied, although results in Goldberg and Segrest (1987) might be extended to shed light on this issue.

The third failure mode is partially addressed by the schema theorem; adjusting crossover rate may help overcome this problem. Representational changes can also

[1] P is a multiset, and an element of $P \cap H$ is regarded as having the multiplicity it had in P.

be useful. Early studies of inversion (a permutation applied to string positions) were done by Bagley (1967), Cavicchio (1970) and Frantz (1972). Holland (1975) discusses inversion as a basic genetic operator. More recently, Whitley (1987) has reported encouraging results with the use of inversion, and Goldberg and Bridges (1990) have considered reordering operators to prevent schemata disruption. Goldberg et. al. (1989 and 1990) introduced messy GAs and cut and splice operators to combat crossover disruption and linkage problems.

The fourth failure mode has been studied from the perspective of Walsh transforms by Bethke (1981) and Goldberg (1988, 1989b). Although the Walsh transform approach allows the construction of deceptive problems, the analysis is performed in the transform space, and lacks some degree of intuitive accessibility. Holland (1989) investigated a (computationally equivalent) hyperplane transform which deals with schemata directly. Bridges and Goldberg (1989) have shown the computational equivalence of the hyperplane transform and the Walsh transform, and extend schema analysis to population estimates.

Our work follows and extends the themes developed in Liepins and Vose (1990) and Vose and Liepins (1990). We directly construct fully deceptive functions whenever the chromosome length n is greater than two. For these same fully deceptive functions, we specify linear transformations that render the functions fully easy. We introduce the concept of basis sets and use it to construct various classes of problems of intermediate deceptiveness.

The second half of our paper goes beyond static schema analysis and models the (expected) genetic algorithm process as a dynamic system. What emerges from this model is a clearer understanding of selection as a "focusing operator" and recombination as a "dispersion operator", and a decomposition of genetic search which makes explicit the transitions between populations, between selection probabilities, and between nonuniform Walsh transforms from one generation to the next.

2 Embedding, Representation and Deceptiveness

A useful point of departure for understanding difficulties with genetic optimization is a commutative diagram (figure 1) that makes explicit the steps involved in their use. Frequently, the function f to be optimized is defined on a real vector space S. Since genetic search explores a finite space of bit strings, some finite subset D of S needs to be selected for investigation. Thereafter, D is regarded as the domain of f and is mapped into bit strings by an invertible map i. We refer to the injection d which maps D into S as the *embedding*, and refer to i as the *representation*. The following diagram induces functions f_d and f_i defined on D and Ω by commutativity.

Let $s^* \in S$ be an optimum of f (i.e., a point which maximizes f). The objective of genetic optimization is to determine a point $x^* \in \Omega$ such that the difference $|f(s^*) - f_i(x^*)|$ is acceptably small.

The success of genetic optimization depends on both the embedding and the representation. For real valued optimization problems, it may be important that the embedding is centered near a function optimum and is of sufficient resolution to reflect the function's variation. One can attempt to achieve this by dynamically changing the embedding. Shaefer (1985, 1987) implements this idea in his ARGOT

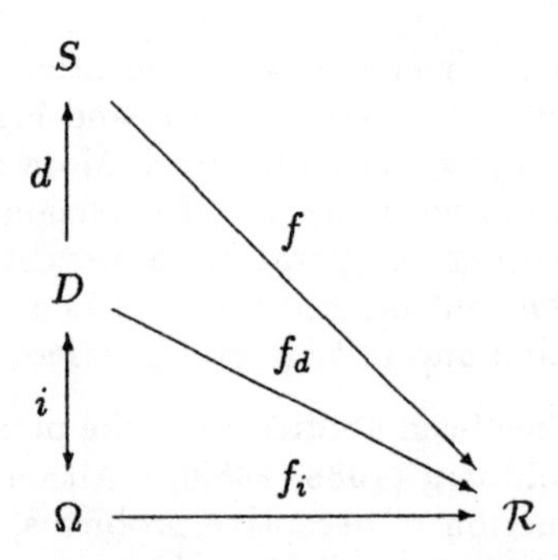

Figure 1: Commutative Diagram for Genetic Optimization

code by varying the parameters of an affine mapping which defines the embedding.

The role of representation is perhaps not as clear. Its importance can be explained in terms of the "building block hypothesis." This hypothesis asserts that GA search proceeds not from individual chromosome to individual chromosome, but rather from high utility schemata with few fixed bits to high utility schemata with many fixed bits. If in the chosen representation the function optima does not lie in the schemata estimated to be of high utility, genetic search may be mislead.

The first published study of deceptiveness was undertaken by Bethke (1981). His analysis made use of the Walsh functions as a basis for the set of real valued functions on Ω. Define the j th Walsh function w_j for $j = <j_1, \ldots, j_n> \in \Omega$ by

$$w_j(x) = (-1)^{j \cdot x}$$

where

$$j \cdot x = \sum_{i=1}^{n} j_i x_i$$

Given a function $f : \Omega \to \mathcal{R}$, define the j th Walsh coefficient $\hat{f}_j$ by

$$\hat{f}_j = 2^{-n/2} \sum_{x \in \Omega} f(x) w_j(x)$$

The inversion formula

$$f(x) = 2^{-n/2} \sum_{j \in \Omega} \hat{f}_j w_j(x)$$

is a special case of the relationship between a function defined on a group and the set of group characters (Rudin, 1967). The utility of the schema H with respect to the function f can be expressed in terms of the Walsh coefficients:

$$\begin{aligned} f_\Omega(H) &= \frac{1}{|H|} \sum_{x \in H} f(x) = |H|^{-1} 2^{-n/2} \sum_{x \in H} \sum_{j \in \Omega} \hat{f}_j w_j(x) \\ &= |H|^{-1} 2^{-n/2} \sum_{j \in \Omega} \hat{f}_j \sum_{x \in H} w_j(x) \end{aligned}$$

Let the order $o(j)$ of a bit string j be the number of 1's in j, and let the order $o(H)$ of a schema H be the number of fixed positions in H. Two schemata are said to be competing if they have the same order, with *'s in the same positions, and different fixed bits.

Bethke's results rest on the observations that:

1. $f(x) = 2^{-n/2}\left(\sum_{j\cdot x \text{ is even}} \hat{f}_j - \sum_{j\cdot x \text{ is odd}} \hat{f}_j\right)$.
2. $\sum_{x\in H} w_j(x) = \mid \{x \in H : x \cdot j \text{ is even}\} \mid - \mid \{x \in H : x \cdot j \text{ is odd}\} \mid$
3. If j contains a 1 in a position where H contains a $*$, then $\hat{f}_j$ does not influence $f_\Omega(H)$.
4. If $o(j) > o(H)$, then $\hat{f}_j$ does not influence $f_\Omega(H)$.

Let H and H' be two competing schemata such that for all fixed positions i, $H'_i = 0 \Rightarrow H_i = 0$. It follows from these observations that if $\hat{f}_j = 0$ when $1 < o(j) \leq o(H)$ and if $\hat{f}_j > 0$ when $o(j) = 1$, then $f_\Omega(H) > f_\Omega(H')$.

Definition 1: Let f be a function with global optima at $\{x^*, \ldots\}$. Then f is deceptive of order m iff there exists $x \notin \{x^*, \ldots\}$ such that when H and H' are competing schemata of order not greater than m,

$$x \in H \implies f_\Omega(H) > f_\Omega(H')$$

We are now ready to provide Bethke's construction. Let $2 \leq d = 2b < n$ be the desired order of deceptiveness. The previous discussion implies the existence of a constant $c_d < 0$ (depending on d) such that the function f defined below in terms of its Walsh coefficients has maximum at $x^* = <1, \ldots, 1>$ and is deceptive of order d (lower order schema containing $<0, \ldots, 0>$ have greater utility):

$$\hat{f}_j = \begin{cases} 1 & \text{if } o(j) = 1 \\ c_d & \text{if } o(j) = d+1 \\ 0 & \text{otherwise} \end{cases}$$

This construction begs several related questions. Do functions exist which are deceptive of all orders $d < n$? Do functions exist which are deceptive of order $d < n-1$, but whose schemata are correctly aligned thereafter? The combinatorics of the Walsh transform analysis quickly become unwieldy, and these questions are better answered in other ways.

Definition 2: Let f be a real valued function on Ω with unique global maximum at x^*, and let x^c be the binary complement of x^*. The function f is fully deceptive iff whenever H and H' are competing schemata of order less than n,

$$x^c \in H \implies f_\Omega(H) > f_\Omega(H')$$

The opposite of fully deceptive functions are fully easy functions.

Definition 3: Let f be a real valued function on Ω with unique optimum at x^*. Then f is fully easy iff whenever H and H' are competing schemata,

$$x^* \in H \implies f_\Omega(H) > f_\Omega(H')$$

Goldberg (1989) constructed an example of a fully deceptive function for $n = 3$. Liepins and Vose (1990) showed that all fully deceptive functions can be expressed as the sum of a fully easy function and a spike function at the optimum point. Furthermore, they constructed fully deceptive functions for string lengths $n > 2$:

$$f(x) = \begin{cases} 1 - 2^{-n} & \text{if } o(x) = 0 \\ 1 - (1 + o(x))/n & \text{if } 0 < o(x) < n \\ 1 & \text{if } o(x) = n \end{cases}$$

Liepins and Vose reported that this class of fully deceptive functions could be transformed into fully easy functions through the transformation $g(x) = f \circ M(x)$, where M is an invertible linear map over Z_2. Regarding binary strings as column vectors, their $n \times n$ matrix M is

$$m_{ij} = \begin{cases} 0 \text{ for } i = j \neq n \\ 1 \text{ otherwise} \end{cases}$$

Holland (1989) noted that neither the original function nor the Walsh transform readily promote schemata analysis. Instead, he proposed the hyperplane transform which depends directly on schemata utilities. Holland's hyperplane transforms are related to a general method of representation based on what we call basis sets. Let $X = \{x_1, \ldots, x_n\}$ be a finite space. To each subset s of X corresponds an incidence vector v defined by

$$v_i = \begin{cases} 1 & \text{if } x_i \in s \\ 0 & \text{otherwise} \end{cases}$$

Definition 4: A collection S of subsets of X is a basis for X iff the associated incidence vectors form a basis for for $\mathcal{R}^n$.

Lemma 1: The collection of all schemata containing $<1\ldots1>$ is a basis for Ω.

Proof: For any schema s, replace the "*"s in s with "0"s and interpret the result as a binary integer. This defines a map h from those schemata containing $<1\ldots1>$ to the set of integers $\{0, \ldots, 2^n - 1\}$. Form a matrix having as rows the incidence vectors corresponding to schemata of S, and order the rows according to increasing values of h. This matrix is upper diagonal and hence of full rank.

Theorem 1: Let $x \in \Omega$. The collection of all schemata containing x is a basis for Ω.

Proof: Let σ be the matrix corresponding to that permutation which sends the j th component of a binary vector to the $j \oplus x$ th position, where $x \in \Omega$ is fixed and $\oplus$ denotes componentwise exclusive-or. If v is the incidence vector for a schema s, it follows that σv is the incidence vector for the schema $x \oplus s$. Therefore the incidence vectors associated with a translation (by x in Ω) of a basis are obtained by mapping the incidence vectors associated with that basis by σ. Since permutation matrices are invertible, they preserve linear independence, and since x was arbitrary, it can translate the basis of Lemma 1 to any point of Ω.

Theorem 2: Let S be a basis for X, and f a real valued function defined on S. There exists a unique function $g : X \to \mathcal{R}$ such that

$$s \in S \Longrightarrow \frac{1}{|s|} \sum_{x \in s} g(x) = f(s)$$

Proof: let M be the matrix having as rows the incidence vectors corresponding to elements of S, let g be the column vector of required function values, and let f be the column vector of given values. The condition to be satisfied is

$$DMg = f$$

where D is a diagonal matrix containing $|s|^{-1}$ for $s \in S$. Hence g is uniquely determined by $g = M^{-1}D^{-1}f$

Definition 5: Let x be a point in Ω. A schema path at x is a nested sequence of schemata $H_0 \subset \cdots \subset H_n$ containing x such that $o(H_i) = i$

Let us now return to the concept of deceptiveness. Intuitively, deceptiveness occurs whenever a "good path" leads to a "bad point" or a "bad path" leads to a "good point".

Definition 6: Let $x \in \Omega$ and let J be a schema path at x. Then f is decreasing at x along J of order (a, b) iff whenever H and H' are two schemata in J,

$$a \leq o(H) < o(H') \leq b \Longrightarrow f_\Omega(H) > f_\Omega(H')$$

If $a = 0$, we shall use the term "decreasing along J of order b". The definitions for increasing along J are defined analogously.

Observe that fully deceptive functions have a unique optimal x^* and are increasing at x^c of order $n-1$ along all schema paths. Fully easy functions have a unique optimal x^* and are increasing at x^* of order n along all schema paths. This follows from the relation

$$F_\Omega(H) = \frac{1}{2}\left(F_\Omega(H_0) + F_\Omega(H_1)\right)$$

where the H_j are schemata obtained from H by replacing a * of H with j.

We can now turn our attention to the existence of classes of functions of intermediate deceptiveness. We assume that the functions of interest have a unique optimal which without loss of generality is at $<0, \ldots, 0>$. We prove each of the following classes are nonempty:

C1. Functions with several schema paths at the optimal; some of them increasing of order n, and others decreasing of order $n-1$.

C2. Functions all of whose schema paths at the optimal are increasing for some order $d < n-1$ and decreasing thereafter (except at order n).

C3. Functions all of whose schema paths at the optimal are decreasing for some order $d < n-1$ and increasing thereafter.

These classes are interesting because real problems could presumably have some paths that are deceptive and other paths that aren't, or could have some regions of deceptiveness either preceded or followed by regions which are nondeceptive. Intuitively, one might expect that the density of nondeceptive paths or the depth of deceptiveness is related to whether a GA discovers an optimum.

The proof that these classes are nonempty follows from the observation that each is defined in terms of schema paths at the single point $<0, \ldots, 0>$. By Theorem 1, the collection of all schemata at a point forms a basis, hence the schemata involved in the definitions of these classes are linearly independent. It follows from Theorem 2 that assigning arbitrary values to any set of linearly independent schemata will induce a fitness function consistent with the given utilities.

3 GAs as Dynamic Systems

In this section we sketch our development of GAs as dynamic systems. Since many of our results have cumbersome proofs, we will report on our progress in this section and refer the reader to Vose and Liepins (1990) for technical details.

We represent recombination (1-point, 2-point, or uniform crossover followed by mutation) as a quadratic operator determined by a fixed non-negative symmetric matrix M in conjunction with a group of permutations on chromosome strings. The matrix M has special properties, many of which result from the commutativity of crossover and mutation with group translation. We formalize selection as multiplication by a diagonal matrix F. Spectral analysis of M and F allows us to draw preliminary conclusions about fixed points and their stability. We also obtain an explicit formula relating the nonuniform Walsh transform to the dynamics of genetic search.

Let Ω be the set of all length n binary strings, and let $N = 2^n$. Thinking of elements of Ω as binary numbers, we identify Ω with the interval of integers $[0, N-1]$. We also regard Ω as the product group

$$Z_2 \times \ldots \times Z_2$$

where Z_2 denotes the additive group of integers modulo 2. The group operation $\oplus$ acts on integers in $[0, N-1]$ via these identifications, and we use $\otimes$ to represent componentwise multiplication.

The t th generation of the genetic algorithm is modeled by a vector $s^t \in \mathcal{R}^N$, where the ith component of s^t is the probability that i is selected for the gene pool. Populations excluding members of Ω are modeled by vectors s^t having corresponding coordinates zero.

Let $p^t \epsilon \mathcal{R}^N$ be a vector with i th component equal to the proportion of i in the t th generation, and let $r_{i,j}(k)$ be the probability that k results from the recombination process based on parents i and j. If $\mathcal{E}$ denotes expectation, then

$$\mathcal{E}\; p_k^{t+1} = \sum_{i,j} s_i^t s_j^t r_{i,j}(k)$$

Let $C(i,j)$ represent the possible results of crossing i and j, and note that $k \oplus \ell \in C(i,j)$ if and only if $k \in C(i \oplus \ell, j \oplus \ell)$. If $X(i)$ represents the result of mutating i for some fixed mutation, then $k \oplus \ell = X(i)$ if and only if $k = X(i \oplus \ell)$. Since recombination is a combination of operations which commute with group translation, it follows that

$$r_{i,j}(k \oplus \ell) = r_{i \oplus k, j \oplus k}(\ell)$$

This allows recombination to be expressed via the matrix M defined by $m_{i,j} = r_{i,j}(0)$. Let F be the nonnegative diagonal matrix with ii th entry $f(i)$, where f is the objective function, and let σ_j be permutations on $\mathcal{R}^N$ given by

$$\sigma_j < y_0, \ldots, y_{N-1} >^T = < y_{j\oplus 0,\ldots,} y_{j\oplus(N-1)} >^T$$

where T denotes transpose. Define operators $\mathcal{M}$, $\mathcal{F}$, and $\mathcal{G}$ by

$$\begin{aligned} \mathcal{M}(s) &= < (\sigma_o s)^T M \sigma_o s, \ldots, (\sigma_{N-1} s)^T M \sigma_{N-1} s >^T \\ \mathcal{F}(s) &= F s \\ \mathcal{G} &= \mathcal{F} \circ \mathcal{M} \end{aligned}$$

and let $\sim$ represent the equivalence relation on $\mathcal{R}^N$ defined by $x \sim y$ if and only if $\exists \lambda > 0 \,.\, x = \lambda y$. It follows that

$$\mathcal{E} s^{t+1} \sim \mathcal{G}(s^t)$$

The expected behavior of a simple GA is therefore determined by matrices M and F. Fitness information appropriate for the selection operator $\mathcal{F}$ is contained in F, while M encodes mixing information appropriate for the recombination operator $\mathcal{M}$. Moreover, the relation

$$s^{t+1} \sim \mathcal{G}(s^t)$$

is an exact representation of the limiting behavior as population size $\longrightarrow \infty$. One natural geometric interpretation of this formalization is to regard F and $\mathcal{M}$ as maps from $\mathcal{S}$, the nonnegative points of the unit sphere in $\mathcal{R}^N$, to $\mathcal{S}$ (since apart from the origin, each equivalence class of $\sim$ has a unique member of norm 1). An initial population then corresponds to a point on $\mathcal{S}$, the progression from one generation to the next is given by the iterations of $\mathcal{G}$, and convergence (of the GA) corresponds to a fixed point of $\mathcal{G}$.

The properties of the operator $\mathcal{F}$ are straightforward to analyze. Regarding $\mathcal{F}$ as a map on S, its fixed points correspond to the eigenvectors of F, i.e., the unit basis vectors $u_0, \ldots, u_{N-1}$. If $f(i) = f(j)$, then by passing to a quotient space (moding out by the linear span of u_i and u_j), the subspace corresponding to i and j is collapsed to a single dimension. Hence we may assume that f is injective by considering a suitable homomorphic image. The basin of attraction of the fixed point u_j is given by the intersection of $\mathcal{S}$ with the (solid) ellipsoid

$$\sum_i \left(s_i \frac{f(i)}{f(j)} \right)^2 < 1$$

Only the fixed points corresponding to the maximal value of the objective function f are in the interior of their basins of attraction. Hence all other fixed points are unstable. This follows from the observation that when $f(j)$ is maximal, no point of $\mathcal{S}$ moves away from u_j since

$$\sum_i \left(s_i \frac{f(i)}{f(j)} \right)^2 \leq \sum_i s_i^2 = 1$$

Intuitively, Theorem 1 is not surprising. Selection is a focusing operator which moves the population towards one containing only the maximally fit individuals which are initially present. The properties of $\mathcal{M}$ are less immediate.

For 1-point crossover with mutation, we explicitly calculate the mixing matrix M for crossover rate χ and mutation rate μ as

$$m_{ij} = \frac{(1-\mu)^n}{2}\left\{\left(\frac{\mu}{1-\mu}\right)^{|i|}\left(1-\chi+\frac{\chi}{n-1}\sum_{k=1}^{n-1}\left(\frac{1-\mu}{\mu}\right)^{\Delta_{i,j,k}}\right)\right.$$

$$\left. + \left(\frac{\mu}{1-\mu}\right)^{|j|}\left(1-\chi+\frac{\chi}{n-1}\sum_{k=1}^{n-1}\left(\frac{\mu}{1-\mu}\right)^{\Delta_{i,j,k}}\right)\right\}$$

Here integers are to be regarded as bit vectors when occurring in $|\cdot|$, division by zero when $\mu \in \{0,1\}$ is to be removed by continuity, and

$$\Delta_{i,j,k} = |\,(2^k-1)\otimes i\,| - |\,(2^k-1)\otimes j\,|$$

Our results concerning fixed points and their stability derive from general properties of M. The most obvious of which are that M is nonnegative, symmetric, and for all i, j satisfies

$$\sum_k m_{i\oplus k, j\oplus k} = \sum_k r_{i\oplus k, j\oplus k}(0) = \sum_k r_{i,j}(k) = 1$$

Associated with the matrix M is a matrix M_* related to the differential of $\mathcal{M}$ whose i,j th entry is $m_{i\oplus j, i}$. We have shown that if M_* is positive and its second largest eigenvalue is less than 1/2, then every fixed point of recombination is asymptotically stable. Several computer runs calculating the spectrum of M_* show that for one point crossover with mutation, if $0 \le \mu \le 0.5$ then

- The second largest eigenvalue of M_* is $\frac{1}{2} - \mu$
- The third largest eigenvalue of M_* is $2\left(1-\frac{\chi}{n-1}\right)\left(\frac{1}{2}-\mu\right)^2$

Thus, when the mutation rate is between 0 and 1/2, the fixed population distributions under 1-point crossover followed by mutation are asymptotically stable. A conjecture of G. R. Belitskii and Yu. I. Lyubich relating to discrete dynamical systems further implies that fixed population would therefore be unique. Using symmetry properties, we can show the uniform population (i.e., all chromosomes represented equally) is fixed by recombination. Hence the uniform population is the only fixed point of recombination. This supports our view of recombination as a "dispersion" operator.

A much less obvious property of M is that conjugation by the symmetric idempotent matrix $\mathcal{W}$ representing the Walsh transform triangulates M_*. Moreover, conjugation by $\mathcal{W}$ also simultaneously diagonalizes the permutation matrices σ_j to ± 1 along the diagonal, and makes the matrix M sparse.

Let Q represent the sparse matrix $\mathcal{W}M\mathcal{W}$, let D_j represent the ± 1 diagonal matrix $\mathcal{W}\sigma_j\mathcal{W}$, let $Q_j = D_j Q D_j$, and define the operator $\mathcal{H}$ by

$$\mathcal{H}(s) = < s^T Q_0\, s, \ldots, s^T Q_{N-1}\, s >^T$$

Note that $\mathcal{H}$ has relatively simple structure; it is a system of quadratic forms having sparse coefficient matrices which differ only in their signs. Moreover, $\mathcal{G}$ is representable in terms of $\mathcal{H}$:

$$\mathcal{G}(s) = F \circ \mathcal{H} \circ \mathcal{W}(s)$$

Let v^t be defined by $v^t = \mathcal{W}s^t = \mathcal{W}Fp^t =$ the nonuniform Walsh transform of f at generation t. Therefore

$$s^{t+1} \sim \mathcal{G}(s^t) \Longrightarrow \mathcal{W}s^{t+1} \sim \mathcal{W} \circ \mathcal{G}(s^t) \Longrightarrow v^{t+1} \sim \mathcal{W} \circ \mathcal{G} \circ \mathcal{W}(v^t)$$

Hence the operator $\mathcal{W} \circ \mathcal{G} \circ \mathcal{W}$ maps the nonuniform Walsh transform from one generation to the next. Moreover we have the simple representation

$$\mathcal{W} \circ \mathcal{G} \circ \mathcal{W} = \mathcal{W} \circ F \circ \mathcal{H}$$

The operators $\mathcal{F}$, $\mathcal{H}$, and $\mathcal{W}$ may be interpreted as *selection*, *mixing* and *transform* respectively. Moreover, their interleaving models the progression of genetic search

$$\cdots \longrightarrow s^t \overset{\mathcal{W}}{\longrightarrow} v^t \overset{\mathcal{H}}{\longrightarrow} p^{t+1} \overset{\mathcal{F}}{\longrightarrow} s^{t+1} \longrightarrow \cdots$$

The transition from one generation to the next may therefore be equivalently regarded as proceeding through selection vectors, population vectors, or nonuniform Walsh transforms.

4 Summary

In this paper we have addressed both static and dynamic properties of the genetic algorithm. The static analysis addressed problem deceptiveness. The dynamic analysis addressed the time evolution of the expected population distribution.

We began our discussion of deceptiveness by explicating four failure modes for the genetic algorithm. Next, we summarized Bethke's results regarding difficult functions. We exhibited fully deceptive functions and invertible linear transformations which transform these functions into fully easy functions. We further introduced basis sets and used them to prove the existence of functions having various intermediate degrees of deceptiveness.

Our modeling of GAs as dynamical systems focused on the expressibility of recombination as quadratic forms in terms of a single, fixed mixing matrix M. We have discovered several special properties of this matrix and have related the spectrum of an associated matrix M_* to the stability of fixed points of recombination. Computer calculations indicate a simple relation between the spectrum of M_* and the crossover and mutation rate which leads to the conclusion that that the only stable fixed population distribution for recombination is uniform. Our model leads to a decomposition of genetic search which makes explicit the transitions between populations, between selection probabilities, and between nonuniform Walsh transforms from one generation to the next.

References

J. D. Bagley. (1967) The Behavior of Adaptive Systems which Employ Genetic and Correlational Algorithms. (Doctoral dissertation, University of Michigan). *Dissertation Abstracts International* **28**(12), 5106B. (University Microfilms No. 68-7556).

A. D. Bethke. (1981) Genetic Algorithms as Function Optimizers. (Doctoral dissertation, University of Michigan). *Dissertation Abstracts International* **41**(9), 3503B. (University Microfilms No. 8106101).

W. W. Bledsoe. (1961, November) *The Use of Biological Concepts in the Analytical Study of Systems.* Paper presented at the ORSA-TIMS National meeting, San Francisco, CA.

C. L. Bridges & D. E. Goldberg. (1989) *A Note on the Non-Uniform Walsh-Schema Transform* (TCGA Report No. 89004). Tuscaloosa: The University of Alabama, The Clearinghouse for Genetic Algorithms.

D. J. Cavicchio. (1970) *Adaptive Search Using Simulated Evolution.* Unpublished doctorial dissertation, University of Michigan, Ann Arbor.

K. A. DeJong. (1975) An Analysis of the Behavior of a Class of Genetic Adaptive Systems. (Doctoral dissertation, University of Michigan). *Dissertation Abstracts International* **36**(10), 5140B. (University Microfilms No. 76-9381).

D. R. Frantz. (1972) Non-Linearities in Genetic Adaptive Search. (Doctoral dissertation, University of Michigan). *Dissertation Abstracts International,* **33**(11), 5240B–5241B. (University Microfilms No. 73-11 116).

D. E. Goldberg & P. Segrest. (1987) Finite Markov Chain Analysis of Genetic Algorithms, *Genetic Algorithms and their Applications: Proceedings of the Second International Conference on Genetic Algorithms*, 1-8.

D. E. Goldberg. (1988) *Genetic Algorithms and Walsh Transforms: Part I, a Gentle Introduction* (TCGA Report No. 88006). Tuscaloosa: University of Alabama, The Clearinghouse for Genetic Algorithms.

D. E. Goldberg. (1989a) *Genetic Algorithms in Search, Optimization, and Machine Learning.* Addison-Wesley.

D. E. Goldberg. (1989b) *Genetic Algorithms and Walsh Functions: Part II, Deception and its Analysis* (TCGA Report No. 89001). Tuscaloosa: University of Alabama, The Clearinghouse for Genetic Algorithms.

D. E. Goldberg, B. Korb, & K. Deb. (1989) *Messy Genetic Algorithms: Motivation, Analysis and First Results* (TCGA Report No. 89003). Tuscaloosa: The University of Albama. The Clearinghouse for Genetic Algorithms.

D. E. Goldberg & C. L. Bridges. (1990) An Analysis of a Reordering Operator on a GA-Hard Problem, *Biological Cybernetics.*

D. E. Goldberg, K. Deb, & B. Korb. (1990) Messy Genetic Algorithms Revisited: Studies in Mixed Size and Scale, *Complex Systems*, 4(4), 415-444.

J. J. Grefenstette. (1984) *A User's Guide to GENESIS* (Technical Report No. CS-84-11). Nashville: Vanderbilt University, Department of Computer Science.

J. J. Grefenstette. (1986) Optimization of Control Parameters for Genetic Algorithms. *IEEE Transactions on Systems, Man, and Cybernetics, SMC* **16**(1) 122–128.

J. H. Holland. (1975) *Adaptation in Natural and Artificial Systems.* Ann Arbor: The University of Michigan Press.

J. H. Holland. (1989) Searching Nonlinear Functions for High Values. *Applied Mathematics and Computation* (to appear).

G. E. Liepins & M. R. Hilliard. (1989) Genetic Algorithms: Foundations and Applications, *Annals of Operations Research*, **21** 31-58.

G. E. Liepins & M. D. Vose. (1990) Representational Issues in Genetic Optimization, *Journal of Experimental and Theoretical Artificial Intelligence*, **2**(2), 4-30.

W. R. Rudin. (1967) *Fourier Analysis on Groups.* Wiley Interscience.

C. G. Shaefer. (1985) Directed Trees Method for Fitting a Potential Function. *Proceedings of an International Conference on Genetic Algorithms and Their Applications* (pp 74 –88). Hillside, NJ: Lawrence Erlbaum Associates.

C. G. Shaefer. (1987) The ARGOT Strategy: Adaptive Representation Genetic Optimizer Technique. *Genetic Algorithms and Their Applications: Proceedings of the Second International Conference on Genetic Algorithms.* 50–58.

M. D. Vose & G. E. Liepins. (1990) Punctuated Equilibria in Genetic Search, to appear in *Complex Systems.*

M. D. Vose & D. Battle. (1990) *Isomorphisms Of Genetic Algorithms* Workshop on the Foundations of Genetic Algorithms & Classifier Systems. Bloomington Indiana, July 1990.

D. Whitley. (1987) Using Reproductive Evaluation to Improve Genetic Search and Heuristic Discovery. *Genetic Algorithms and Their Applications: Proceedings of the Second International Conference on Genetic Algorithms.* Lawrence Erlbaum Associates.

PART 2

SELECTION AND CONVERGENCE

An Extension To the Theory of Convergence and a Proof of the Time Complexity of Genetic Algorithms

Carol A. Ankenbrandt
Center for Intelligent and Knowledge-based Systems
Department of Computer Science
Tulane University
New Orleans, La. 70118

Abstract

Genetic algorithms, search procedures modelled after the mechanics of natural selection, are difficult to analyze because their time complexity is related to characteristics of the application problem domain. The standard recurrence relation for binary and nonbinary genetic algorithms is solved using an inductive proof and a proof using linear finite difference equations. This result is used to estimate the average and worst case time complexity for genetic algorithms. The resulting equations define the relationship between estimated run time, problem size, fitness ratio and the complexity of the domain specific evaluation function for domains with constant fitness ratio for binary and nonbinary GAs. The fitness ratio is a characteristic of the problem domain reflecting the relative merit of some candidate solutions over others.

Keywords: convergence, complexity, implementation parameters, fitness ratio

1. Introduction and Background

Genetic algorithms (GAs) are search procedures modelled after the mechanics of natural selection and originally developed by John Holland [Holland, 1975]. Candidate solutions are termed organisms and are represented as an ordered lists of values. Each specific value, or allele, represents a partial solution. Induction is performed via an analog of natural selection rather than deduction by a simulated reasoning process. Sets of organisms are grouped into populations. A genetic algorithm evaluates a population based on application dependent criteria and generates a new one iteratively. Each successive population is called a generation. The GA creates an initial generation, G(0), and for each generation G(t), the genetic algorithm generates a new one, G(t+1). An abstract view of a serial genetic algorithm is given in Figure 1.

```
generate initial population, G(0);
evaluate G(0);
t := 1;
repeat
        generate G(t) using G(t-1);
        evaluate G(t);
        t := t + 1;
until solution is found (convergence).
```

Figure 1. An Abstract View of a Serial Population Driven Genetic Algorithm

GAs differ from traditional search techniques in several ways. First, genetic algorithms are designed to optimize the tradeoff between exploring new search points and exploiting the information discovered so far. This was proved using an analogy with the two armed bandit problem [Holland, 1975, pp. 75-88]. Second, GAs have the property of implicit parallelism. Implicit parallelism means that the effect of the genetic algorithm is equivalent to an extensive search of the schemas of the domain search space. Schemas are lower order spaces in the domain search space, denoted as an ordered list of symbols. Each symbol is either a member of the same alphabet used to representation the organisms or a wildcard symbol, *, which matches any value. The order of a schema is a measure of its specificity, the number of symbols in the schema that are not wildcard symbols. For a GA with an optimal population size m, the number of short schemas sampled is on the order of m^3, as proved by [Holland, 1975] and explained in [Goldberg, 1989a, pp. 40-41]. This proof has been modified, to prove that the number of low order schemas sampled is also on the order of m^3, for a GA with an optimal population size m[Ankenbrandt, 1990a]. Third, genetic algorithms are randomized algorithms, as GAs use operators with probabilistic results. The results for any given operation are based on the value of a random number. Fourth, GAs operate on several solutions simultaneously, gathering information from many current points to direct the search. This factor makes GAs less susceptible to the problems of local maxima and noise.

From a mechanistic view, GAs are an iterative process where each iteration has two steps, generate and evaluate. The generate step includes a selection operator and modification operators. In the selection operation, domain information is used to guide the selection of new candidates for following iterations. High quality candidates are given more emphasis in following iterations because they are selected more often. Domain information is used to determine the fitness of a candidate, a measure of its quality. The fitness function, or evaluation function, maps a candidate solution into the nonnegative real numbers. The fitness ratio is the average fitness of candidates with a particular allele value over the average fitness of all other candidates in the generation. The fitness ratio may be more difficult to determine for some domains than others. The concept of the fitness ratio is new, so methods to determine fitness ratios are not well developed. In the modification operation, mixing is performed by crossover operators, which combine candidates selected to form new candidates. The GA converges when there is no change in the candidates from iteration to iteration. At convergence, often all the candidates are exactly alike.

As with many heuristic software algorithms, the problem of characterizing the performance of genetic algorithms in various domains is complex. The performance of a genetic algorithm varies with the application domain as well as with the implementation parameters. Previous researchers have empirically attempted to find reasonable settings for GA implementation parameters over several domains [Schaffer, et. al, 1989, Grefenstette, 1986, De Jong, 1975]. One implementation parameter that is varied in these studies is the population size. A separate implementation parameter is the cardinality of the alphabet chosen to represent the organism. The results presented here extend the theoretical basis for choices of some of these implementation parameters. Specifically, the predicted performance of a GA is shown to be a function of the population size, the cardinality chosen for representation, the complexity of the evaluation function, and the fitness ratio. The results here can be used to guide implementation choices. Some examples of implementation options are increasing population size at the cost of the increased number of generations until convergence, using a higher versus a lower cardinality of representation for organisms, and increasing in the complexity of the evaluation function for faster convergence at the price of the increased resources needed for that more complex function.

The results presented here address a separate problem, identifying which domains are suitable for genetic algorithms. These results may be used as a basis for identifying domain characteristics of problems suitable for GAs by predicting the behavior of GAs in terms of the fitness ratio for that domain and the specified evaluation function. If the domain has a suitable representation, and an associated evaluation function that provide a given fitness ratio, then the GA performance is predicted by the relationship proved in this paper. Note that although the analysis is presented for fitness ratios that do not vary with time, the results here can be applied to domains where the fitness ratio does vary with time. There are several methods possible, one of which is given in [Ankenbrandt, 1990a], where the minimum fitness ratio for the one max problem is used to approximate the total performance of the GA. The minimum fitness ratio was the best predictor of GA performance in that study, since the GA is not converged until it is converged along all allele positions. Fitness ratios can also be measured online, as an indicator of the progress a GA is making toward a solution.

The probabilistic average case complexity and worst case complexity estimates are presented, and proved by induction and by finite difference methods. Genetic algorithms with proportional selection have average and worst case time complexity on the order of (evaluate * m ln m)/ln r, where m is the population size, r is the fitness ratio, and evaluate is the complexity of the domain dependent evaluation function. The average case analysis presented is based on the assumption that all allele values are equally likely in the initial population. The worst case analysis presented assumes only one member of the population has the needed allele value in the initial population. The analysis here is a probabilistic analysis because the formula used as the basis is a derivation of the schema theorem for GAs using proportional selection, as discussed in [Goldberg, 1989b]. Selection methods have been successfully implemented that closely approximate this formula, so this source of error is negligible. Because the basis formula is a derivation of the schema theorem, it can be applied to each individual allele position. This is reasonable in this study, since the focus is not on quantifying implicit parallelism, where crossover disruption factors would be important. Crossover operators disrupt schemas with order of at least two. Since this study is addressing the question of the time it takes a GA to converge, and the GA is converged when each single allele position is converged, crossover disruption is not a factor. For an analysis of convergence using other selection methods, see [Goldberg and Deb, 1990]. Goldberg and Deb state that the results presented here are consistent with their results.

2. The Solution of the Recurrence Relation By Inspection and An Inductive Proof for Binary Genetic Algorithms

Let t denote time, in $\{0,1,2,...\}$. Let $P_{(i)}$ represent the proportion of alleles set to the value one at time t = i for a particular allele position j. Let $P_{(0)}$ represent P at time 0. Let f_1 represent the fitness of all organisms sampled with allele value one in a particular position j. Similarly, let f_0 represent the fitness of all organisms sampled with allele value zero in position j. Let r represent the fitness ratio, where $r = f_1/f_0$. Assume that r is constant over time. Start with the recurrence relation given in equations 1 and 2. Next, the terms for $P_{(1)}$ to $P_{(3)}$ are developed, solving the recurrence relation by inspection.

$$P_{(t+1)} = \frac{f_1 P_{(t)}}{f_1 P_{(t)} + [1 - P_{(t)}] f_0} \tag{1}$$

$$P_{(t+1)} = \frac{r P_{(t+0)}}{1 + (r-1) P_{(t+0)}} \tag{2}$$

$$P_{(1)} = \frac{r P_{(0)}}{1 + (r-1) P_{(0)}} \tag{3}$$

$$P_{(2)} = \frac{r P_{(1)}}{1 + (r-1) P_{(1)}} \tag{4}$$

$$P_{(2)} = \frac{r\left[\dfrac{r\,P_{(0)}}{1 + (r-1)\,P_{(0)}}\right]}{1 + (r-1)\left[\dfrac{r\,P_{(0)}}{1 + (r-1)\,P_{(0)}}\right]} \tag{5}$$

$$P_{(2)} = \frac{r^2 P_{(0)}}{1 + (r - 1)P_{(0)} + (r - 1)rP_{(0)}} \tag{6}$$

$$P_{(2)} = \frac{r^2 P_{(0)}}{1 - P_{(0)} + r^2 P_{(0)}} \tag{7}$$

$$P_{(3)} = \frac{rP_{(2)}}{1 + (r - 1)P_{(2)}} \tag{8}$$

$$P_{(3)} = \frac{r\left[\dfrac{r^2\,P_{(0)}}{1 - P_{(0)} + r^2\,P_{(0)}}\right]}{1 + (r-1)\left[\dfrac{r^2\,P_{(0)}}{1 - P_{(0)} + r^2\,P_{(0)}}\right]} \tag{9}$$

$$P_{(3)} = \frac{r^3 P_{(0)}}{1 - P_{(0)} + r^2 P_{(0)} + (r - 1)\,r^2 P_{(0)}} \tag{10}$$

$$P_{(3)} = \frac{r^3 P_{(0)}}{1 - P_{(0)} + r^3 P_{(0)}} \tag{11}$$

Equation 12 is found by inspection, and proved by induction.

$$P_{(t)} = \frac{r^t P_{(0)}}{(1 - P_{(0)}) + r^t P_{(0)}} \tag{12}$$

2.1. Proof By Induction

1. Test the base case, where t = 0.

$$P_{(t+1)} = \frac{r\,P_{(t+0)}}{1 + (r-1)\,P_{(t+0)}} \tag{13}$$

$$P_{(1)} = \frac{r\,P_{(0)}}{1 + (r-1)\,P_{(0)}} \tag{14}$$

$$P_{(1)} = \frac{r^1 P_{(0)}}{(1 - P_{(0)}) + r^1 P_{(0)}} \tag{15}$$

$$P_{(t)} = \frac{r^t P_{(0)}}{(1 - P_{(0)}) + r^t P_{(0)}} \tag{16}$$

2. Assume $P_{(i)}$, then show $P_{(i+1)}$.

$$P_{(i)} = \frac{r^i P_{(0)}}{(1 - P_{(0)}) + r^i P_{(0)}} \quad ==> \quad P_{(i+1)} = \frac{r^{(i+1)}\,P_{(0)}}{(1 - P_{(0)}) + r^{(i+1)}\,P_{(0)}} \tag{17}$$

$$P_{(i)} = \frac{r^i P_{(0)}}{(1 - P_{(0)}) + r^i P_{(0)}} \tag{18}$$

$$P_{(i+1)} = \frac{r\,P_i}{1 + (r-1)P_{(i)}} \tag{19}$$

$$P_{(i+1)} = \frac{r\left[\dfrac{r^i P_{(0)}}{(1 - P_{(0)}) + r^i P_{(0)}}\right]}{1 + (r-1)\left[\dfrac{r^i P_{(0)}}{(1 - P_{(0)}) + r^i P_{(0)}}\right]} \tag{20}$$

$$P_{(i+1)} = \frac{r^{(i+1)}\,P_{(0)}}{(1 - P_{(0)}) + r^i\,P_{(0)} + (r-1)(r^i\,P_{(0)})} \tag{21}$$

$$P_{(i+1)} = \frac{r^{(i+1)} P_{(0)}}{(1 - P_{(0)}) + (1 + r - 1)\, r^{i} P_{(0)}} \quad (22)$$

$$P_{(i+1)} = \frac{r^{(i+1)} P_{(0)}}{(1 - P_{(0)}) + r^{(i+1)} P_{(0)}} \quad (23)$$

3. Perform induction.

$$P_{(t)} = \frac{r^{t} P_{(0)}}{(1 - P_{(0)}) + r^{t} P_{(0)}} \quad (24)$$

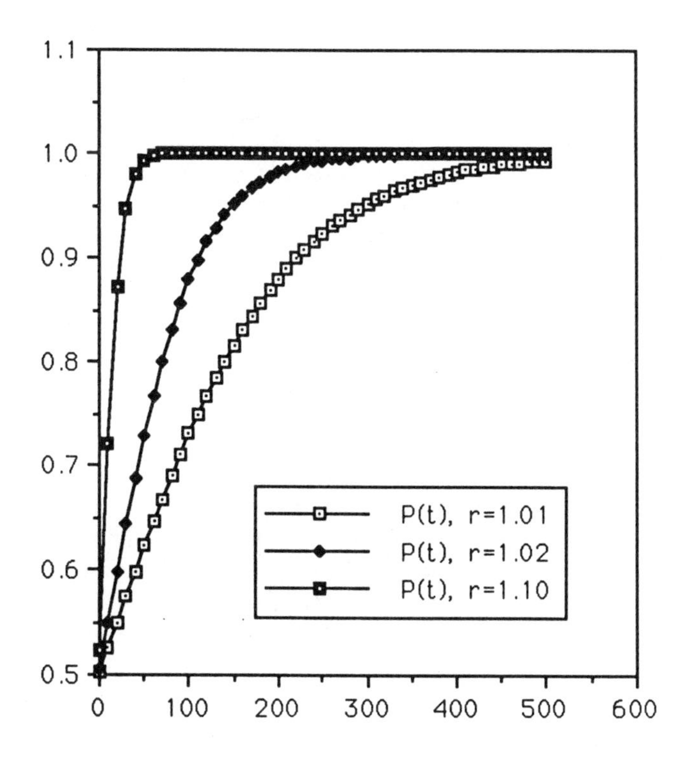

Figure 2. Time, in Generations versus Predicted Proportion of Allele Convergence $P_{(t)}$ for Various Fitness Ratios

The recurrence relation is generalized to accomodate nonbinary genetic algorithms in the following section, so this result holds in the nonbinary case as well. Equation 24 is plotted in Figure 2, showing the predicted proportion of convergence $P_{(t)}$ for several fitness ratios. Figure 2 assumes that the $P_{(0)} = 0.5$, which is reasonable for binary genetic algorithms with a population of reasonable size, or over many runs. (Populations that are very small can only be addressed by probabilistic complexity analysis as presented in this paper if one is concerned with the average GA behavior over many runs. For example, in such pathologically small populations such as those with three members, each organism accounts for one third of the proportion of convergence.)

3. The Recurrence Relation for Convergence for Nonbinary Genetic Algorithms

Consider the recurrence relation governing all genetic algorithms, showing $P_{(t+1)}$ to be a function of $P_{(t)}$, where the function $P_{(t)}$ is the proportion of candidates with a particular allele position set to a specified allele value at time t. For the nonbinary case, the fitness ratio must be redefined. Let the fitness ratio, r, denote the fitness of those organisms with alleles set to a specified allele value for a particular allele position over the fitness of the remaining organisms. For binary GAs, it has been shown that the number of generations to convergence can be estimated by deriving a formula for the derivative, dP/dt [Goldberg,1989b]. As shown in the previous section, an exact solution can be found for $P_{(t)}$ for the binary case, using inductive proof. In this section, the analogous recurrence relation for the formula $P_{(t)}$ for nonbinary GAs is solved directly using methods from linear difference equations. This result is used to derive an improved estimate for the number of generations to convergence.

3.1. The Solution of the Recurrence Relation $P_{(t)}$ for Nonbinary Genetic Algorithms

Let t denote time, in {0,1,2,...}. Let $P_{(i,a)}$ represent the proportion of alleles set to some arbitrary value a at time t = i for a particular allele position j, where the value at position j is defined over any arbitrary alphabet {a,b,...}. Let card(j) denote the cardinality of this alphabet. Let $P_{(0)}$ represent P at time 0. Let f_a represent the fitness of all organisms sampled with allele value a in a particular position j. Similarly, use subscript notation to represent the fitness of all organisms sampled for all other legal alleles in position j. For example, let f_b represent the fitness of all organisms sampled with allele b in position j. In the nonbinary case, consider the analogous recurrence relation given in equation 25.

$$P_{(t+1,a)} = \frac{f_a P_{(t,a)}}{f_a P_{(t,a)} + f_b P_{(t,b)} + f_c P_{(t,c)} + \ldots} \tag{25}$$

Note that the sum of P over all possible allele values in {a,b,...} must equal one, as per equation 26.

$$1 = P_{(t,a)} + P_{(t,b)} + P_{(t,c)} + \ldots \tag{26}$$

$$1 - P_{(t,a)} = P_{(t,b)} + P_{(t,c)} + \ldots \tag{27}$$

Let f_{min} represent the minimum in the set $\{f_b, f_c, \ldots\}$. Let f_{max} represent the maximum in the set $\{f_b, f_c, \ldots\}$. Given that the fitness function is nonnegative, and P is defined in the interval [0,1], then there is some value f in the interval $[f_{min}, f_{max}]$ such that equation 28 holds. Equation 29 follows from substituting this value f into the original recurrence relation in equation 25. Equation 30 follows from substituting the result from equation 27 for the sum in the denominator in equation 29. Let r represent the fitness ratio for the nonbinary case, where $r = f_a/f$. Assume that r is constant over time. Equation 31 follows from dividing equation 30 by f and substituting in the value of the fitness ratio r.

$$f[P_{(t,b)} + P_{(t,c)} + \ldots] = f_b P_{(t,b)} + f_c P_{(t,c)} + \ldots \tag{28}$$

$$P_{(t+1,a)} = \frac{f_a P_{(t,a)}}{f_a P_{(t,a)} + f[P_{(t,b)} + P_{(t,c)} + \ldots]} \tag{29}$$

$$P_{(t+1,a)} = \frac{f_a P_{(t,a)}}{f_a P_{(t,a)} + f[1 - P_{(t,a)}]} \tag{30}$$

$$P_{(t+1,a)} = \frac{r P_{(t,a)}}{1 + (r-1) P_{(t,a)}} \tag{31}$$

This recurrence relation can be solved by inspection and inductive proof, similar to the method used for binary GAs in a previous section of this paper. This recurrence relation can also be solved using a linear difference equation approach, as follows. First, the formula in equation 31 for $P_{(t+1,a)}$ is inverted and a change of variables is introduced, as per equations 33 and 34. This change of variables will allow direct solution, since it will put the equation in a standard form, where the solution to the linear difference equation is found in [Braun, 1983].

$$\frac{1}{P_{(t+1,a)}} = \left[\frac{1}{r}\right] \frac{1}{P_{(t,a)}} + \left[\frac{r-1}{r}\right] \tag{32}$$

$$v_{(t+1)} = \frac{1}{P_{(t+1,a)}}; \quad v_{(t)} = \frac{1}{P_{(t,a)}}; \quad A = \frac{1}{r}; \quad B = \left[\frac{r-1}{r}\right] \tag{33}$$

$$v_{(t+1)} = Av_{(t)} + B \tag{34}$$

A direct solution of this recurrence relation is found in [Braun, 1983], and is sketched here. Let $z_{(0)} = v_{(0)}$ and let equation 35 hold. Let equation 38 introduce the variable C, to simplify equation 37. Letting the subscript notation for C represent time, the simple solution for z is as given in equation 40. Equation 41 follows from substituting for our particular value of C, as given in equation 38, into equation 40, and simplifying.

$$z_{(t)} = \frac{v_{(t)}}{A^{(t)}} \tag{35}$$

$$z_{(t+1)} = \frac{v_{(t+1)}}{A^{(t+1)}} = \frac{Av_{(t)} + B}{A^{(t+1)}} = \frac{v_{(t)}}{A^{(t)}} + \frac{B}{A^{(t+1)}} \tag{36}$$

$$z_{(t+1)} = z_{(t)} + \frac{B}{A^{(t+1)}} \tag{37}$$

$$C = \frac{B}{A^{(t+1)}} \tag{38}$$

$$z_{(t+1)} = z_{(t)} + C \tag{39}$$

$$z_{(t+1)} = z_{(0)} + C_0 + C_1 + \dots + C_t = z_{(0)} + \sum_{m=0}^{(t-1)} C_m \tag{40}$$

$$z_{(t)} = z_{(0)} + \sum_{m=0}^{(t-1)} \frac{B}{A^{m+1}} = v_{(0)} + \frac{B}{A}\left[\frac{1 - \left[\frac{1}{A}\right]^t}{1 - \frac{1}{A}}\right] \tag{41}$$

$$z_{(t)} = v_{(0)} + \frac{B}{A-1}\left[1 - \left[\frac{1}{A}\right]^t\right] \tag{42}$$

Equation 43 follows from equations 42 and 35. Equation 44 follows from substituting from equation 33 into equation 43.

$$v_{(t)} = A^t\left[v_{(0)} + \frac{B}{A-1}\left[1 - \left[\frac{1}{A}\right]^t\right]\right] \tag{43}$$

$$\frac{1}{P_{(t,a)}} = [1/r]^t \left[\frac{1}{P_{(0,a)}} + \frac{\left[\frac{r-1}{r}\right]}{[1/r] - 1} \left[1 - \left[\frac{1}{[1/r]} \right]^t \right] \right] \tag{44}$$

$$\frac{1}{P_{(t,a)}} = [1/r]^t \left[\frac{1}{P_{(0,a)}} + \left[\frac{r-1}{1-r}\right](1 - r^t) \right] = [1/r]^t \left[\frac{1}{P_{(0,a)}} + (r^t - 1) \right] \tag{45}$$

$$P_{(t,a)} = \frac{r^t}{\frac{1}{P_{(0,a)}} + (r^t - 1)} = \frac{P_{(0,a)}\, r^t}{1 + P_{(0,a)}(r^t - 1)} \tag{46}$$

$$P_{(t,a)} = \frac{r^t P_{(0,a)}}{(1 - P_{(0,a)}) + r^t P_{(0,a)}} \tag{47}$$

4. The Solution for the Time to Convergence t_c for Binary and Nonbinary Genetic Algorithms

Let t_c be the value of t when the system reaches convergence. Given $P_{(t,a)}$, t_c can be found. When the system reaches convergence, some allele position j was the last to converge, so if t_c is solved for the slowest individual allele, this is the total t_c for the entire system. To simplify our notation, the allele value subscript a is dropped from P, leaving only the time subscript. Let P_f be the value of P at convergence.

$$P_{(t)} = \frac{r^t P_{(0)}}{(1 - P_{(0)}) + r^t P_{(0)}} \tag{48}$$

$$P_f = \frac{r^{t_c} P_{(0)}}{(1 - P_{(0)}) + r^{t_c} P_{(0)}} \tag{49}$$

$$P_f\left((1 - P_{(0)}) + r^{t_c} P_{(0)} \right) = r^{t_c} P_{(0)} \tag{50}$$

$$P_f (1 - P_{(0)}) = r^{t_c} P_{(0)} (1 - P_f) \tag{51}$$

$$r^{t_c} = \frac{P_f (1 - P_{(0)})}{P_{(0)} (1 - P_f)} \tag{52}$$

Equation 53 follows from assuming r is constant with respect to time.

$$t_c \ln r = \ln \left[\frac{P_f\,(1 - P_{(0)})}{P_{(0)}\,(1 - P_f)}\right] \tag{53}$$

$$t_c = \frac{\ln\left[\dfrac{P_f\,(1 - P_{(0)})}{P_{(0)}\,(1 - P_f)}\right]}{\ln r} \tag{54}$$

4.1. Probabilistic Worst Case Analysis

Any analysis of the behavior of genetic algorithms based on the formula $P_{(t)}$ is a probabilistic case analysis. If all other decisions are made in the most conservative way, this approximates a worst case analysis, as follows. Let n represent the length of the organism. Let m represent the size of the population. Assume no mutation. At least one organism in the population has the allele in consideration set to the value a, if the population is to converge to the value a. So, a minimum value for $P_{(0)}$ is 1/m. Let convergence be defined with a tolerence of γ, where the population is said to have converged to the value a for this allele when $P_{(f)} = 1 - \gamma$.

$$t_c = \frac{\ln\left[\dfrac{(1 - \gamma)\quad(1 - \frac{1}{m})}{\frac{1}{m}(1 - (1 - \gamma))}\right]}{\ln r} \tag{55}$$

$$t_c = \frac{\ln\left[\dfrac{(1 - \gamma)\quad(m - 1)}{\gamma}\right]}{\ln r} \tag{56}$$

A conservative estimate for γ is 1/m. Equation 57 follows from assuming $\gamma = 1/m$.

$$t_c = \frac{\ln\left[\dfrac{(1 - \frac{1}{m})(m - 1)}{\frac{1}{m}}\right]}{\ln r} \tag{57}$$

$$t_c = \frac{\ln\left((m - 1)^2\right)}{\ln r} \tag{58}$$

In the worst case, GA theory estimates m, the population size, to be exponential relative to the length of the organism, where the length of the organism is related to the domain size. For example, in a binary domain of size 2^x, the organism length is usually x. Typically, as GAs are currently implemented, the population size m is set to a constant value between 25 and 100. In either case, our estimate for the complexity of the main driving loop of the entire genetic algorithm is still of low order for domains with moderate fitness ratios r. The intuitive meaning of the fitness ratio r is that r is a measure of the relative difference in quality that alternate solutions pose for that domain.

The main driving loop of the entire genetic algorithm contains a domain dependent procedure, termed evaluate in our abstract representation of a genetic algorithm. Let the term order(evaluate) represent the time complexity of this domain dependent procedure. The other procedures within the main loop have time complexity estimates on the order of the population size m. Then, the worst case complexity of the entire genetic algorithm is of order (order(evaluate) * m ln m^2)/ln r.

4.2. Probabilistic Average Case Analysis for Binary Genetic Algorithms

Let n represent the length of the organism. Let m represent the size of the population. Assume no mutation. On the average, half the organisms in the population have the allele in consideration set to one, if the population is to converge to one for that allele. So, an average value for $P_{(0)}$ is 0.5. Let convergence be defined with a tolerence of γ, where the population is said to have converged to one for this allele when $P_{(f)} = 1 - \gamma$.

$$t_c = \frac{\ln\left[\dfrac{(1-\gamma)\quad(1-0.5)}{0.5\,(1-(1-\gamma))}\right]}{\ln r} \tag{59}$$

$$t_c = \frac{\ln\left[\dfrac{(1-\gamma)}{\gamma}\right]}{\ln r} \tag{60}$$

A conservative estimate for γ is 1/m. Equation 61 follows from assuming that $\gamma = 1/m$.

$$t_c = \frac{\ln\left[\dfrac{(1-\frac{1}{m})}{\frac{1}{m}}\right]}{\ln r} \tag{61}$$

$$t_c = \frac{\ln\,(m-1)}{\ln r} \tag{62}$$

The complexity of the entire genetic algorithm in the average case for binary domains is of (order(evaluate) * m ln m)/ln r.

4.3. Probabilistic Average Case Analysis for Nonbinary Genetic Genetic Algorithms

Let n represent the length of the organism. Let m represent the size of the population. Assume no mutation. On the average, if all allele values have an equal probability, 1/card(j) of the the organisms in the population have the allele j set to the value a, if the population is to converge to the value a for that allele position. So, an conservative estimate for the average value for $P_{(0,a)}$ is 1/card(j). Let convergence be defined with a tolerence of γ, where the population is said to have converged to one for this allele when $P_{(f)} = 1 - \gamma$. A conservative estimate for γ is 1/m. In equation 65, let $\gamma = 1/m$. The time complexity of the entire genetic algorithm in the average case is of the order (order(evaluate) * m ln (card (j) * m))/ln r.

$$t_c = \frac{\ln\left[\dfrac{(1-\gamma)\left[1-\dfrac{1}{\mathrm{card}(j)}\right]}{\dfrac{1}{\mathrm{card}(j)}\,(1-(1-\gamma))}\right]}{\ln r} \tag{63}$$

$$t_c = \frac{\ln\left[\dfrac{(1-\gamma)(\mathrm{card}(j)-1)}{\gamma}\right]}{\ln r} \tag{64}$$

$$t_c = \frac{\ln\left[\dfrac{\left[1-\dfrac{1}{m}\right](\mathrm{card}(j)-1)}{\dfrac{1}{m}}\right]}{\ln r} \tag{65}$$

$$t_c = \frac{\ln\left[(m-1)(\mathrm{card}(j)-1)\right]}{\ln r} \tag{66}$$

5. Examples and Extensions

Recall

$$t_c = \frac{\ln\left[\dfrac{P_{(f)}\,(1-P_{(0)})}{P_{(0)}\,(1-P_{(f)})}\right]}{\ln r} \tag{67}$$

For a problem with a population size of 20, $P_{(f)} = 0.95$ and $P_{(0)} = 0.05$ and $r = 1.1$, using the above equation we find t_c is approximately 62. This result can also be found by plotting $P_{(t)}$ as in Figure 2, and reading the value t_c directly from the graph. This analysis is useful when the fitness ratio, r, can be estimated with a bound. For a complete example, which calculates the the predicted time to convergence using the results in this study and compares this with the actual time to convergence for the one max problem, see [Ankenbrandt, 1990a].

In a related work, a formal approach to validating the use of a particular crossover operator is developed, called the theory of recombination operators, and is demonstrated by formally validating the uniform crossover operator [Ankenbrandt, 1991, Ankenbrandt, 1990a]. To show that a recombination operator is theoretically sound, three conditions have to be met. First, the operator must process enough schemas so that implicit parallelism is preserved. This is normally not a problem. Second, a disruption factor for the schema theorem must be given. Third, abstract qualities of the operator must be identified, such as positional bias and the amount and distribution of information that is inherited from the parents. The work presented here is related because the formulas developed can be used in the disruption factors in the schema theorem. For example, the disruption factor for the uniform crossover operator can be given as a function of the proportion of convergence [Ankenbrandt, 1991, Ankenbrandt, 1990a]. These disruption factors are no longer a function of the defining length of the schemas, but rather a function of the order of the schemas. This supports the theoretical development of GAs using low order schemas as building blocks, rather than focussing exclusively on short defining length building blocks.

6. Conclusions

These results extend the theoretical basis for genetic algorithms, providing an estimate for the run time complexity of a genetic algorithm in terms of the problem size, the cardinality of the representation, the complexity of the evaluation function chosen, and the fitness ratio. Probabilistic average case complexity and worst case complexity estimates are found, and proved by induction as well as finite difference methods. Genetic algorithms with proportional selection have an average and a worst case time complexity on the order of (evaluate * m ln m)/ln r, where m is the population size, r is the fitness ratio, and evaluate is the complexity of the domain dependent evaluation function. These results can be used to predict GA performance, and to choose between implementation options, such as varying population size, selecting the cardinality of representation, and selecting between various domain dependent evaluation functions. A separate problem this result addresses is the problem of identifying which domains are suitable for genetic algorithms. Our result may be used as a basis for identifying domain characteristics of problems suitable for GAs by predicting the behavior of GAs in those domains in terms of the fitness ratio for that domain. These results can also be used in disruption factors for crossover operators in the schema theorem, as seen in [Ankenbrandt, 1991, Ankenbrandt, 1990a]. Extensions to this work include refining the probabilistic estimates given for the time complexity, and developing fitness ratio estimates across various domains.

References

Ankenbrandt, C.A. (1990a). The Time Complexity of Genetic Algorithms and the Theory of Recombination Operators. Doctoral Dissertation, Department of Computer Science, Tulane University, New Orleans, LA.

Ankenbrandt, C.A., B. P. Buckles, and F. E. Petry. (1990b). Scene Recognition Using Genetic Algorithms With Semantic Nets. Pattern Recognition Letters (11), pp. 285-293.

Ankenbrandt, C. A., B. P. Buckles, F. E. Petry, and M. Lybanon. (1989). Ocean Feature Recognition Using Genetic Algorithms With Fuzzy Fitness Functions (GA/F3), in Third Annual Workshop on Space Operations Automation and Robotics (SOAR'89), pp. 679-685, July 25-27, 1989.

Ankenbrandt, C. A. (1991). The Theory of Recombination Operators as Demonstrated with the Uniform Crossover Operator. In progress.

Braun, M. (1983) Differential Equations and Their Applications, 3rd ed, pp. 91 - 92, New York, NY: Springer-Verlag.

De Jong, K. E. (1975) Analysis of the Behavior of a Class of Genetic Adaptive Systems, Doctoral Dissertation, Department of Computer and Communication Sciences, University of Michigan, Ann Arbor, MI.

Goldberg, D. E. (1989a). Genetic Algorithms in Search, Optimization, and Machine Learning. Reading, MA: Addison-Wesley.

Goldberg, D. E. (1989b). Sizing Populations for Serial and Parallel Genetic Algorithms. Proceedings of the Third International Conference on Genetic Algorithms, pp. 70-79. San Mateo, CA: Morgan Kaufman.

Goldberg, D. E. and K. Deb. (1990). A Comparative Analysis of Selection Schemes Used in Genetic Algorithms. The Clearinghouse for Genetic Algorithms Tech Report No. 90007, The University of Alabama Department of Engineering Mechanics, Tuscaloosa, AL.

Grefenstette, J. J. (1986). Optimization of Control Parameters for Genetic Algorithms, IEEE Transactions on Systems, Man & Cybernetics SMC-16, 1, pp. 122-128.

Holland, J. H. (1975). Adaption in Natural and Artificial Systems. Ann Arbor, MI: University of Michigan Press.

Schaffer, J. D., R. A. Caruana, L. J. Eshelman, and R. Das. (1989). A Study of Control Parameters Affecting Online Performance of Genetic Algorithms for Function Optimization. Proceedings of the Third International Conference on Genetic Algorithms, pp. 51-60. San Mateo, CA: Morgan Kaufman.

A Comparative Analysis of Selection Schemes Used in Genetic Algorithms

David E. Goldberg and Kalyanmoy Deb
Department of General Engineering
University of Illinois at Urbana-Champaign
117 Transportation Building
104 South Mathews
Urbana, IL 61801-2996

Abstract

This paper considers a number of selection schemes commonly used in modern genetic algorithms. Specifically, proportionate reproduction, ranking selection, tournament selection, and Genitor (or "steady state") selection are compared on the basis of solutions to deterministic difference or differential equations, which are verified through computer simulations. The analysis provides convenient approximate or exact solutions as well as useful convergence time and growth ratio estimates. The paper recommends practical application of the analyses and suggests a number of paths for more detailed analytical investigation of selection techniques.

Keywords: proportionate selection, ranking selection, tournament selection, Genitor, takeover time, time complexity, growth ratio.

1 Introduction

Many claims and counterclaims have been lodged regarding the superiority of this or that selection scheme in genetic algorithms (GAs), but most of these are based on limited (and uncontrolled) simulation experience; surprisingly little analysis has been performed to understand relative expected fitness ratios, convergence times, or the functional forms of selective convergence. This paper seeks to partially alleviate this dearth of quantitative information by comparing the expected performance of four commonly used selection schemes:

1. proportionate reproduction;
2. ranking selection;
3. tournament selection;
4. Genitor (or "steady state") selection.

Specifically, deterministic finite difference equations are written that describe the change in proportion of different classes of individual, assuming fixed and identical objective function values within each class. These equations are solved explicitly or approximated in time using integrable ordinary differential equations. These solutions are shown to agree well with computer simulations, and linear ranking (Baker, 1985) and binary tournament selection (Brindle, 1981) are shown to give identical performance in expectation. Moreover, ranking and tournament selection are shown to maintain strong growth under normal conditions, while proportionate selection without scaling is shown to be less effective in keeping a steady pressure toward convergence. Whitley's (1989) Genitor or "steady state" (Syswerda, 1989) selection mechanism is also examined and found to be a simple combination of block death and birth via ranking. Analysis of this overlapping population scheme shows that the convergence results observed by Whitley may be most easily explained by the unusually high growth ratio Genitor achieves as compared to other schemes on a generational basis. The analysis also suggests that the premature convergence caused by imposing such high growth ratios is one of the reasons Genitor requires other fixes such as large population sizes or multiple populations.

In the remainder, the fundamental equation of population dynamics—the birth, life, and death equation—is described, and specific equations are written, solved, and compared to simulation results for each of the selection schemes. Different schemes are then compared and contrasted, and the use of these analyses in practical implementations is discussed. The paper concludes by briefly recommending more detailed stochastic analyses and another look at what the k-armed bandit problem can teach us about selection.

2 Selection: A Matter of Birth, Life, and Death

The derivation of Holland's (1975) schema theorem starts by calculating the expected number of copies of a schema under selection alone. The calculation is essentially a continuity or conservation of individuals' relationships, where the sources and sinks of a particular class of individual are all accounted. Under selection alone, individuals can only do one of three things: they may be born, they may live, or they may die. If we consider these events to be moved along synchronously, time step by time step, the following general *birth, life*, and *death* equation may be written:

$$m_{i,t+1} = m_{i,t} + m_{i,t,b} - m_{i,t,d}, \tag{1}$$

where m is the number of individuals, the subscript i identifies the class of individual with common objective function value f_i, the subscript t is a time index (individual or generational), the subscript b signifies individuals being born, the subscript d signifies dying individuals, and the lack of a b or a d subscript signifies living individuals. In the usual nonoverlapping population model, the number of individuals dying in a generation is assumed to equal the number of living individuals, $m_{i,t,d} = m_{i,t}$, and the whole matter hinges around the number of births: $m_{i,t+1} = m_{i,t,b}$. Careful consideration of birth, life, and death will become more important when we analyze an overlapping population model.

The analysis may also be performed by calculating the expected proportions $P_{i,t+1}$ rather than absolute numbers $m_{i,t+1}$:

$$P_{i,t+1} = P_{i,t} + P_{i,t,b} - P_{i,t,d}, \tag{2}$$

where the proportion P is obtained by dividing the class count m by the total number of individuals in the population at that time. Here the subscripts b and d are used as before to denote birth and death respectively.

In the sections that follow, specific equations are written and solved for each of the selection schemes mentioned above.

3 Proportionate Reproduction

The name proportionate reproduction describes a group of selection schemes that choose individuals for birth according to their objective function values f. In these schemes, the probability of selection p of an individual from the ith class in the tth generation is calculated as

$$p_{i,t} = \frac{f_i}{\sum_{j=1}^{k} m_{j,t} f_j}, \tag{3}$$

where k classes exist and the total number of individuals sums to n. Various methods have been suggested for sampling this probability distribution, including Monte Carlo or roulette wheel selection (De Jong, 1975), stochastic remainder selection (Booker, 1982; Brindle, 1981), and stochastic universal selection (Baker, 1987; Grefenstette & Baker, 1989). As we are uninterested here in stochastic differences, these schemes receive identical analytical treatment when we calculate their expected performance.

If we consider a nonoverlapping population of constant size n and assume that n selections are made each generation according to the distribution of equation 3, it is a straightforward matter to calculate the expected number of copies of the ith class in the next generation:

$$\begin{aligned} m_{i,t+1} &= m_{i,t} \cdot n \cdot p_{i,t}; \\ &= m_{i,t} \frac{f_i}{\bar{f}_t}, \end{aligned} \tag{4}$$

where $\bar{f}_t = \sum m_{i,t} f_i / n$ is the average function value of the current generation. This equation may be written in proportion form by dividing by the population size:

$$P_{i,t+1} = P_{i,t} \frac{f_i}{\bar{f}_t}. \tag{5}$$

This equation is solved explicitly in the next section.

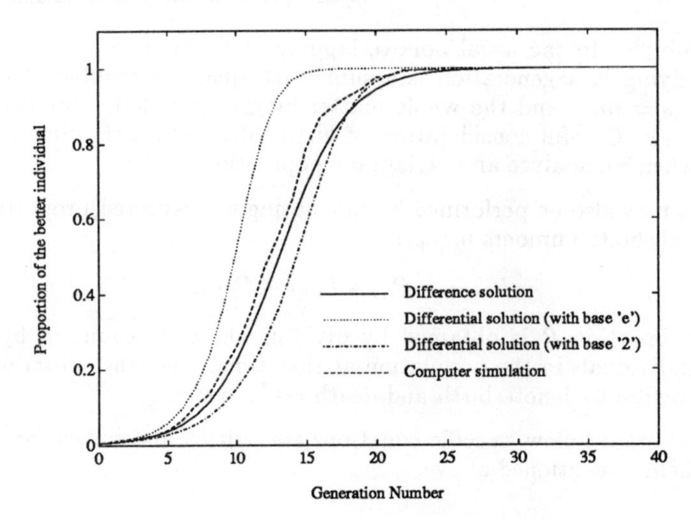

Figure 1: Various differential and difference equation solutions agree well with a representative computer simulation of proportionate reproduction using stochastic universal selection with two alternatives ($r = f_1/f_2 = 1.5$).

3.1 Solving the proportionate reproduction equation

The proportionate reproduction equation (equation 5) may be solved quite directly after an interesting fact is noted. Imagine that a population of individuals grows according to the uncoupled, exponential growth equations: $m_{i,t+1} = m_{i,t}f_i$. If the growth of the proportion of individuals is calculated by dividing through by the total population size at generation $t+1$, we note that the resulting equation for the proportions is identical to that used under the assumption of a fixed population size:

$$P_{i,t+1} = \frac{f_i m_{i,t}}{\sum_j m_{j,t+1}} = \frac{f_i m_{i,t}}{\sum_j f_j m_{j,t}} = \frac{f_i P_{i,t}}{\sum_j f_j P_{j,t}} \tag{6}$$

Since the uncoupled equations may be solved directly ($m_{i,t} = f_i^t m_{i,0}$) the implied proportion equations can also be solved without regard for coupling. Substituting the expression for m at generation t and dividing numerator and denominator through by the total population size at that time yields the exquisitely simple solution:

$$P_{i,t} = \frac{f_i^t P_{i,0}}{\sum_j f_j^t P_{j,0}}. \tag{7}$$

The solution at some future generation goes as the computation over a single generation except that power functions of the objective function values are used instead of the function values themselves. This solution agrees with Ankenbrandt's (1990) solution for $k = 2$, but the derivation of above is more direct and applies to k alternatives without approximation.

In a previous paper (Goldberg, 1989b), the solution to a differential equation approximation of equation 5 was developed for the two-alternative case. That solution, a solution of the same functional form using powers of 2 instead of e, the solution of equation 7, and a representative computer simulation are compared in figure 1. A population size of $n = 200$ and fitness ratio $r = f_1/f_0 = 1.5$ are used, and the simulation and solutions are initiated with a single copy of the better individual. The exact difference equation solution, the approximate solution with powers of 2, and the simulation result agree quite well; the approximate solution with powers of e converges too quickly, although all solutions are logistic as expected.

The exact solution (equation 7) may be approximated in space by treating the alternatives as though they existed over a one-dimensional continuum x, relating positions in space to objective function values with a function $f(x)$.[1] Thus, we may solve for the proportion $P_{I,t}$ of individuals between specified x values $I = \{x : a \leq x < b\}$ at time t as follows:

$$P_{I,t} = \frac{\int_a^b f^t(x) p_0(x) dx}{\int_{-\infty}^{\infty} f^t(x) p_0(x) dx}, \tag{8}$$

where $p_0(x)$ is an appropriate initial density function.

3.2 Two cases: a monomial and an exponential

In general, equation 8 is difficult to integrate analytically, but several special cases are accessible. Limiting consideration to the unit interval and restricting the density function to be uniformly random yields $p_0(x) = 1$. Thus, if $\int f^t(x)dx$ can be integrated analytically a time-varying expression for the proportion may be obtained. We consider two cases, $f(x) = x^c$ and $f(x) = e^{cx}$.

Consider the monomial first. Under the previous assumptions, equation 8 may be integrated with $f(x) = x^c$ and upper and lower limits of x and $x - 1/n$:

$$P_{I,t} = x^{ct+1} - (x - 1/n)^{ct+1}. \tag{9}$$

These limits parameterize individual classes on the variable x, where $x = 1$ is the best individual and $x = 0$ is the worst, thereby permitting an approximation of the growth of an individual with specified rank in a population of size n. This space-continuous solution, the exact solution to the difference equation, and a representative computer simulation are compared in figure 2 for the linear objective function $f(x) = x$. The simulation and the discrete solution use $n = k = 256$ alternatives with one of each alternative at the start. It is interesting to note that the solutions to the difference equation and its space-continuous approximation are virtually identical, and both compare well to the representative simulation shown in the figure.

This analysis may be used to calculate the takeover time for the best individual. Setting $x = 1$ in equation 9, yields a space-continuous solution for the growth of

[1] The ordering of the f values is unimportant in this analysis. In what follows, a number of monotonically increasing functions are considered, and these may be viewed as representative of many other objective functions with similar image densities. Alternatively they may be viewed as scaling functions used on functions of relatively uniform image density: $g(f(x))$ with $f(x)$ linear.

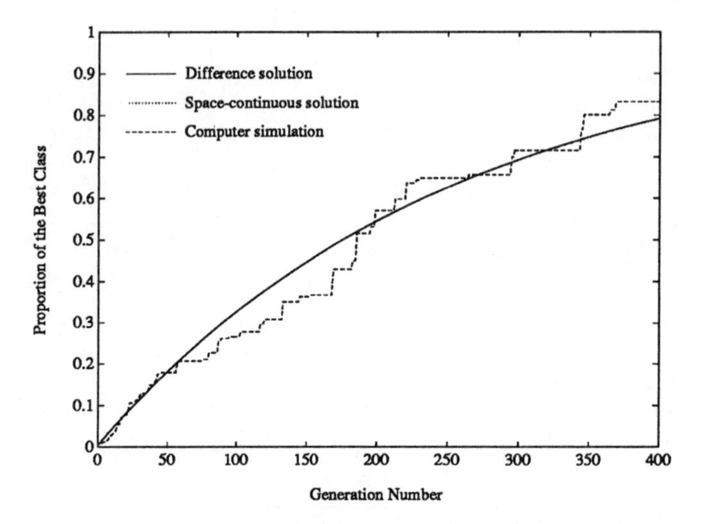

Figure 2: A comparison of the discrete difference equation solution, the approximate continuous solution, and a representative simulation of SUS proportionate reproduction for the function $f(x) = x$ shows substantial agreement between simulation and either model. The exact solution to the difference equation and the space-continuous solution are virtually identical.

the best class:

$$P_t^* = 1 - \left(\frac{n-1}{n}\right)^{ct+1} \tag{10}$$

Setting this proportion equal to $\frac{n-1}{n}$, we calculate the time when the population contains $n-1$ best individuals, the takeover time t^*:

$$ct^* + 1 = \frac{1}{1 - \frac{\log(n-1)}{\log n}}. \tag{11}$$

As the exponent on the monomial increases, the takeover time decreases correspondingly. This helps explain why a number of investigators have adopted polynomial scaling procedures to help speed GA convergence (Goldberg, 1989a). The expression $\frac{1}{1-\frac{\log(n-1)}{\log n}}$ may be simplified at large n. Expanding $\log(n-1)$ in a Taylor series about the value n, keeping the first two terms, and substituting into the expression yields $\frac{1}{1-\frac{\log(n-1)}{\log n}} \approx n \ln n$, the approximation improving with increasing n. Using this approximation, we obtain the takeover time approximation

$$t^* = \frac{1}{c}(n \ln n - 1) \tag{12}$$

Thus, the takeover time for a polynomially distributed objective function is $O(n \log n)$. It is interesting to compare this takeover time to that for an exponentially distributed (or exponentially scaled) function.

An exponential objective function may be considered similarly. Under the previous assumptions, equation 8 may be integrated using $f(x) = e^{cx}$ and the same limits of integration as before:

$$P_{I,t} = \frac{e^{cxt}(1 - e^{-ct/n})}{e^{ct} - 1}. \tag{13}$$

Considering the best group (setting $x = 1$) and solving for the takeover time (the time when the proportion of the best group equals $\frac{n-1}{n}$) yields the approximate equation as follows:

$$t^* = \frac{1}{c} n \ln n. \tag{14}$$

It is interesting that under the unit interval consideration, both a polynomially distributed function and an exponentially distributed function have the same computational speed of convergence.

3.3 Time complexity of proportionate reproduction

The previous estimates give some indication of how long a GA will continue until it converges substantially. Here, we consider the time complexity of the selection algorithm itself per generation. We should caution that it is possible to place too much emphasis on the efficiency of implementation of a set of genetic operators. After all, in most interesting problems the time to evaluate the function is much greater than the time to iterate the genetics, $t_f \gg t_{ga}$, and fiddling with operator time savings is unlikely to pay off. Nonetheless, if a more efficient operator can be used without much bother, why not do so?

Proportionate reproduction can be implemented in a number of ways. The simplest implementation (and one of the earliest to be used) is to simulate the spin of a weighted roulette wheel (Goldberg, 1989a). If the search for the location of the chosen slot is performed via linear search from the beginning of the list, each selection requires $O(n)$ steps, because on average half the list will be searched. Overall, roulette wheel selection performed in this method requires $O(n^2)$ steps, because in a generation n spins are required to fill the population. Roulette wheel selection can be hurried somewhat, if a binary search (like the bisection method in numerical methods) is used to locate the correct slot. This requires additional memory locations and an $O(n)$ sweep through the list to calculate cumulative slot totals, but overall the complexity reduces to $O(n \log n)$, because binary search requires $O(\log n)$ steps per spin and n spins.

Proportionate reproduction can also be performed by stochastic remainder selection. Here the expected number of copies of a string is calculated as $m_i = \frac{f_i}{\bar{f}}$, and the integer portions of the count are assigned deterministically. The remainders are then used probabilistically, to fill the population. If done without replacement, each remainder is used to bias the flip of a coin that determines whether the structure receives another copy or not. If done with replacement, the remainders are used to size the slots of a roulette wheel selection process. The algorithm without replacement is $O(n)$, because the deterministic assignment requires only a single pass (after the calculation of $\bar{f}$, which is also $O(n)$), and the probabilistic assignment is likely to terminate in $O(1)$ steps. On the other hand, the algorithm when performed with replacement takes on the complexity of the roulette wheel, because $O(n)$ of the individuals are likely to have fractional parts to their m values.

Stochastic universal selection is performed by sizing the slots of a weighted roulette wheel, placing equally spaced markers along the outside of the wheel, and spinning the wheel once; the number of copies an individual receives is then calculated by counting the number of markers that fall in its slot. The algorithm is $O(n)$, because only a single pass is needed through the list after the sum of the function values is calculated.

4 Ranking Selection

Baker (1985) introduced the notion of ranking selection to genetic algorithm practice. The idea is straightforward. Sort the population from best to worst, assign the number of copies that each individual should receive according to a non-increasing assignment function, and then perform proportionate selection according to that assignment. Some qualitative theory regarding such schemes was presented by Grefenstette and Baker (1989), but this theory provides no help in evaluating expected performance. Here we analyze the performance of ranking selection schemes somewhat more quantitatively. A framework for analysis is developed by defining assignment functions and these are used to obtain difference equations for various ranking schemes. Simulations and various difference and differential solutions are then compared.

4.1 Assignment functions: a framework for the analysis of ranking

For some ranking scheme, we assume that an assignment function α has been devised that satisfies three conditions:

1. $\alpha(x) \in R$ for $x \in [0,1]$.
2. $\alpha(x) \geq 0$.
3. $\int_0^1 \alpha(\eta) d\eta = 1$.

Intuitively, the product $\alpha(x)dx$ may be thought of as the proportion of individuals assigned to the proportion dx of individuals who are currently ranked a fraction x below the individual with best function value (here $x = 0$ will be the best and $x = 1$ will be the worst to connect with Baker's formulation, even though this convention is the opposite of the practice adopted in section 3).

With this definition, the cumulative assignment function β may be defined as the integral of the assignment from the best ($x = 0$) to a fraction x of the current population:

$$\beta(x) = \int_0^x \alpha(\eta) d\eta. \tag{15}$$

Analyzing the effect of ranking selection is now straightforward. Let P_i be the proportion of individuals who have function value better than or equal to f_i and let Q_i be the proportion of individuals who have function value worse than that same value. By the definitions above, the proportion of individuals assigned to the proportion $P_{i,t}$ in the next generation is simply the cumulative assignment value of the current proportion:

$$P_{i,t+1} = \beta(P_{i,t}). \tag{16}$$

The complementary proportion may be evaluated as well:

$$Q_{i,t+1} = 1 - \beta(P_{i,t}) = 1 - \beta(1 - Q_{i,t}). \tag{17}$$

In either case, the forward proportion is only a function of the current value and has no relation to the proportion of other population classes. This contrasts strongly to proportionate reproduction, where the forward proportion is strongly influenced by the current balance of proportions and the distribution of the objective function itself. This difference is one of the attractions of ranking methods in that an even, controllable pressure can be maintained to push for the selection of better individuals. Analytically, the independence of forward proportion makes it possible to calculate the growth or decline of individuals whose objective function values form a convex set. For example, if P_1 represents the proportion of individuals with function value greater than or equal to f_1 and Q_2 represents the proportion of individuals with proportion less than f_2, the quantity $1 - P_1 - Q_2$ is the proportion of individuals with function value between f_1 and f_2.

4.2 Linear assignment and ranking

The most common form of assignment function is linear: $\alpha(x) = c_0 - c_1 x$. Requiring a non-negative function with non-increasing values dictates that both coefficients be greater than zero and that $c_0 \geq c_1$. Furthermore, the integral condition requires that $c_1 = 2(c_0 - 1)$. Integrating α yields $\beta(x) = c_0 x - (c_0 - 1)x^2$. Substituting the cumulative assignment function into equation 16 yields the difference equation

$$P_{i,t+1} = P_{i,t}\left[c_0 - (c_0 - 1)P_{i,t}\right]. \tag{18}$$

The equation is the well known logistic difference equation; however, the restrictions on the parameters preclude any of its infamous chaotic behavior, and its solution must stably approach the fixed point $P_i = 1$ as time goes on.

In general, equation 18 has no convenient analytical solution (other than that obtained by iterating the equation), but in one special case a simplified solution can be derived. When $c_0 = c_1 = 2$, the complementary equation simplifies as follows:

$$Q_{i,t+1} = 1 - \beta(1 - Q_{i,t}) = Q_{i,t}^2. \tag{19}$$

Solving for Q at generation t yields the following:

$$Q_{i,t} = Q_{i,0}^{2^t}. \tag{20}$$

Since $Q = 1 - P$, the solution for P may be obtained directly as

$$P_{i,t} = 1 - (1 - P_{i,0})^{2^t}. \tag{21}$$

Calculating the takeover time by substituting initial and final proportions of $\frac{1}{n}$ and $\frac{n-1}{n}$ respectively and simplifying yields the approximate equation $t^* = \log n + \log(\ln n)$, where log is taken base 2 and ln is the usual natural logarithm.

Other cases of linear ranking may be evaluated by turning to the type of differential equation analysis used elsewhere (Goldberg, 1989b). Approximating the finite difference by its derivative in one step yields the logistic differential equation

$$\frac{dP_i}{dt} = cP_i(1 - P_i), \tag{22}$$

where $c = c_0 - 1$. Solving by elementary means, we obtain the solution

$$P_{i,t} = \frac{1}{1 + \frac{1-P_{i,0}}{P_{i,0}} e^{-ct}}. \tag{23}$$

The solution overpredicts proportion early on, because of the error made by approximating the difference by the derivative. This error can be corrected approximately by using 2 in place of e in equation 23. In either approximation the takeover time may be calculated in a straightforward manner:

$$t^* = \frac{2}{c} \log(n - 1), \tag{24}$$

where the logarithm should be taken base e in the case of the first approximation and base 2 in the case of the second.

The two differential equation solutions, the exact solution to the difference equation, and a representative simulation using stochastic universal selection are shown in figure 3 for the case of linear ranking with $c_0 = c_1 = 2$. Here a population of size $n = 256$ is started with a single copy of the best individual. The difference equation solution and the simulation are very close to one another as expected. The differential equation approximations have the correct qualitative behavior, but the solution using e converges too rapidly, and the solution using 2 agrees well early on but takes too long once the best constituents become a significant proportion of the total population.

4.3 Time complexity of ranking procedures

Ranking is a two-step process. First the list of individuals must be sorted, and next the assignment values must be used in some form of proportionate selection. The calculation of the time complexity of ranking requires the consideration of these separate steps.

Sorting can be performed in $O(n \log n)$ steps, using standard techniques. Thereafter, we know from previous results that proportionate selection can be performed in something between $O(n)$ and $O(n^2)$. Here, we will assume that a method no worse than $O(n \log n)$ is adopted, concluding that ranking has time complexity $O(n \log n)$.

5 Tournament Selection

A form of tournament selection attributed to unpublished work by Wetzel was studied in Brindle's (1981) dissertation, and more recent studies using tournament schemes are found in a number of works (Goldberg, Korb, & Deb, 1989; Muhlenbein, 1990; Suh & Van Gucht, 1987). The idea is simple. Choose some number of individuals randomly from a population (with or without replacement), select the

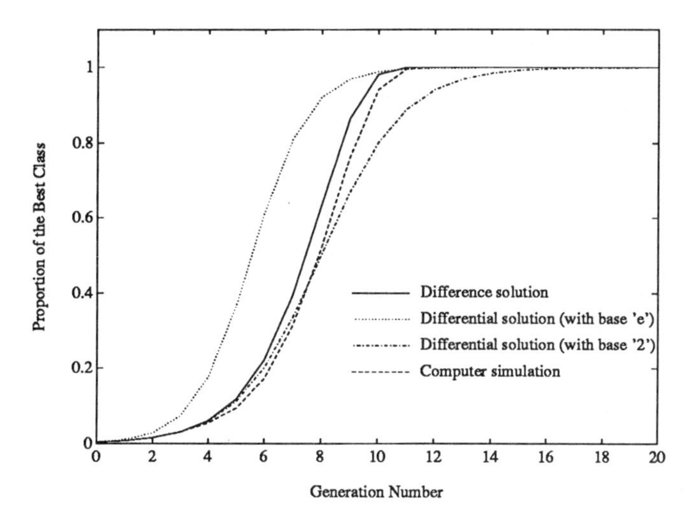

Figure 3: The proportion of individuals with best objective function value grows as a logistic function of generation under ranking selection. A representative simulation using linear ranking and stochastic universal selection agrees well with the exact difference equation solution ($c_0 = c_1 = 2$). The differential equation approximations are too rapid or too slow depending upon whether exponentiation is performed base e or base 2.

best individual from this group for further genetic processing, and repeat as often as desired (usually until the mating pool is filled). Tournaments are often held between pairs of individuals (tournament size $s = 2$), although larger tournaments can be used and may be analyzed. We start our analysis by considering the binary case and later extend the analysis to general s-ary tournaments.

5.1 Binary tournaments: $s = 2$

Here we analyze the effect of a probabilistic form of binary tournament selection.[2] In this variant, two individuals are chosen at random and the better of the two individuals is selected with fixed probability p, $0.5 < p \leq 1$. Using the notation of section 4, we may calculate the proportion of individuals with function value better than or equal to f_i, the proportion at the next generation quite simply:

$$P_{i,t+1} = p[2P_{i,t}(1 - P_{i,t}) + P_{i,t}^2] + (1 - p)P_{i,t}^2. \tag{25}$$

Collecting terms and simplifying yields the following:

$$P_{i,t+1} = 2pP_{i,t} - (2p - 1)P_{i,t}^2. \tag{26}$$

[2]The probabilistic variation was brought to our attention by Donald R. Jones (personal communication, April 20, 1990) at General Motors Research Laboratory. We analyze this variant, because the deterministic version is a special case and because the probabilistic version can be made to agree in expectation with ranking selection regardless of c_0.

Letting $2p = c_0$ and comparing to equation 18, we note that the two equations are identical. This is quite remarkable and says that binary tournament selection and linear ranking selection are identical in expectation. The solutions of the previous section all carry forward to the case of binary tournament selection as long as the coefficients are interpreted properly ($2p = c_0$ and $c = 2p - 1$).

Simulations of tournament selection agree well with the appropriate difference and differential equation solutions, but we do not examine these results here, because the analytical models are identical to those used for linear ranking, and the tournament selection simulation results are very similar to those presented for linear ranking. Instead, we consider the effect of using larger tournaments.

5.2 Larger tournaments

To analyze the performance of tournament selection with any size tournament, it is easier to consider the doughnut hole (the complementary proportion) rather than the doughnut itself (the primary proportion). Considering a deterministic tournament[3] of size s and focusing on the complementary proportion Q_i, a single copy will be made of an individual in this class only when all s individuals in a competition are selected from this same lowly group:

$$Q_{i,t+1} = Q_{i,t}^{s}, \tag{27}$$

from which the solution follows directly:

$$Q_{i,t} = Q_{i,0}^{s^t}. \tag{28}$$

Recognizing that $P_i = 1 - Q_i$, we may solve for the primary proportion of individuals as follows:

$$P_{i,t} = 1 - (1 - P_{i,0})^{s^t}. \tag{29}$$

Solving for the takeover time yields an asymptotic expression that improves with increasing n:

$$t^* = \frac{1}{\ln s}[\ln n + \ln(\ln n)]. \tag{30}$$

This equation agrees with the previous calculation for takeover time in the $c_0 = 2$ solution to linear ranking selection when $s = 2$. Of course, binary tournament selection and linear ranking selection ($c_0 = 2$) are identical in expectation, and the takeover time estimates should agree.

The difference equation model and a representative computer simulation are compared in figure 4 for a tournament of size $s = 3$. As before, a solution and a representative simulation are run with $n = k = 256$, starting with a single copy of each alternative. The representative computer simulation shown in the figure matches the difference equation solution quite well.

Figure 5 compares the growth of the best individual starting with a proportion $\frac{1}{256}$ using tournaments of sizes $s = 2, 4, 8, 16$. Note that as the tournament size increases, the convergence time is cut by the ratio of the logarithms of the tournament sizes as predicted.

---

[3] Here we consider a deterministic competition, because the notion of a probabilistic tournament does not generalize from binary to s-ary tournaments easily.

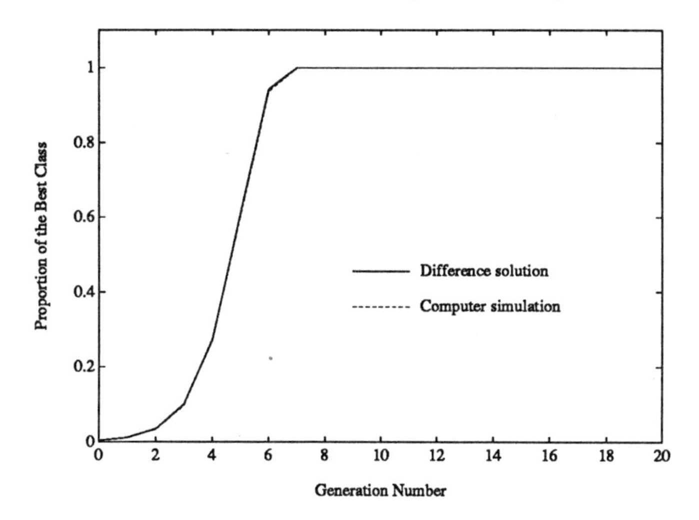

Figure 4: A comparison of the difference equation solution and a representative computer simulation with a ternary tournament ($s = 3$) demonstrates good agreement.

5.3 Time complexity of tournament selection

The calculation of the time complexity of tournament selection is straightforward. Each competition in the tournament requires the random selection of a constant number of individuals from the population. The comparison among those individuals can be performed in constant time, and n such competitions are required to fill a generation. Thus, tournament selection is $O(n)$.

We should also mention that tournament selection is particularly easy to implement in parallel. All the complexity estimates given in this paper have been for operation on a serial machine, but all the other methods discussed in the paper are difficult to parallelize, because they require some amount of global information. Proportionate selection requires the sum of the function values. Ranking selection (and Genitor, as we shall soon see) requires access to all other individuals and their function values to achieve global ranking. On the other hand, tournament selection can be implemented locally on parallel machines with pairwise or s-wise communication between different processors the only requirement. Muhlenbein (1989) provides a good example of a parallel implementation of tournament selection. He also claims to achieve niching implicitly in his implementation, but controlled experiments demonstrating this claim were not presented nor were analytical results given to support the observation. Some caution should be exercised in making such claims, because the power of stochastic errors to cause a population to drift is quite strong and is easy to underestimate. Nonetheless, the demonstration of an efficient parallel implementation is useful in itself.

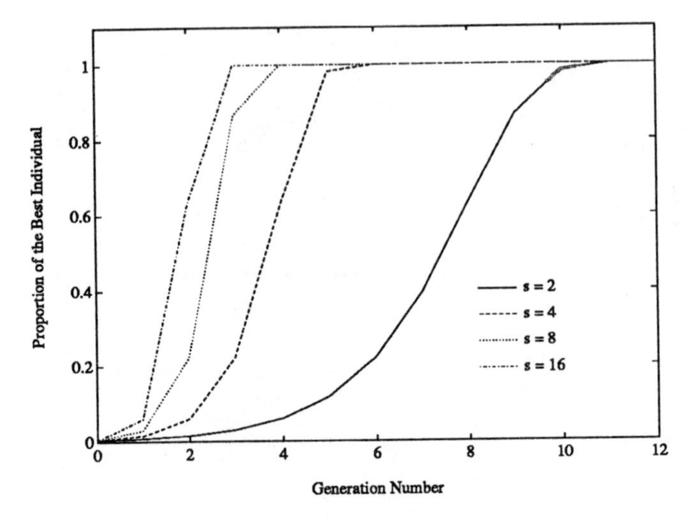

Figure 5: Growth of the proportion of best individual versus generation is graphed for a number of tournament sizes.

6 Genitor

In this section, we analyze and simulate the selection method used in Genitor (Whitley, 1989). Our purpose is twofold. First, we would like to give a quantitative explanation of the performance Whitley observed in using Genitor, thereby permitting comparison of this technique to others commonly used. Second, we would like to demonstrate the use of the analysis methods of this paper in a somewhat involved, overlapping population model, thereby lighting a path toward the analysis of virtually any selection scheme.

Genitor works individual by individual, choosing an offspring for birth according to linear ranking, and choosing the currently worst individual for replacement. Because the scheme works one by one it is difficult to compare to generational schemes, but the comparison can and should be made.

6.1 An analysis of Genitor

We use the symbol τ to denote the individual iteration number and recognize that the generational index may be related to τ as $t = \tau/n$. Under individual-wise linear ranking the cumulative assignment function β is the same as before, except that during each assignment we only allocate a proportion $1/n$ of the population (a single individual). For block death, the worst individual (the individual with rank between $\frac{n-1}{n}$ and one) will lose a proportion $\frac{1}{n}$ of his current total. Recognizing that the best individual never loses until he dominates the population, it is a straightforward

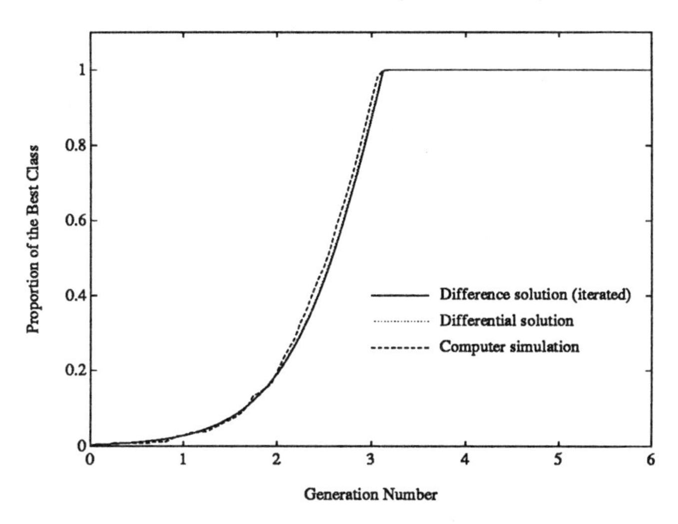

Figure 6: Comparison of the difference equation solution, differential equation solution, and computer simulations of Genitor for the function $f(x) = x$, $n = k = 256$. Linear ranking with $c_0 = 2$ is used, and the individual iteration number (τ) has been divided by the population size to put the computations in terms of generations. Solutions to the difference and differential equations are so close that they appear as a single line on the plot, and both compare well to the representative computer simulation shown.

matter to write the birth, life, and death equation for an iteration of the ith class:

$$P_{i,\tau+1} = \begin{cases} P_{i,\tau} + \beta(P_{i,\tau})/n, & \text{if } P_{i,\tau} \le \frac{n-1}{n}; \\ P_{i,\tau} + \beta(P_{i,\tau})/n - (P_{i,\tau} - \frac{n-1}{n}), & \text{otherwise.} \end{cases} \tag{31}$$

A simplified exact solution of this equation (other than by iteration) is nontrivial. Therefore, we approximate the solution by subtracting the proportion at generation τ from both sides of the equation, thereafter approximating the finite difference by a time derivative. The resulting equation is logistic in form and has the following solution:

$$P_t = \frac{c_0 P_0 e^{c_0 t}}{c_0 + (c_0 - 1)P_0(e^{c_0 t} - 1)}. \tag{32}$$

Note that the class index i has been dropped and that the solution is now written in terms of the generational index t, enabling direct comparisons to other generational schemes. The difference equation (iterated directly), the differential equation solution, and a representative computer simulation are compared in figure 6, a graph of the proportion of the best individual ($n = k = 256$) versus generation. It is interesting that the solution appears to follow exponential growth that is terminated when the population is filled with the best individual. The solution is logistic, but its fixed point is $P = \frac{c_0}{1-c_0}$, which can be no less than 2 ($1 < c_0 \le 2$). Thus, by the time any significant logistic slowing in the rate of convergence occurs, the solution has already crashed into the barrier at $P = 1$.

To compare this scheme to other methods, it is useful to calculate the free or early growth rate. Considering only linear terms in the difference equation, we obtain $P_{\tau+1} = (1+\frac{c_0}{n})P_\tau$. Over a generation n individual iterations are performed, obtaining $P_n = (1+\frac{c_0}{n})^n P_0$, which approaches $e^{c_0} P_0$ for moderate to large n. Thus we note an interesting thing. Even if no bias is introduced in the ranked birth procedure (if $c_0 = 1$), Genitor has a free growth factor that is no less than e. In other words, unbiased Genitor pushes harder than generation-based ranking or tournament selection, largely a result of restricting death to the worst individual. When biased ranking ($c_0 > 1$) is used, Genitor pushes very hard indeed. For example, with $c_0 = 2$, the selective growth factor is $e^2 = 7.389$. Such high growth rates can cover a host of operator losses, recalling that the net growth factor γ is the product of the growth factor obtained from selection alone ϕ and the schema survival probability obtained by subtracting operator losses from one:

$$\gamma = \phi\,[1-\epsilon], \tag{33}$$

where $\epsilon = \sum_\omega \epsilon_\omega$, the sum of the operator disruption probabilities. For example, with $c_0 = 2$ and $\phi = 7.389$, Genitor can withstand an operator loss of $\epsilon = 1 - \frac{1}{7.389} = 0.865$; such an allowable loss would permit the growth of building blocks with defining lengths roughly 87% of string length. Such large permissible errors, however, come at a cost of increased premature convergence, and we speculate that it is precisely this effect that motivated Whitley to try large population sizes and multiple populations in a number of simulations. Large sizes slow things down enough to permit the growth and exchange of multiple building blocks. Parallel populations allow the same thing by permitting the rapid growth of the best building blocks within each subpopulation, with subsequent exchanges of good individuals allowing the cross of the best bits and pieces from each subpopulation. Unfortunately, neither of these fixes is general, because codings can always be imagined that make it difficult to cut and splice the correct pieces. Thus, it would appear that there still is no substitute for the formation and exchange of tight building blocks.

Moreover, we find no support for the hypothesis that there is something special about overlapping populations. This paper has demonstrated conclusively that high growth rates are acting in Genitor; this factor alone can account for the observed results, and it should be possible to duplicate Whitley's results through the use of any selection scheme with equivalent duplicative horsepower. We have not performed these experiments, but the results of this paper provide the analytical tools necessary to carry out a fair comparison. Exponential scaling with proportionate reproduction, larger tournaments, or nonlinear ranking should give results similar to Genitor, if similar growth ratios are enforced and all other operators and algorithm parameters are the same.

6.2 Genitor's takeover time and time complexity

The takeover time may be approximated. Since Genitor grows exponentially until the population is filled, the takeover time may be calculated from the equation $\frac{n-1}{n} = \frac{1}{n}e^{c_0 t^*}$. Solving for the takeover time yields the following equation:

$$t^* = \frac{1}{c_0}\ln(n-1). \tag{34}$$

Table 1: A Comparison of Three Growth Ratio Measures

SCHEME	ϕ_t	ϕ_e	ϕ_l
Proportionate	$\frac{f_1}{f_t}$	$\frac{f_1}{f}$	$\frac{f_1}{f_1+f_2}$
Linear ranking	$c_0-(c_0-1)P$	c_0	$\frac{c_0+1}{2}$
Tournament, p	$2p-(2p-1)P$	$2p$	$\frac{2p+1}{2}$
Tournament, s	$\sum_{i=1}^{s}\binom{s}{i}P^{i-1}(1-P)^{s-i}$	s	$2(1-2^{-s})$
Genitor	no closed form in t	e^{c_0}	$e^{0.5(c_0+1)}$

The time complexity of Genitor may also be calculated. Once an initial ranking is established, Genitor does not need to completely sort the population again. Each generated individual is simply inserted in its proper place; however, the search for the proper place requires $O(\log n)$ steps if a binary search is used. Moreover, the selection of a single individual from the ranked list can also be done in $O(\log n)$ steps. Since both of these steps must be performed n times to fill an equivalent population (for comparison with the generation-based schemes), the algorithm is clearly $O(n \log n)$.

Next, we cross-compare different schemes on the basis of early and late growth ratios, takeover times, and time complexity computations for the selection algorithms themselves.

7 Selection Procedures Head to Head

In this section, we gather the growth ratios, takeover times, and time complexity calculations for each of the selection mechanisms to permit a side-by-side comparison.

7.1 Growth ratios due to selection

The analyses of the previous section permit the comparison of the selection methods on the basis of their expected growth ratio for members of the best class. Specifically, we compare three values: the growth ratio at generation t, $\phi_t = P^*_{t+1}/P^*_t$, the *early* (or *free*) ratio ϕ_e defined as the growth ratio when the proportion of individuals is insubstantial (when $P^*_t \approx 0$), and the *late* (or *constrained*) ratio ϕ_l defined as the growth ratio when the best occupy 50% of the population and the remainder of the population is occupied by second-best structures. Table 1 compares each of the schemes on the basis of these three growth ratio measures. As we can see, proportionate selection is dependent on the objective function used, and the early growth ratio is likely to be quite high, and the late growth ratio is likely to be quite low. It is exactly this effect that has caused researchers to turn to scaling techniques and ranking methods. The results presented in section 3 using an exponential function are interesting and seem to recommend the use of an exponential scaling function to control the degree of competition. The c coefficient may be used in a manner similar to the inverse of temperature in simulated annealing to control the

Table 2: A Comparison of Takeover Time Values

SCHEME	t^*
Proportionate x^c	$\frac{1}{c}(n \ln n - 1)$
Proportionate e^{cx}	$\frac{1}{c} n \ln n$
Linear ranking $c_0 = 2$	$\log n + \log(\ln n)$
Linear ranking (diff. eq.)	$\frac{2}{c_0 - 1} \log(n-1)$,
Tournament p	same as linear ranking with $c_0 = 2p$
Tournament s	$\frac{1}{\ln s}[\ln n + \ln(\ln n)]$
Genitor	$\frac{1}{c_0} \ln(n-1)$

accentuation of salient features (Goldberg, 1990).

As was mentioned earlier, linear ranking and binary tournament selection agree in expectation, both allowing early growth ratios of between one and two, depending on the adjustment of the appropriate parameter (c_0 or p). Both achieve late ratios between 1 and 1.5. Tournament selection can achieve higher growth ratios with larger tournament sizes; the same effect can be achieved in ranking selection with nonlinear ranking functions, although we have not investigated these here. Genitor achieves early ratios between e and e^2, and it would be interesting to compare Genitor selection with tournament selection or ranking selection with appropriate tournament size or appropriate nonlinear assignment function.

7.2 Takeover time comparison

Table 2 shows the takeover times calculated for each of the selection schemes. Other than the proportionate scheme, the methods compared in this paper, all converge in something like $O(\log n)$ generations. This, of course, does not mean that real GAs converge to global optima in that same time. In the setting of this paper, where we are doing nothing more than choosing the best from some fixed population of structures, we get convergence to the best. In a real GA, building blocks must be selected and juxtaposed in order to get copies of the globally optimal structure, and the variance of building block evaluation is a substantial barrier to convergence. Nonetheless, the takeover time estimates are useful and will give some idea how long a GA can be run before mutation becomes the primary mechanism of exploration.

7.3 Time complexity comparison

Table 3 gathers the time complexity calculations together. The best of the methods are $O(n)$, and it is difficult to imagine how fewer steps can be used since we must select n individuals in some manner. Of the $O(n)$ methods, tournament selection is the easiest to make parallel, and this may be its strongest recommendation, as GAs cry out for parallel implementation, even though most of us have had to make do with serial versions. Whether paying the $O(n \log n)$ price of Genitor is worth its somewhat higher later growth ratio is unclear, and the experiments recommended earlier should be performed. Methods with similar early growth ratios and not-too-

Table 3: A Comparison of Selection Algorithm Time Complexity

SCHEME	TIME COMPLEXITY
Roulette wheel proportionate	$O(n^2)$
RW proportionate w/binary search	$O(n \log n)$
Stochastic remainder proportionate	$O(n)$
Stochastic universal proportionate	$O(n)$
Ranking	$O(n \ln n)+$ time of selection
Tournament Selection	$O(n)$
Genitor w/binary search	$O(n \log n)$

different late growth ratios should perform similarly. Any such comparisons should be made under controlled conditions where only the selection method is varied, however.

8 Selection: What Should We Be Doing?

This paper has taken an unabashedly descriptive viewpoint in trying to shed some analytical light on various selection methods, but the question remains: how should we do selection in GAs? The question is a difficult one, and despite limited empirical success in using this method or that, a general answer remains elusive.

Holland's connection (1973, 1975) of the k-armed bandit problem to the conflict between exploration and exploitation in selection still stands as the only sensible theoretical abstraction of the general question, despite some recent criticism (Grefenstette & Baker, 1989). Grefenstette and Baker challenge the k-armed model by posing a partially deceptive function, thereafter criticizing the abstraction because the GA does not play the deceptive bits according to the early function value averages. The criticism is misplaced, because it is exactly such deceptive functions that the GA must play as a higher-order bandit (in a 3-bit deceptive subfunction, the GA must play the bits as an eight-armed bandit) and the schema theorem says that it will do so if the linkage is sufficiently tight. In other words, GAs will play the bandit problems at as high a level as they can (or as high a level as is necessary), and it is certainly this that accounts at least partially for the remarkable empirical success that many of us have enjoyed in using simple GAs and their derivatives.

Moreover, dismissing the bandit model is a mistake for another reason, because in so doing we lose its lessons about the effect of noise on schema sampling. Even in easy deterministic problems—problems such as $\sum_i a_i x_i + b$, $a_i, b \in R$, and $x_i \in \{0,1\}$—GAs can make mistakes, because alleles with small contribution to objective function value (alleles with small a_i) get fixed, a result of early spurious associations with other highly fit alleles or plain bad luck. These errors can occur, because the variation of other alleles (the sampling of the *'s in schemata such as **1**) is a source of noise as far as getting a particular allele set properly is concerned. Early on this noise is very high (estimates have been given in Goldberg, Korb, & Deb, 1989), and only the most salient building blocks dare to become fixed. This fixation reduces the variance for the remaining building blocks, permitting less salient alleles or allele combinations to become fixed properly. Of course, if along the way down this salience ladder, the correct building blocks have been lost somehow (through spurious linkage or cumulative bad luck), we must wait for mutation to restore them. The waiting time for this restoration is quite reasonable for low-order schemata but grows exponentially as order increases.

Thinking of the convergence process in this way suggests a number of possible ways to balance or overcome the conflict between exploration and exploitation:

- Use slow growth ratios to prevent premature convergence.
- Use higher growth ratios followed by building block rediscovery through mutation.
- Permit localized differential mutation rates to permit more rapid restoration of building blocks.
- Preserve useful diversity temporally through dominance and diploidy.
- Preserve useful diversity spatially through niching.
- Eliminate building block evaluation noise altogether through competitive templates.

Each of these is examined in somewhat more detail in the remainder of the section.

One approach to obtaining correct convergence might be to slow down convergence enough so that errors are rarely made. The two-armed bandit convergence graphs presented elsewhere (Goldberg, 1989a) suggest that using convergence rates tuned to building blocks with worst function-difference-to-noise-ratio is probably too slow to be practical, but the idea of starting slowly and gradually increasing the growth ratios makes some sense in that salient building blocks will be picked off with a minimum of pressure on not-so-salient allele combinations. This is one of the fundamental ideas of simulated annealing, but simulated annealing suffers from its lack of a population and its lack of interesting discovery operators such as recombination. The connection between simulated annealing and GAs has become clearer recently (Goldberg, 1990) through the invention of *Boltzmann tournament selection*. This mechanism stably achieves a Boltzmann distribution across a population of strings, thereby allowing a controllable and stable distribution of points to be maintained across both space and time. More work is necessary, but the use of such a mechanism together with well designed annealing schedules should be helpful in controlling GA convergence. As was mentioned in the paper, similar mechanisms can also be implemented under proportionate selection through the use of exponential scaling and sharing functions.

The opposite tack of using very high growth ratios permits good convergence in some problems by grabbing those building blocks you can get as fast as you can, thereafter restoring the missing building blocks through mutation (this appears to be the mechanism used in Genitor). This works fine if the problems are easy (if simple mutation can restore those building blocks in a timely fashion), and it also explains why Whitley has turned to large populations or multiple populations when deceptive problems were solved (L. Darrell Whitley, personal communication, September, 1989). The latter applications are suspect, because waiting for high-order schemata to be rediscovered through mutation or waiting for crossover to splice together two intricately intertwined deceptive building blocks are both losing propositions (they are low probability events), and the approach is unlikely to be practical in general.

It might be possible to encourage more timely restoration of building blocks by having mutation under localized genic control, however. The idea is similar to that used in Bowen (1986), where a set of genes controlled a chromosome's mutation and crossover rates, except that here a large number of mutation-control genes would be added to give differential mutation rates across the chromosome. For example, a set of genes dictating high ($p_m \approx 0.5$) or low ($p_m \approx 0$) mutation rate could be added to control mutation on function-related genes (a fixed mutation rate could be used on the mutation-control genes). Early on salient genes could achieve highest function value by fixing the correct function-related allele and fixing the associated mutation-control allele in the low position. At the same time, poor alleles would be indifferent to the value of their mutation allele, and the presence of a number of mutation-control genes set to the high allele would ensure the generation of a significant proportion of the correct function-related alleles when those poorer alleles become salient.

This mechanism is not unlike that achieved through the use of dominance and diploidy as has been explored elsewhere (Goldberg & Smith, 1987; Smith, 1988). Simply stated, dominance and diploidy permit currently out-of-favor alleles to remain in abeyance, sampling currently poorer alleles at lower rates, thereby permitting them to be brought out of abeyance quite quickly when the environment is favorable. Some consideration needs to be given toward recalling groups of alleles together, rather than on the allele-by-allele basis tried thus far (the same comment applies to the localized mutation scheme suggested in the previous paragraph), but the notion of using the temporal recall of dominance and diploidy to handle the nonstationarity of early building block sampling appears sound.

The idea of preserving useful diversity temporally helps recall the notion of diversity preservation spatially (across a population) through the notion of *niching* (Deb, 1989; Deb & Goldberg, 1989; Goldberg & Richardson, 1987). If two strings share some bits in common (those salient bits that have already been decided) but they have some disagreement over the remaining positions and are relatively equal in overall function value, wouldn't it be nice to make sure that both get relatively equal samples in the next and future generations. The schema theorem says they will (in expectation), but small population selection schemes are subject to the vagaries of *genetic drift* (Goldberg & Segrest, 1987). Simply stated, small stochastic errors of selection can cause equally good alternatives to converge to one alternative or another. Niching introduces a pressure to balance the subpopulation sizes in accordance with niche function value. The use of such niching methods can form an effective pressure to maintaining useful diversity across a population, allowing that diversity to be crossed with other building blocks, thereby permitting continued exploration.

The first five suggestions all seek to balance the conflict between exploration and exploitation, but the last proposal seeks to eliminate the conflict altogether. The elimination of building block noise sounds impossible at first glance, but it is exactly the approach taken in *messy genetic algorithms* (Goldberg, Deb, & Korb, 1990; Goldberg & Kerzic, 1990; Goldberg, Korb, & Deb, 1990). Messy GAs (mGAs) grow long strings from short ones, but so doing requires that missing bits in a problem of fixed length be filled in. Specifically, partial strings of length k (possible building blocks) are overlaid with a *competitive template*, a string that is locally optimal at the level $k-1$ (the competitive template may be found using an mGA at the lower level). Since the competitive template is locally optimal, any string that gets a value in excess of the template contains a k-order building block by definition. Moreover, this evaluation is without noise (in deterministic functions), and building blocks can be selected deterministically without fear; simple binary tournament selection has been used as one means of conveniently doping the population toward the best building blocks. Some care must be taken to compare related building blocks to one another, lest errors be made when subfunctions are scaled differently. Also, some caution is required to prevent hitchhiking of wrong (parasitic) incorrect bits that agree with the template but later can prevent expression of correct allele combinations. Reasonable mechanisms have been devised to overcome these difficulties, however, and in empirical tests mGAs have always converged to global optima in a number of provably deceptive problems. Additionally, mGAs have been shown to converge in time that grows as a polynomial function of the number of decision variables on a serial machine and as a logarithmic function of the number of decision variables on a parallel machine. It is believed that this convergence is correct (the answers are global) for problems of bounded deception. More work is required here, but the notion of strings that grow in complexity to more completely solve more difficult problems has a nice ring to it if we think in terms of the way nature has filled this planet with increasingly complex organisms.

In addition to trying these various approaches toward balancing or overcoming the conflict of exploration and exploitation, we must not drop the ball of analysis. The methods of this paper provide a simple tool to better understand the expected behavior of selection schemes, but better probabilistic analyses using Markov chains (Goldberg & Segrest, 1987), Markov processes, stochastic differential and difference equations, and other techniques of the theory of stochastic processes should be tried with an eye toward understanding the variance of selection. Additionally, increased study of the k-armed bandit problem might suggest practical strategies for balancing the conflicts of selection when they arise. Even though conflict can apparently be sidestepped in deterministic problems using messy GAs, eventually we must return to problems that are inherently noisy, and the issue once again becomes germane.

9 Conclusions

This paper has compared the expected behavior of four selection schemes on the basis of their difference equations, solutions to those equations (or related differential equation approximations), growth ratio estimates, and takeover time computations. Proportionate selection is found to be significantly slower than the other three types. Linear ranking selection and a probabilistic variant of binary tournament selection have been shown to have identical performance in expectation, with binary tournament selection preferred because of its better time complexity. Genitor selection, an overlapping population selection scheme, has been analyzed and compared to the others and tends to show a higher growth ratio than linear ranking or binary tournament selection performed on a generation-by-generation basis. On the other hand, tournament selection with larger tournament sizes or nonlinear ranking can give growth ratios similar to Genitor, and such apples-to-apples comparisons have been suggested.

Additionally, the larger issue of balancing or overcoming the conflict of exploration and exploitation inherent in selection has been raised. Controlling growth ratios, localized differential mutation, dominance and diploidy, niching, and messy GAs (competitive templates) have been discussed and will require further study. Additional descriptive and prescriptive theoretical work has also been suggested to further understanding of the foundations of selection. Selection is such a critical piece of the GA puzzle that better understanding at its foundations can only help advance the state of genetic algorithm art.

Acknowledgments

This material is based upon work supported by the National Science Foundation under Grant CTS-8451610. Dr. Goldberg gratefully acknowledges additional support provided by the Alabama Research Institute, and Dr. Deb's contribution was performed while supported by a University of Alabama Graduate Council Research Fellowship.

References

Ankenbrandt, C. A. (1990). *An extension to the theory of convergence and a proof of the time complexity of genetic algorithms* (Technical Report CS/CIAKS-90-0010/TU) New Orleans: Center for Intelligent and Knowledge-based Systems, Tulane University.

Baker, J. E. (1985). Adaptive selection methods for genetic algorithms. *Proceedings of an International Conference on Genetic Algorithms and Their Applications*, 100-111.

Baker, J. E. (1987). Reducing bias and inefficiency in the selection algorithm. *Proceedings of the Second International Conference on Genetic Algorithms,* 14-21.

Booker, L. B. (1982). Intelligent behavior as an adaptation to the task environment. (Doctoral dissertation, Technical Report No. 243, Ann Arbor: University of Michigan, Logic of Computers Group). *Dissertation Abstracts International, 43*(2), 469B. (University Microfilms No. 8214966)

Bowen, D. (1986). A study of the effects of internally determined crossover and mutation rates on genetic algorithms. Unpublished manuscript, University of Alabama, Tuscaloosa.

Brindle, A. (1981). *Genetic algorithms for function optimization* (Doctoral dissertation and Technical Report TR81-2). Edmonton: University of Alberta, Department of Computer Science.

De Jong, K. A. (1975). An analysis of the behavior of a class of genetic adaptive systems. (Doctoral dissertation, University of Michigan). *Dissertation Abstracts International, 36*(10), 5140B. (University Microfilms No. 76-9381)

Deb, K. (1989). *Genetic algorithms in multimodal function optimization* (Master's thesis and TCGA Report No. 88002). Tuscaloosa: University of Alabama, The

Clearinghouse for Genetic Algorithms.

Deb, K., & Goldberg, D. E. (1989). An investigation of niche and species formation in genetic function optimization. *Proceedings of the Third International Conference on Genetic Algorithms*, 42-50.

Goldberg, D. E. (1989a). *Genetic algorithms in search, optimization, and machine learning.* Reading, MA: Addison-Wesley.

Goldberg, D. E. (1989b). Sizing populations for serial and parallel genetic algorithms. *Proceedings of the Third International Conference on Genetic Algorithms*, 70-79.

Goldberg, D. E. (1990). A note on Boltzmann tournament selection for genetic algorithms and population-oriented simulated annealing. *Complex Systems, 4*, 445-460.

Goldberg, D. E., Deb, K. & Korb, B. (1990). Messy Genetic Algorithms Revisited: Nonuniform Size and Scale. *Complex Systems, 4*, 415-444.

Goldberg, D. E., & Kerzic, T. (1990). *mGA1.0: A Common Lisp implementation of a messy genetic algorithm* (TCGA Report No. 90004). Tuscaloosa: University of Alabama, The Clearinghouse for Genetic Algorithms.

Goldberg, D. E., Korb, B., & Deb, K. (1990). Messy genetic algorithms: Motivation, analysis, and first results. *Complex Systems, 3*, 493-530.

Goldberg, D. E., & Richardson, J. (1987). Genetic algorithms with sharing for multimodal function optimization. *Proceedings of the Second International Conference on Genetic Algorithms*, 41-49.

Goldberg, D. E., & Segrest, P. (1987). Finite Markov chain analysis of genetic algorithms. *Proceedings of the Second International Conference on Genetic Algorithms*, 1-8.

Goldberg, D. E., & Smith, R. E. (1987). Nonstationary function optimization using genetic algorithms with dominance and diploidy. *Proceedings of the Second International Conference on Genetic Algorithms*, 59-68.

Grefenstette, J. J. & Baker, J. E. (1989). How genetic algorithms work: A critical look at implicit parallelism. *Proceedings of the Third International Conference on Genetic Algorithms*, 20-27.

Holland, J. H. (1973). Genetic algorithms and the optimal allocations of trials. *SIAM Journal of Computing, 2(*2), 88-105.

Holland, J. H. (1975). *Adaptation in natural and artificial systems.* Ann Arbor, MI: University of Michigan Press.

Muhlenbein, H. (1989). Parallel genetic algorithms, population genetics and combinatorial optimization. *Proceedings of the Third International Conference on Genetic Algorithms*, 416-421.

Smith, R. E. (1988). *An investigation of diploid genetic algorithms for adaptive search of nonstationary functions* (Master's thesis and TCGA Report No. 88001). Tuscaloosa: University of Alabama, The Clearinghouse for Genetic Algorithms.

Suh, J. Y. & Van Gucht, D. (1987). *Distributed genetic algorithms* (Technical Report No. 225). Blooomington: Indiana University, Computer Science Department.

Syswerda, G. (1989). Uniform crossover in genetic algorithms. *Proceedings of the Third International Conference on Genetic Algorithms*, 2-9.

Whitley, D. (1989). The Genitor algorithm and selection pressure: Why rank-based allocation of reproductive trials is best. *Proceedings of the Third International Conference on Genetic Algorithms*, 116-121.

A Study of Reproduction in Generational and Steady-State Genetic Algorithms

Gilbert Syswerda
BBN Laboratories
BBN Systems and Technologies Corporation
10 Moulton Street
Cambridge, MA 02138
syswerda@bbn.com

Abstract

Two techniques of population control are currently used in the field of serial genetic algorithms: generational and steady state. Although they have been used somewhat interchangeably in the past, it has become apparent that the two techniques are actually quite different. In this paper, I study the behavior of each with regard to reproduction, and show that while each can be made similar with respect to the schema theorem, in practice their behavior is quite different.

Keywords: reproduction, steady-state reproduction, generational reproduction, population

1 INTRODUCTION

Genetic algorithms (GAs) modify a population of potential solutions during the course of a run, using both the application of operators such as crossover and mutation, and the application of a reproductive technique. There are currently two reproductive techniques in general use. The first, which is probably the most widely used, we will call *generational* reproduction, and the second *steady state* reproduction [Syswerda 1989]. Briefly, generational reproduction replaces the entire population at once, while steady-state reproduction replaces only a few members at a time.

The two techniques have been used in the past as if though they were interchangeable, with research results being presented without acknowledgment of which type of algorithm was used. It has been becoming apparent, however, that the two techniques are actually quite different and that the choice of which to use can dramatically affect the performance of a genetic algorithm.

According to the schema theorem [Holland 1975], the performance of a genetic algorithm is highly dependent on its reproductive behavior. The schema theorem states that the rate of increase of above average schemata in a population is directly dependent on the reproductive rate of individual population members. In this paper, I begin a comparison of the two techniques by examining how each affects the reproduction of fit members in a population. This is done by first computing ideal performance curves and then comparing actual performance against ideal performance.

Note that the horizontal axis of all graphs in this paper are proportional to the number of evaluations that would occur during the running of a GA. This allows direct comparison of the graphs, since each represents the same number of evaluations.

2 Ideal Reproductive Performance

The ideal performance of a reproductive technique is defined to be the performance of the technique when used with an infinite population. An infinite population can be simulated by allowing fractional members to exist in the population.

All experiments presented in this paper used a population of size 100, with 10 members having a value of 10 (the best), 10 with value 9, etc. Various reproductive strategies were run, and the population percentages of each fitness value plotted against the number of generations.

2.1 Generational Reproduction

Generational reproduction replaces the entire population with a new population. This is done by repeatedly selecting a member from the old population, according to fitness and with replacement, and adding that member to the new population. Since selection is biased by fitness, members having a fitness value greater than the population average will on average be represented in the new population to a greater extent. Likewise, members with below-average fitness will decrease in the population according to the same ratio.

To compute the ideal curves for generational reproduction, I allow partial members to be represented, simulating in effect an infinite population. For each generation, the number of members for each fitness value was increased or decreased in the population according to its fitness ratio as follows:

$$n_{t+1} = n_t \frac{f}{f_{avg}}$$

The results for 20 generations are presented in figure 1a. Since we are using a population size of 100, each generation involves 100 reproductive events, for a total of 2000 evaluations. As can be seen, after 20 generations the population has nearly converged to members with value 10.

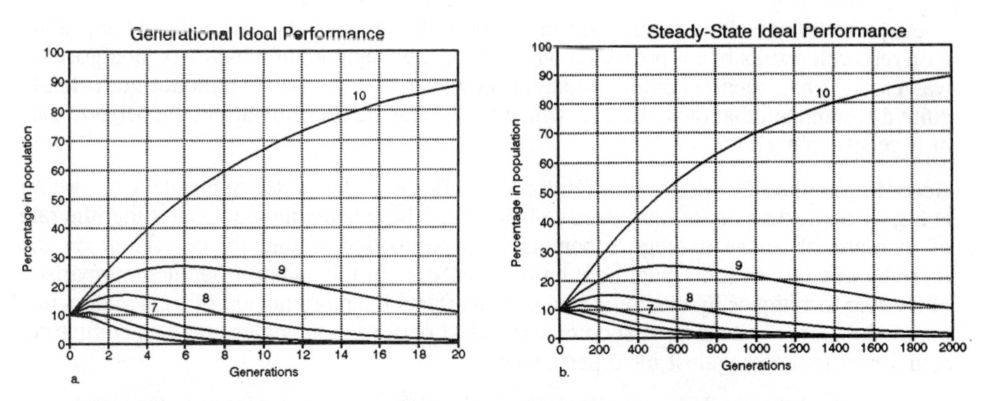

Figure 1: Ideal reproduction curves for generational and steady-state GAs.

2.2 Steady-State Reproduction

The steady-state technique replaces only a few members during a generation, usually whatever number is required for an operator. In these studies, one member was reproduced in each generation. This was done by selecting a member of the population according to its fitness, just as with generational reproduction, and making a copy. However, in order to insert the copy into the population, room must be made for it, which means we must also have a method for selecting a member to be deleted. To make steady-state reproduction most like generational, members were deleted randomly. We will come back to this issue, since random deletion is not the method of choice.

To compute the ideal values, each member in the population was increased in number according to its chance of being selected for reproduction, and decreased according its chance of being deleted.

$$n_{t+1} = n_t + n_t(\frac{f}{sum} - \frac{1}{size}),$$

where *sum* is the sum of the fitness values of the population members, and *size* is the size of the population.

The results are presented in figure 1b. Note that for the same number of reproductive events to occur, 100 generations of a steady-state GA must occur for each generation of a generational GA.

The graphs for steady-state and generational reproduction are nearly identical. This demonstrates that with respect to the schema theorem, either algorithm is equivalent as far as the required reproductive behavior is concerned.

3 Actual Behavior

It is instructive to examine how each algorithm performs in practice. To do this, an actual *finite* population of size 100 was constructed with the same distribution of fitness values as above, and each reproductive algorithm run on it. At the end of each generation, the population percentages were recorded. The results are presented in figures 2 and 3.

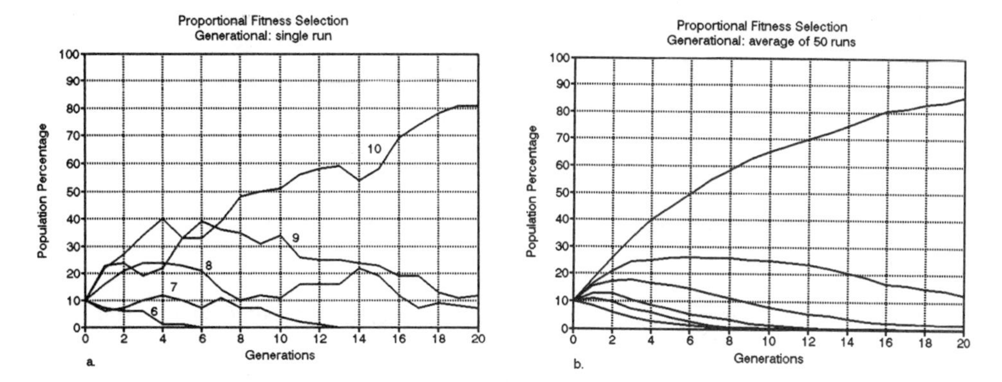

Figure 2: Simulated reproductive behavior of generational GAs.

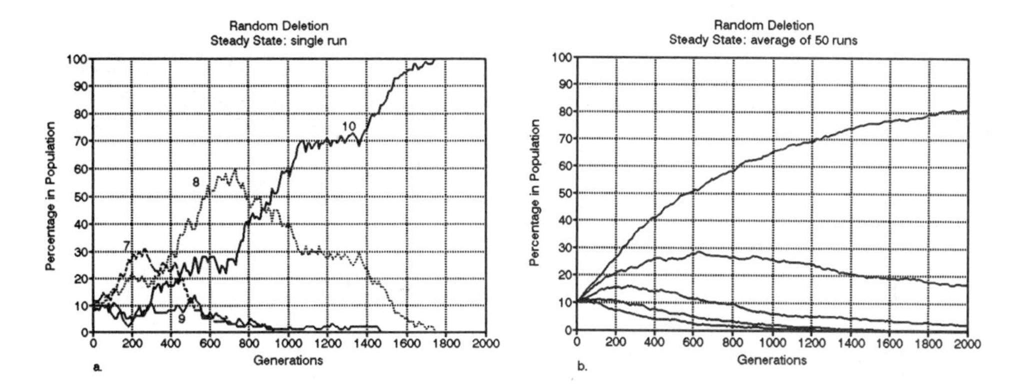

Figure 3: Simulated reproductive behavior of steady-state GAs.

I have presented both a typical single run, and the average of 50 runs. If we look first at the average runs, and compare them with the ideal numbers expected, we can see that on average, each algorithm performs close to the ideal. However, for any single run, the results are far from ideal. Both steady-state and generational reproduction exhibit a large amount of variance.

However, users of each technique rarely run their algorithms as described above. Generational practitioners typically use Baker's multi-arm spinner selection method [Baker 1987]. Basically, Baker proposed using a spinner that has as many equally-spaced arms as there are members in the population. Each member gets a piece of the spinner pie sized according to its fitness value. The spinner is spun, and each arm then points to a member that will be reproduced in the new population.

Baker's selection method works very well. Each individual run results in nearly ideal results. A typical single run and the average of 50 runs are presented in figure 4.

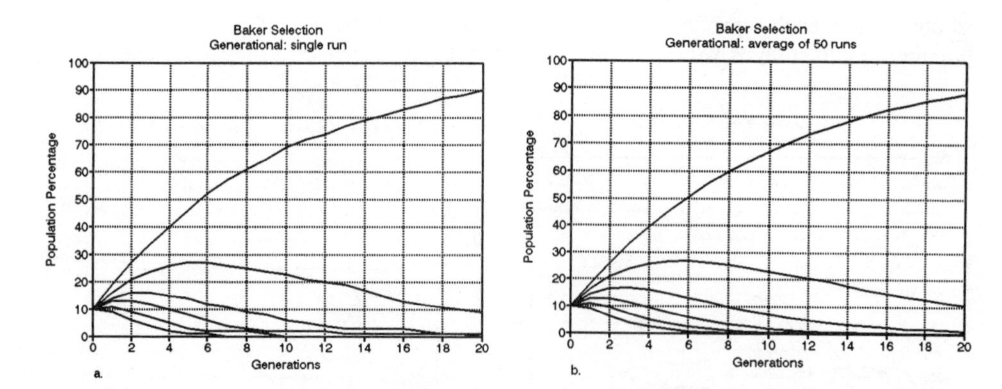

Figure 4: Simulated reproductive behavior of generational GAs using Baker's selection method.

Most steady-state GA practitioners do not use random deletion. Instead, they either delete the worst member in the population, or use some sort of ranking scheme. Below, I present the performance graphs for three alternative deletion methods.

- **Delete least fit.** Figure 5 presents the results of simply deleting the least fit member of the population. The argument given by its defenders is that since they have little chance of being selected for reproduction, they are the best candidates for deletion.

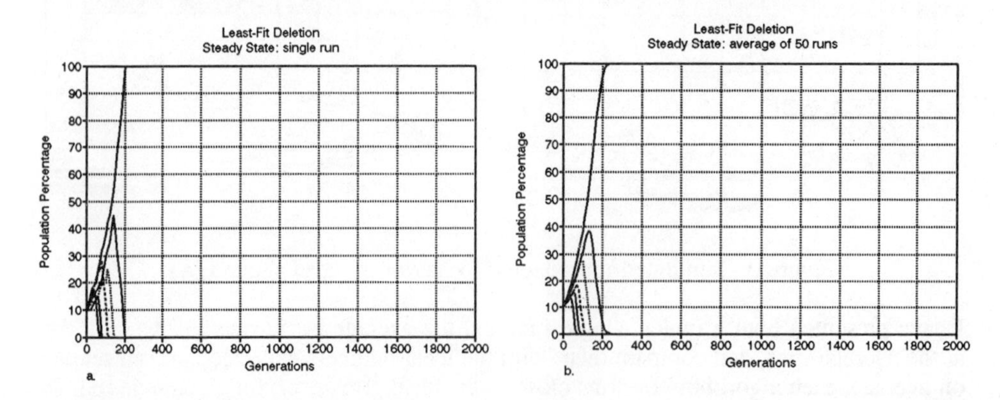

Figure 5: Simulated reproductive behavior of steady-state GAs using the least fit deletion method.

- **Exponential ranking.** This is method I typically use. The worst member has some chance of being deleted, say $1/10$. If it is not selected, then the next to the last also has a 0.1 chance, and so on. For a population of 100, the best member has a very low $0.9^{99} = 0.00003$ chance of being deleted. Results are presented in figure 6.
- **Reverse fitness.** Each member has a chance of being deleted according to how bad it is. In this case, a "badness" value is computed by subtracting a member's value from 11. Results are presented in figure 7.

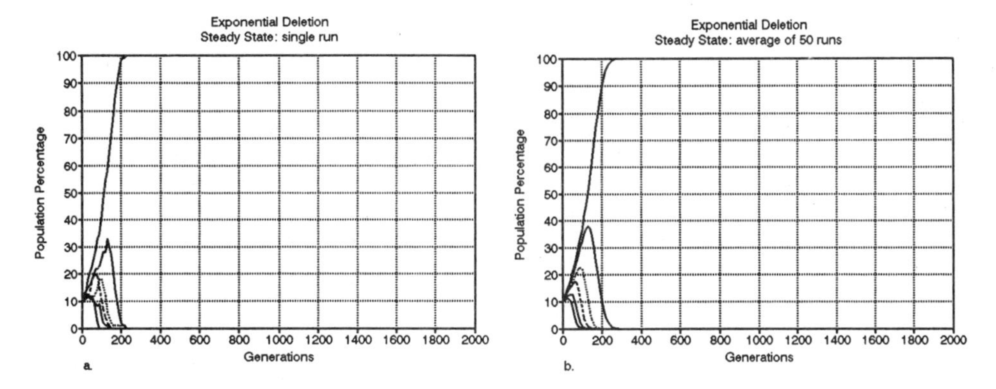

Figure 6: Simulated reproductive behavior of steady-state GAs using exponential ranking as the deletion method.

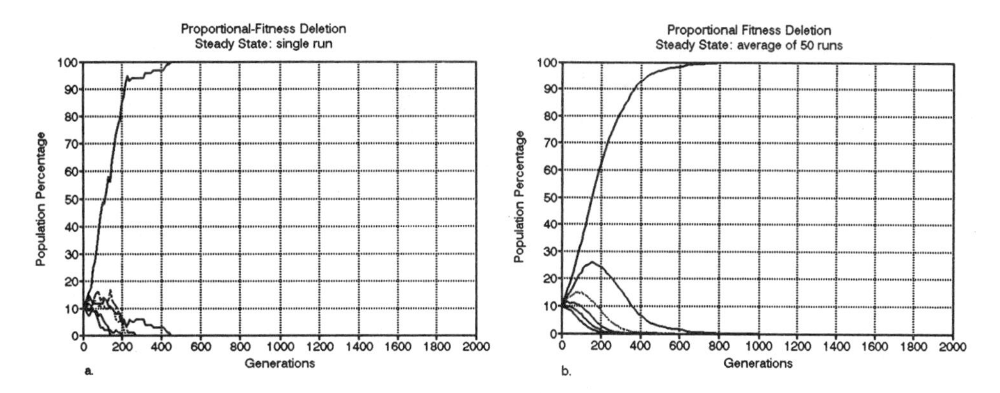

Figure 7: Simulated reproductive behavior of steady-state GAs using reverse fitness as the deletion method.

I have found exponential ranking to be better than deleting the least fit. Random deletion is worse than either (although this conclusion is not backed up by much data). I have yet to try reverse fitness.

4 Discussion

The steady-state reproductive technique has already been widely used. It is commonly used in the GA component of the Michigan approach to classifier systems [Holland 1986, page 148]. Whitley has made it an explicit component of his Genitor algorithm [Whitley 1988, 1989]. Others have confessed to using the steady-state population model without being aware that most other people used a generational model.

I have shown that steady-state and generational genetic algorithms can be made to look similar as far as the schema theorem is concerned. This was done by using random selection as the deletion technique for steady-state GAs.

However, few people actually use steady-state GAs in this manner. Instead, least-fit or exponential ranking are the preferred selection methods for deletion, and these methods dramatically change the population percentage figures. Compare figures 4b and 5b. These figures are most representative of the way each algorithm is typically used. One might imagine that the graph of steady-state reproduction is a compressed version of the graph of generational reproduction, achieving in one tenth the time the results that the generational approach obtains. Interestingly, informal testing and comparison of the two approaches indicate that at least for some problems, steady-state GAs do find as good or better solutions in much less time than generational GAs. A careful benchmark of the two approaches is warranted.

There are other issues to consider when comparing steady-state and generational GAs. Users of generational GAs often provide a guarantee that the best member in the current population will be present in the next. This is not necessary with steady-state GAs, since the deletion methods most often used automatically grant elitist status to *all* good members of the population.[1]

Also, when exploration is added in the form of crossover and mutation, new population members are created. If a good new member is created in a steady-state GA, it is immediately available for use. In generational GAs, a good new member is not used until the next generation, which in our example may be 50 operations away. By the time generational GAs start using new good material, steady-state GAs may have already thoroughly integrated the material into the population.[2]

Finally, steady-state GAs allow us to make another change that significantly improves performance: one can impose the condition that only one copy of any chromosome exist in the population at any one time. This simple change has, on some problems, resulted in a large improvement in the speed at which a steady-state GA converges on the correct solution [Syswerda 1989, Davis 1991]. It also allows the GA to play the end game better, since the population never fully converges.

References

Baker, E. J. Reducing bias and inefficiency in the selection algorithm. In *Proceedings of the Second Int'l. Conference on Genetic Algorithms*, 14-21, John J. Grefenstette, ed., Morgan Kaufmann, June 1987.

Davis, L. (ed.), (1991). *Handbook of Genetic Algorithms.* New York, New York: Van Nostrand Reinhold.

Goldberg, D. E. (1989). *Genetic Algorithms in Search, Optimization, and Machine Learning.* Addison-Wesley Publishing Company, Inc.

Holland, J. H. (1975). *Adaptation in Natural and Artificial Systems.* University of Michigan Press.

Holland, J. H., Holyoak, K. J., Nisbett, R. E., and Thagard, P. R. (1987). *Induction:*

[1]Darrell Whitley has examined this effect in the context of preservation of long schemata [Whitley 1988].

[2]This was pointed out to me by Lawrence Davis.

Processes of Inference, Learning, and Discovery. MIT Press.

Syswerda, G. Uniform crossover in genetic algorithms. In *Proceedings of the Third Int'l. Conference on Genetic Algorithms and their Applications*, 2-9, J. D. Schaffer, ed., Morgan Kaufmann, June 1989.

Whitley, D., The genitor algorithm and selection pressure: why rank-based allocation of reproductive trials is best. In *Proceedings of the Third Int'l. Conference on Genetic Algorithms and their Applications*, 116-121, J. D. Schaffer, ed., Morgan Kaufmann, June 1989.

Whitley, D., Kauth, J. (1988). *Genitor: A Different Genetic Algorithm*. Technical Report CS-88-101, Colorado State University.

Spurious Correlations and Premature Convergence in Genetic Algorithms

J. David Schaffer **Larry J. Eshelman** **Daniel Offutt**

Philips Laboratories,
North American Philips Corporation,
345 Scarborough Road,
Briarcliff Manor, New York 10510

Abstract

What distinguishes genetic algorithms (GAs) from other search methods is their inherent exploitive sampling ability known as implicit parallelism. We argue, however, that this exploitive behavior makes GAs sensitive to spurious correlations between schemata that contribute to performance and schemata that are parasitic. If not combatted, this can lead to premature convergence. Among crossover operators, some are more disruptive than others, and traditional arguments have held that less disruption is better for implicit parallelism. To explore this issue we examine the behavior of two crossover operators, two-point and uniform crossover, on a problem contrived to contain specific spurious correlations. The more disruptive operator, uniform crossover, is more effective at combatting the spurious correlations at the expense of also more disruption of the effective schemata. Elitist selection procedures are shown to be able to ameliorate this somewhat, suggesting that research into ways of dealing with the effects of the inevitable sampling errors may lead to generally more robust algorithms.

Keywords: implicit parallelism, schema sampling, spurious correlation, premature convergence, uniform crossover, two-point crossover, elitist selection

1 Introduction

Traditional genetic algorithm (GA) theory has shown how many schemata may be sampled according to their observed fitnesses by the combined actions of survival-of-the-fittest selection and recombination. This is called implicit parallelism and is the major distinction of GAs. Nevertheless, GAs have sometimes been observed to converge prematurely to suboptimal solutions. To help explain this observation, we draw attention to spurious correlations which must inevitably arise when one attempts to exploit a number of schemata that is much larger than the number of samples (independent observations).

Recently Grefenstette and Baker (1989) have illustrated that the observed fitnesses that are exploited may not be a schema's true average fitness (as implied by the k-armed bandit analogy), but rather may be better than average samples within a schema. While this observation may spell trouble for the k-armed bandit analogy, it nevertheless is an example of what we would call proper search behavior; the better schemata are being properly sampled. We focus on an example of improper search behavior, in which some schemata (call them parasites) may be sampled at rates that are not justified by their true fitnesses. This is likely to occur when, through sampling error, all individuals that contain some valuable schema (a host) also contain the parasite(s). If the GA's recombination operator is not sufficiently vigorous in separating parasites from hosts, then the GA may converge to a population that includes the parasites. Note that this phenomenon is not an under sampling of observed schemata, but rather a failure to sample observed schemata in new contexts; a sacrifice of exploration in favor of exploitation.

This realization may help explain some recent results that suggest uniform crossover (defined later) may be more successful than the traditional 1- or 2-point crossover operator for some (perhaps many) problems (Ackley 1987, Syswerda 1989). GA traditionalists (e.g., Goldberg 1989, p. 119) have argued against this operator as being too disruptive to allow the propagation of the good schemata once they are found. Our consideration of spurious correlations has led us to reexamine this operator in light of its ability to disrupt them. We find that when uniform crossover is combined with a selection procedure that provides protection for previously located schemata through some form of elitism, then one may be able to have the benefits of vigorous exploration for disrupting spurious correlations without excessive sacrifice of the exploitation. We note that Ackley (1987) and Syswerda (1989) both used some form of elitism in their experiments.

After introducing the necessary concepts from schema sampling theory, we describe an artificial problem that allows us to examine the fate of deliberately introduced spurious correlations under the influence of different crossover and selection operators. While we do not claim that this illustration proves the general superiority of any one combination of operators, we believe that sampling errors are an inherent problem with implicit parallelism that forces a tradeoff between exploration and exploitation. Our illustration is designed to clarify the nature of these tradeoffs.

2 Schema Sampling Theory and Sampling Error

Before we can begin to analyze the expected behavior of a GA, we need to define schemata and introduce some of their features. Schemata are combinations of alleles which can be represented as strings over the same symbols used in the chromosomes, plus a "don't care" symbol (#). For example, if the chromosomes are bit strings (the alleles are 0 and 1) of length 4, then these three strings are schemata:

$$H_a = 1\ \#\ \#\ \#$$
$$H_b = 1\ \#\ \#\ 0$$
$$H_c = \#\ 0\ 0\ 1$$

Schema H_a represents the set of all the 4-bit strings that begin with a 1, H_b represents all the 4-bit strings that begin with 1 and end with 0; H_a is a superset of H_b and is independent of (i.e. shares no defined loci with) schema H_c. Schemata can be thought of as defining hypersurfaces (hence the symbol H) of the L-dimensional hypercube that represents the space of all bit strings of length L.

There are two features of schemata that are sometimes employed in the analysis of the behavior of GAs, schema order and schema defining length. The order of a schema is simply the number of non-# symbols in it, and defining length is the difference between the locus of the first and last non-# symbols (also the number of places within the schema where crossover might cut the string). The order and defining length of the example schemata are:

$$O(H_a) = 1 \quad D(H_a) = 0$$
$$O(H_b) = 2 \quad D(H_b) = 3$$
$$O(H_c) = 3 \quad D(H_c) = 2$$

Note that every chromosome is a member of 2^L different schemata. Hence, a population of M chromosomes will contain at least 2^L and at most $M2^L$ schemata. Even for modest values of L and M, these numbers can be very large so there will be a large number of schemata available for processing in the population. The inherent property of genetic search called *implicit parallelism* may be stated as follows:

Schema Sampling Theorem: *a large number of the schemata in a population will be sampled in future trials exponentially over time, where the growth rate is their fitness ratio.*

By fitness ratio (FR), we mean the ratio of the average fitness of the chromosomes in the current population that are members of the schema to the average fitness of the whole population. Thus, we expect the above average (FR > 1) schemata to increase in the gene pool and the below average (FR < 1) to decrease, both exponentially. Furthermore, this effect occurs without our having explicitly to compute these ratios; it occurs as a result of the combined action of survival-of-the-fittest selection and genetic recombination. It has been estimated that the "large number" referred to in the theorem is at least on the order of M^3 for some specified level of processing error (Goldberg 1989, p. 40).

The notion of processing error in this M^3 analysis refers to the disruption of schemata by the crossover operator, but there is another type of error that needs to be considered, sampling error. Anyone familiar with sampling error will immediately be suspicious of a claim that M^3 schemata can be reliably exploited based on only M observations. It is highly likely that at least some of the observed correlations will be spurious (i.e., result from sampling error) and an algorithm that overexploits them is likely to converge to a suboptimal solution.

3 Empirical Evidence

In this section we illustrate some of the aspects of spurious correlations as they appear during genetic searches. To create a sharp illustration, we have created a problem in which there is one highly influential schema H_1 that confers considerable survival advantage on any individual who possesses it. We also include other schemata that are much less influential, although not without influence. This seems to be a situation that is not unlikely to occur in real search problems. If, for instance, in a parameter optimization problem, there are some parameters whose high order bits strongly determine function value and there are other parameters, or some lower order bits, with less impact, then something like this situation is likely to be present.

To build a problem with this characteristic, we begin with a simple problem called One Max that has been studied by Ackley (1987), Syswerda (1989) and others. The One Max function simply returns the number of ones in any bit string given to it. One Max has a single global optimum and there is no epistasis (i.e. each locus makes an independent contribution to the fitness of the chromosome). As Ackley and Syswerda both observed, this problem is solved more effectively by a GA using uniform crossover than the traditional one-point (or the slightly better, two-point) crossover. Uniform crossover (UX) crosses the alleles independently at each locus with some probability (we use 0.5). Two-point crossover (2X), on the other hand, selects two loci at random and crosses over the string segment between them. We will use these two crossover operators for our illustrations.

To provide a baseline for comparison, we tested a 100-bit One Max Problem and our results are summarized in Table 1. The figures reported are those for the best solution found during each search. Note that this solution is not necessarily present in the final population; if a good solution was found and lost, then it is still reported as the best found. For this and all the tables below, we present statistics from 100 independent searches [1] with each operator. Since our GA minimizes, the 100-bit chromosomes were given a score of 100 minus the number of ones in the string. The global optimum of zero was found only once out of 100 tries by a GA using 2X, whereas UX failed to find it only four times, each time missing it by only a single bit. All of these searches terminated when the gene pool was converged. The fact that many of the searches using 2X converged on solutions with some "zero" alleles in them is evidence for improper exploitation of spurious correlations between these zeros and the ones in the better than average individuals.

One Max by itself is an uninteresting problem for studying variations on the GA because it has no epistasis (i.e. the contribution to fitness of each bit is independent of all others), and it is epistasis that frustrates simpler search methods. A simple hillclimber, for example, can solve One Max more quickly than a GA. We added a controlled amount of epistasis to the One Max problem by prepending a 10-bit header to an otherwise One Max string of 100 bits. The contribution to fitness of this header segment is equal to the maximum for the rest of the string (100 points) but is given only if all 10 bits are zero. Since our GA minimizes, we subtract each string's score from 200 giving the global optimum score, 0, for a string with 10 zeros followed by

[1] Each search used a population size of 50, 100% crossover, no mutation or inversion and Baker's unbiased selection procedure (Baker 1987). The fitness of each chromosome was scaled to the worst individual in the population, called *dynamic linear fitness scaling* by Grefenstette and Baker (1989). Each search was terminated when the population converged.

Table 1: Results for the One Max Problem

Crossover	best solution				trials to convergence	
	mean	sem[a]	min[b]	max	mean	sem
2X	4.23	0.20	0(1)	9	1640	16
UX	0.04	0.02	0(96)	1	1824	8

[a] Standard error of the mean.

[b] The number in parentheses is the number of times the minimum score of zero was found.

100 ones. This problem has one very influential schema of order 10 (H_1) and many other schemata of lesser import. O(H_1) is high enough that it is unlikely to be randomly introduced into a population ($p = 2^{-10}$) and it has no building blocks (i.e. lower order schemata that contribute to fitness and that can be combined to form H_1). Furthermore, its defining length (9 for a header of 10 contiguous bits) can be varied by placing the defined loci wherever we wish along the chromosome. We present one instance of increasing this defining length later. In all the experiments that we report, we seeded an otherwise random initial population with one instance of an all-zero string. This individual is very likely to be the most fit since it scores 100 and the expected score for its random neighbors is 150 (i.e., 200 - 50). By doing this, we have deliberately introduced a correlation between H_1 and poor alleles in the rest of the string.

How might we expect the two crossover operators to cope with this problem? Two-point crossover is known to possess considerable positional bias which means that two schemata are less likely to be disrupted (tested separately in the offspring) if they are close together on the chromosome than if they are far apart (Eshelman et al. 1989). This is the property that causes it to perform worse than uniform crossover on the One Max problem; the individuals with more ones than average propagate their schemata in the gene pool, taking some spurious zeros along with the ones. On the other hand, the highly disruptive nature of UX will make it hard to maintain H_1 while it vigorously searches for the maximum number of ones in the rest of the string. In other words, we expect 2X to tend to preserve both the host and the parasites, while UX will tend to disrupt them both. This example highlights the exploration/exploitation tradeoff that is at the heart of GA theory; the more "successful" algorithm will be the one that makes the more effective tradeoff. Given that there is a heavy penalty for disrupting the header bits — half the possible score — we would expect 2X to do much better than UX.

Before presenting the performance results, we illustrate the difference in disruptive behavior (positional bias) between these operators in Figure 1. Each curve presents the number of ones along the chromosome averaged for each final population and summed for 100 searches. It can be seen that for 2X, the "zero" alleles (spurious correlations) in the One Max part of the string are more likely to survive into the converged population when they are located close to H_1. Note that "close to H_1 applies to both ends of the string; 2X treats the chromosome as a ring. No such positional bias is present for UX. Also apparent is the disruption of H_1 by UX (there are some ones in the H_1 segment) and not by 2X.

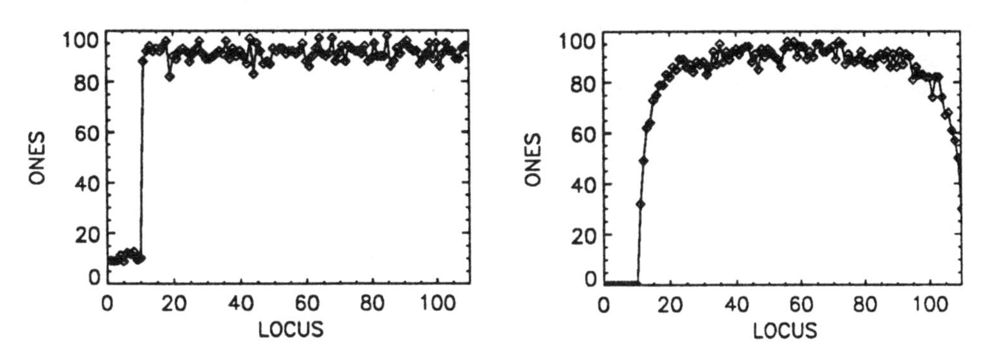

Figure 1. The number of ones along the chromosome averaged for each final population and summed for 100 searches with two-point and uniform crossover operators for the One Max problem with a 10-bit header.

The performance results, summarized in Table 2, confirm our hypothesis, at least on average.

Table 2: Results for the One Max with 10-bit Header Problem

Crossover	best solution					trials to convergence	
	mean	sem	median	min	max	mean	sem
2X	15.05	0.35	15	6	23	1755	20
UX[a]	24.62	3.32	12	0(10)	100(14)	11725	1876

[a] Of these 100 runs, 17 failed to converge and were halted after 50000 trials. These include the 14 runs that lost H_1.

In 14 of its searches, UX lost the H_1 schema (i.e. the best solution found was the seeded individual in the initial population) whereas 2X never lost it. Clearly 2X's lower disruption rate paid off. On the other hand, there were 26 searches using UX whose best result was better than the best results found by 2X, and UX found the global optimum on ten of these. So sometimes UX's increased exploration (via more vigorous recombination) pays off. The question we address next is whether there is any way that we can get the benefits of increased exploration without having to pay the full price of the consequent increase in disruption.

In a previous paper (Eshelman et al. 1989) we considered one strategy for doing this, using a crossover operator that is more disruptive than 2X but less so than UX, e.g., eight-point crossover. In the remainder of this paper, however, we shall explore an alternative strategy — what has been known in the traditional GA as the elitist selection strategy (Grefenstette 1986). The elitist strategy insures that the structure with the best performance always survives intact into the next generation. Although the elitist strategy has commonly been used with traditional one-point or two-point crossover, one would expect that it would benefit UX more than 2X, since it would counteract the tendency for UX to disrupt and lose the H_1 schema. This expectation was confirmed (again, on average) by repeating the above experiment using the elitist strategy. The results are given in Table 3.

Table 3: Results for the One Max with 10-bit Header Problem Using Elitism

Crossover	best solution					trials to convergence	
	mean	sem	median	min	max	mean	sem
2X	18.82	0.39	19	10	33	1446	20
UX	17.02	0.65	17	3	32	2036	20

Elitism significantly improved the average performance of UX and reduced its variance while making the average performance of 2X worse. Furthermore, UX's worst performance improved (it never lost H_1 since elitism guarantees this), but its best performance suffered (the global optimum was not located on any of the runs). Both searches tended to converge more rapidly with elitism. While elitism prevents the loss of H_1 (which 2X didn't need), it also biases the schema sampling in the gene pool by favoring the other schemata in the best individual, including any zero alleles (spurious correlations) in the One Max portion of the chromosome and so worsens the performance of 2X.

We have seen that the epistasis provided by the 10-bit header tends to give 2X an advantage over UX, but there is no reason to suppose that in a real search problem a highly epistatic schema will consist of contiguous bits. We can illustrate the impact of increasing D(H_1) by breaking H_1 into two segments of 5 bits each with a One Max segment between. This will increase 2X's probability of disrupting it. Figure 2 shows the distribution of alleles in the solutions when the string is still 110 bits, but the gap is 50 bits. Behavior similar to Figure 1 is seen again [2]. Table 4 shows that the performance of UX on this problem is essentially the same as it was on the previous one, but 2X does significantly worse when its positional bias is no longer an advantage.

Although using the elitist strategy significantly improves the mean performance of UX, it hurts the performance of 2X. The reason seems to be that it biases selection in favor of individuals containing the H_1 schema. One effect of this is that zeros outside this critical schema are also propagated throughout the population. One would expect that this would also happen with UX, but its undesirable effects are counteracted by the guaranteed preservation of the H_1 schema. This undesirable effect, however, can be countered by a more even-handed elitism — instead of preserving the single best individual seen so far, a population-elitist strategy can be adopted. Every individual mates every generation without regard for its fitness. Then all the offspring are pooled

[2] The effect of the elitist strategy (not shown) is to shift these curves down by a similar amount in both the header and One Max segments.

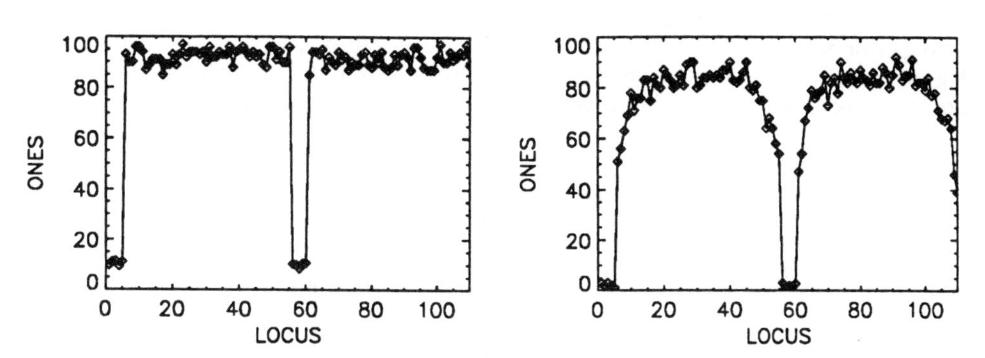

Figure 2. The number of ones along the chromosome averaged for each final population and summed for 100 searches with two-point and uniform crossover operators for the One Max problem with a 50-bit gap in H_1 (without elitism).

Table 4: Results for 50-bit Gap Problem

Cross	elitist	best solution					trials to conv.	
		mean	sem	median	min	max	mean	sem
2X	yes	26.34	0.60	26	14	44	1389	21
2X	no	25.41	1.56	23	11	100(2)	2182	484
UX	yes	16.54	0.66	16	4	37	2057	20
UX	no	23.93	3.15	12	0(13)	100(11)	11588	1862

with all the parents and the best 50% are selected to yield the next generation [3]. The performances using this form of selection are shown in Table 5. Comparing with Table 2, we see that both crossovers' performances were improved with more conservative selection, although UX realized the greater benefit. This form of selection seems to more successfully maintain population diversity, as reflected in the larger number of trials to convergence. This in turn gives both crossover operators

[3] This is the strategy used by CHC (Eshelman 1991). Syswerda (1989) and Whitley (1989) have used a selection strategy somewhat similar to this. However, their procedures differ in that they both produce offspring and introduce them into the population one or two at a time instead of in batches (generations); they both apply a performance bias in parent selection; and they do not compare the performance of the offspring to that of the string it replaces. Ackley (1987) used a variation in which an offspring replaced a string in the population chosen at random from among those worse than the mean. The Evolution Strategy work in Germany (Hoffmeister and Bäck 1990) first used this type of selection (called (μ+λ)-ES), but later opted for the selection scheme of the traditional GA (called (μ,λ)-ES) to be able to cope better with non-stationary and noisy tasks.

more time to build up good schemata from the building blocks in the gene pool, while working against the spurious correlations (they propagate more slowly). The very good performance of UX combined with population elitist selection seems to show that one can exploit the benefits of the vigorous search from a highly disruptive crossover operator by combining it with a conservative selection operator.

Table 5: Results for One Max with 10-bit Header Problem with Population Elitist Selection

Crossover	best solution					trials to convergence	
	mean	sem	median	min	max	mean	sem
2X	11.81	0.36	11	5	23	2267	22
UX	4.78	1.44	1	0(32)	100(2)	2435	45

4 Conclusions

A GA is able to search many schemata in parallel because the operator it uses to generate new trials — crossover — disrupts relatively few schemata. Unfortunately, this same property is highly susceptible to exploiting spurious correlations between schemata that contribute to performance and schemata that are parasitic. This phenomenon tends to make GAs subject to premature convergence on suboptimal solutions.

We have argued that since spurious correlations are inherent to GAs, they cannot be eliminated; however, we have explored ways in which their effect might be ameliorated. We have shown that on some problems where there is little epistasis, a highly disruptive recombination operator like uniform crossover may be more of an asset than a liability. Furthermore, we have argued that even when a highly epistatic element is introduced into the problem, the disruptive nature of uniform crossover can sometimes be compensated for by introducing an elitist strategy which preserves the single best individual found so far. It may be noteworthy that the recently published studies (with which we are familiar) (Ackley 1987, Caruana et al. 1989, Eshelman et al. 1989, Schaffer and Morishima 1987) showing an improvement over the traditional GA with more disruptive crossover operators have all used some form of elitism. We have noted, however, that the individual-elitist strategy, by repeatedly reintroducing the same individual, also favors spurious correlations. Finally, we have suggested that this second effect can be ameliorated by switching to a population-elitist strategy in which new individuals are introduced into the population only by replacing the worst members of the population. This selection scheme, by giving all members of the current generation the same number of offspring and making replacement contingent upon fitness, delays convergence and gives crossover more time to build up superior schemata.

It should be stressed that it has not been our purpose to demonstrate the superiority of any particular combination of operators and selection strategies. The problem we used through out this paper was contrived to demonstrate the conflicting demands of preserving a critical schema and avoiding spurious correlations. For any interesting problem, some tradeoff will be called for and will be critical in determining the outcome. We do feel that the GA community, by focusing on the power of implicit parallelism, has tended to overlook the negative effects of spurious correlations and consequently ignored potential ways of ameliorating this effect.

If one may characterize traditional thinking as seeing fitness proportional reproduction as the source of implicit parallelism, then crossover is viewed as a source of loss (by schemata disruption). We suggest an inversion of this thinking by seeing crossover as the source of search and the limited population size as a source of loss (by forcing some previous results to be forgotten). Population-elitist selection foregoes fitness proportional reproduction at each generation in order to minimize this loss. And, as Eshelman points out, fitness proportional reproduction does occur with population-elitist selection, but over many generations rather than at every generation (Eshelman 1991) since the number of matings is proportional to lifespan. This seems more similar to what we observe in Nature.

References

D. H. Ackley. (1987) *A Connectionist Machine for Genetic Hillclimbing.* Boston, MA: Kluwer Academic Publishers.

J. E. Baker. (1987) Reducing Bias and Efficiency in the Selection Algorithm. *Genetic Algorithms and Their Applications: Proceedings of the Second International Conference on Genetic Algorithms*, 14-21. Hillsdale, NJ: Lawrence Erlbaum Associates.

R. A. Caruana, L. J. Eshelman and J. D. Schaffer. (1989) Representation and Hidden Bias II: Eliminating Defining Length Bias in Genetic Search via Shuffle Crossover. *Proceedings of the 11th International Joint Conference on Artificial Intelligence*, San Mateo, CA: Morgan Kaufmann.

L. J. Eshelman, R. A. Caruana and J. D. Schaffer. (1989) Biases in the Crossover Landscape. *Proceedings of the Third International Conference on Genetic Algorithms*, 10-19. San Mateo, CA: Morgan Kaufmann.

L. J. Eshelman. (1991) The CHC Adaptive Search Algorithm: How to Have Safe Search When Engaging in Nontraditional Genetic Recombination. In G. J. E. Rawlins (ed.), *Foundations of Genetic Algorithms and Classifier Systems.* San Mateo, CA: Morgan Kaufmann.

D. E. Goldberg. (1989) *Genetic Algorithms in Search, Optimization, and Machine Learning.* Reading, MA: Addison Wesley.

J. J. Grefenstette. (1986) Optimization of Control Parameters for Genetic Algorithms. *IEEE Transactions on Systems, Man & Cybernetics*, **SMC-16**(1): 122-128.

J. J. Grefenstette and J.E. Baker. (1989) How Genetic Algorithms Work: A Critical Look at Implicit Parallelism. *Proceedings of the Third International Conference on Genetic Algorithms*, 20-27. San Mateo, CA: Morgan Kaufmann.

F. Hoffmeister and T. Bäck. (1990) Genetic Algorithms and Evolution Strategies: Similarities and Differences. *Proceedings of the First International Workshop on Parallel Problem Solving from Nature*. Dortmund, Germany: University of Dortmund.

J. D. Schaffer and A. Morishima. (1987) An Adaptive Crossover Distribution Mechanism for Genetic Algorithms. *Genetic Algorithms and Their Applications: Proceedings of the Second International Conference on Genetic Algorithms*, 36-40. Hillsdale, NJ: Lawrence Erlbaum Associates.

G. Syswerda. (1989) Uniform Crossover in Genetic Algorithms. *Proceedings of the Third International Conference on Genetic Algorithms*, 2-9. San Mateo, CA: Morgan Kaufmann.

D. Whitley. (1989) The GENITOR Algorithm and Selection Pressure: Why Rank-Based Allocation of Reproductive Trials is Best. *Proceedings of the Third International Conference on Genetic Algorithms*, 116-121. San Mateo, CA: Morgan Kaufmann.

PART 3

CLASSIFIER SYSTEMS

Representing Attribute-Based Concepts in a Classifier System

Lashon B. Booker
Artificial Intelligence Technical Center
The MITRE Corporation
7525 Colshire Drive
McLean, VA 22102

Abstract

Legitimate concerns have been raised about the expressive adequacy of the classifier language. This paper shows that many of those concerns stem from the inadequacies of the binary encodings typically used with classifier systems, not the classifier language per se. In particular, we describe some straightforward binary encodings for attribute-based instance spaces. These encodings give classifier systems the ability to represent ordinal and nominal attributes as expressively as most symbolic machine learning systems, without sacrificing the building blocks required by the genetic algorithm.

Keywords: Classifier systems, concept learning, feature manifolds, binary encodings, classifier representations.

1 Introduction

Classifier system rules are fixed length strings in the alphabet { `0,1,#`} . The left hand side of each rule is a simple conjunctive expression having a limited number of terms. Consequently, a single classifier cannot be used to express arbitrary, general relationships among structured attributes. The simplicity of this language is part of a deliberate tradeoff of the descriptive power of individual rules for overall adaptive efficiency (Booker, Goldberg, & Holland,1989). A simple syntax facilitates learning under the genetic algorithm by offering a large repertoire of structural regularities (or *building blocks*) that are easily extracted, evaluated, and recombined in useful ways. This is a powerful substrate for learning that, it is claimed, gives classifier systems a robust ability to adapt. Moreover, it is claimed that the sacrifice of descriptive power in individual rules does not severely limit the capabilities of classifier systems that use *sets* of rules to represent knowledge.

Though sets of classifiers are sufficient in principle to represent complex concepts and relationships, legitimate concerns have been raised about the expressive adequacy of the classifier language (Belew & Forrest,1988; Carbonell,1989). For example, one can easily produce concept descriptions in the classifier language that are difficult to interpret or have weak semantics (Belew & Forrest,1988). Moreover, the classifier language seems to be — at first glance anyway — too "low level" (Grefenstette, 1989) to support straightforward requirements for understanding and communicating the knowledge a system has learned. These requirements are easily met using a "high level" rule language. It is therefore reasonable to ask if the descriptive power of the classifier language is sufficient in any practical sense for realistic learning problems.

This paper points out that, to a large extent, these difficulties stem from the inadequacies of the binary encodings typically used with classifier systems, not the classifier language per se. That is why it is a mistake to conclude (eg. Carbonell (1989)) that the classifier language is incapable of supporting the kinds of symbolic concept descriptions that are easily communicated to humans or interpreted by other software systems. Binary encodings may be "low level" syntactically, but that does not mean they lack significant expressive capabilities. In fact, given an adequate set of encodings, the descriptive power of the classifier language compares favorably with the description languages used in many symbolic paradigms to learn concepts in attribute-based instance spaces. The primary contribution of this paper is to introduce some of these more expressive encodings.

We begin with a review of the standard binary encodings used in classifier systems, and discuss their strengths and weaknesses. In subsequent sections we describe more powerful binary encodings and indicate how they can be used to produce characteristic concept descriptions of the kind found in "high level" symbolic learning systems.

2 Standard Binary Encodings

By far the most widely used encodings in classifier systems are the simple ones that order a set of properties or objects, then map them directly into consecutive bits in a string , or into consecutive numbers represented as bit strings. Two kinds of simple encodings are prevalent: feature lists and linear orderings.

Simple Feature Lists In the feature list encoding each bit corresponds to a single property, feature-value pair, or predicate. Bit i in the bit string representation is `1` if feature i is present, `0` otherwise. The corresponding classifier syntax for concept descriptions uses a `1` for each feature that is present in all concept instances, `0` for each feature that is absent in all instances, and `#` for features that have no consistent value among instances and are therefore irrelevant. The conjunction of common feature values and irrelevant features serves as a definition of the pattern class. This encoding is useful for learning pattern classes defined by a set of critical features whose presence or absence is mandatory for class membership. Pattern classes defined by a disjunction of these simple conjunctive expressions can be represented by a set of classifiers, each classifier representing one disjunct.

Simple Linear Orderings In the linear orderings, a string of bits is used to represent parameter values belonging to a linearly ordered set of mutually exclusive values. Examples include integer–valued parameters, continuous parameters represented to some fixed precision, and symbolic parameters that can be mapped under similarity–preserving transformations onto a set of numbers. The bit string representation is usually either a binary coding or a Gray coding[1]. Gray coding schemes are sometimes preferred for genetic algorithms because they preserve distance relationships between adjacent values (Caruana & Schaffer,1988). Under these simple encodings, classifiers are generalizations that correspond to hyperplanes in the space $\{0,1\}^l$ where l is the length of the bit strings. Hyperplanes are a rich, but not exhaustive, collection of generalizations over this representation space that are especially helpful for learning pattern classes defined by periodicities in the parameter space. Arbitrary generalizations can be represented by using sets of classifiers.

Goldberg's (1985) gas pipeline control model is a good example of how to properly use these simple encodings in an application. The input message from the environment is composed of several parameters. The ranges of continuous parameters like pipeline pressure and flow are partitioned into subintervals. The values of these parameters are represented by binary integers that identify which subinterval a value belongs to. Parameters like "time of year" are adequately represented by a simple predicate that distinguishes between two states such as "winter" and "summer". A more complex example of how to use the simple encodings is given by Forrest (1985). Her work shows how the knowledge contained in a symbolic semantic network description (eg. KL-ONE or NETL) can be mapped into a set of classifiers that support the same information retrieval operations. This mapping demonstrates that, even using simple encodings, classifier systems can implement powerful knowledge representation constructs.

Though simple encodings are powerful in principle, it is clear that for many problems they are not sufficiently expressive. One can easily produce classifiers that are difficult to interpret or have weak semantics. This happens, for instance, whenever the structural similarities among concept instances are not all preserved under the binary encoding; or, whenever the binary representation admits bits or bit combinations that do not correspond to any meaningful set of attributes. The simple encodings also make it difficult to achieve certain common kinds of generalizations. For example, it is awkward to use them for nominal parameters that have unordered values, or for making graded – rather than all or none – comparisons between two numbers. Attempts to use simple encodings for these purposes can quickly lead to unsatisfactory classifier system implementations (eg. Schuurmans & Schaeffer,1989).

---

[1] One simple way to generate a Gray coded string $g_0 g_1 \ldots g_n$ is to start with the binary coded string $b_0 b_1 \ldots b_n$ and use the relationships $g_0 = b_0$, $g_i = b_i \,\hat{}\, b_{i-1}$ for $i > 0$, where ^ is the exclusive–OR operator.

The observations of Schuurmans and Schaeffer are especially relevant here since they discuss the weaknesses of simple encodings in considerable detail. Four specific issues are identified:

1. Search bias — The standard encodings unavoidably make some generalizations easy to learn and other generalizations difficult, if not impossible, to learn. The severity of this bias is reduced considerably if we use several classifiers to represent a concept. In that case, though, questions are raised about how *succinctly* each concept can be expressed.
2. Limited disjunction — Disjunctive combinations are an important class of abstractions, and most disjunctions cannot be represented using a single classifier condition. Here again, sets of classifiers provide an answer but expressive adequacy is still a concern.
3. Positional semantics — It is very easy to design representations in which important dimensions of a problem are implicitly encoded in the length of the string, the juxtaposition of two substrings, etc. None of these aspects of the representation are amenable to generalization. In order to avoid position dependent semantics, one often has to solve a parameterization problem (see below), or use rule conditions capable of encoding either a disjunctive combination or a range of attribute values.
4. Parameterization — There is no mechanism in classifier representations to bind selected message bits to match #'s in identical positions over separate classifier conditions. Consequently, classifier systems cannot easily parameterize solutions which require some agreement among several distinct classifiers about matched messages.

Schuurmans and Schaeffer argue that the root cause of these problems is the lack of expressiveness in classifier representations. Their remedy is to move the classifier system framework closer to traditional symbolic paradigms by adding more complex syntactic constructs like variables to the classifier language.

In considering the issues Schuurmans and Schaeffer raise, the point about search bias can be quickly dismissed because it is based on naive assumptions about the role of bias in learning. It has long been recognized that inductive bias is *necessary* in any generalization scheme that makes inductive leaps beyond the data that has been observed (Mitchell, 1980). Furthermore, there is no "universal" bias that facilitates learning in all domains (Dietterich, 1989). An effective bias must be provided by the system designer and chosen in a domain dependent manner. The fact that Schuurmans and Schaeffer can identify a bias in the classifier language is therefore not surprising. Search biases are a fundamental characteristic of all symbolic approaches to generalization, not some particular weakness of the classifier system methodology. The real issue is whether or not classifier systems allow us to implement biases that are helpful for a given task. That point is closely related to the other issues Schuurmans and Schaeffer raise.

The parameterization issue cannot be dismissed so easily. Parameterization and variable binding are problematic in classifier systems for many of the same reasons they pose difficulties for connectionist systems. It is not clear, though, that the way to address this problem is to introduce variables into the classifier language as

Schuurmans and Schaeffer suggest. If the classifier language is made more complex so that traditional programming methods can be easily used, the efficient matching operations and the well-chosen building blocks the language provides may be lost. It is more desirable to consider representation schemes that reduce the need for variables! Indexical representations (Agre, 1988) are a promising alternative account of representation designed to do just that. These representations are grounded in perception and action, thereby avoiding most of the overhead associated with reasoning about arbitrary, abstract variables. A limited number of *markers* with pre-defined semantics are implemented and dynamically bound to objects in the world by a sensory-motor subsystem. Internal representations only need to refer to and reason about the properties of marked objects. This approach drastically reduces the size of the parameterization problem, and makes variable binding — to the extent that it is needed at all — an explicit *action* that occurs in the context of some specific situated activity. Indexical representations have been used successfully in connectionist networks (Agre & Chapman, 1988) and in reinforcement learning systems (Whitehead & Ballard, 1990) to manage the parameterization problem. They are a promising option for handling variable binding in classifier systems as well.

In the next section we address the remaining two issues raised by Schuurmans and Schaeffer, paying particular attention to the technical problem of using the classifier language to represent both disjunctive generalizations and intervals of parameter values.

3 More Expressive Encodings

Because the simple encodings have been used almost exclusively in most classifier system implementations, the weaknesses of these encodings have been misconstrued as weaknesses of the classifier language. However, it has been known for some time (Booker,1982) that expressive encodings for classifier system messages can be derived from Hayes-Roth's (1976) work on pattern classification. These encodings increase the expressiveness of individual rules without any change in the syntax of the classifier language. In this section we review Hayes-Roth's ideas and show how they can be used in a classifier system.

3.1 Feature Manifolds

Hayes-Roth (1976) developed a system for pattern classification that used bit strings to represent both pattern exemplars and pattern classes. He advocated the use of bit string encodings because they afford efficient methods for classification and abstraction whenever the bits correspond to characteristic features. The bitwise logical product of two exemplar strings yields the set of features common to both exemplars. The match between an exemplar string and a pattern string can be determined by simple bitwise masking operations. The key to the power and applicability of this approach rests on an ability to properly organize the features associated with a pattern classification problem. If the features can be organized in a manner that reveals a (possibly new) set of discrete features that facilitate the required generalizations, then the bit string approach can be used to solve the problem. Hayes-Roth calls such an organization of features a *feature manifold.* Two manifold types he devised can be easily adapted for use in classifier systems[2].

3.1.1 Ordinal Manifolds

An ordinal parameter is one whose range is a linearly ordered set of mutually exclusive values. Ordinal feature manifolds are designed to facilitate graded comparisons of ordinal, continuous, or noisy data. These manifolds are analogous to the organization of receptive fields in visual perception, and are similar to the coarse coding techniques commonly used in connectionist systems. The basic idea is that the range of an ordinal parameter X is covered by a finite set of adjacent, overlapping intervals $\{[x_1-r_1, x_1+r_1], \ldots, [x_n-r_n, x_n+r_n]\}$. Each number r_i is called a *radius of generalization* for the manifold. The intervals can be viewed as receptive fields in the sense that the occurence of any value x within radius r_i of x_i simultaneously implies the occurence of x_i. Each x_i is considered a *property* in the manifold, and all data values x are described in terms of the presence or absence of these properties. The bit string representation for values of X is a string of length n where bit i indicates whether or not the value belongs to the receptive field centered at x_i. The properties shared by two values can be computed using a simple bitwise logical product of their bit string representation. The manifold radii r_i determine the level of discrimination possible between values of X by specifying the amount overlap between adjacent receptive fields. Ordinal manifolds preserve some information about distance in the sense that the number of properties shared by two values is a monotonic function of the distance between those two values. Two values will share no properties if they differ by more than $r + d$ where d is the amount of separation between the centers of adjacent fields and r is the maximum of the r_i.

For example, suppose the range of X is the continous interval $[0, 10]$ and we choose the eleven integers in the interval as centers for receptive fields of radius $r_i = 1.5$, $i = 0, \ldots, 10$. In this manifold, the value 2.9 is represented by the string 00111000000 and the value 5.2 is represented by the string 00001110000. The bitwise logical product of these two strings is 00001000000, indicating that the receptive field centered at 4 is the only field containing both values. This encoding directly supports inferences about the range of ordinal data that is characteristic of the underlying pattern class. The generalization radii can be interpreted either as typical measurement errors to be accounted for when comparing values and generalizing; or, as the parameters of discriminability that determine which values are assigned equivalent representations.

3.1.2 Complementary Manifolds

Complementary feature manifolds are structured to facilitate the discovery of disjunctive regularities using conjunctive bit string representations. The organization of complementary feature manifolds is roughly analogous to the way visual perception systems are organized to represent mutually exclusive features. Perceptual systems often represent mutually exclusive features using mutually inhibitory neural structures whose receptive fields have complementary states. In an analogous way, representations in a complementary manifold are composed of sets of paired, mu-

---

[2]These manifolds could also be helpful when using a genetic algorithm to solve an optimization or search problem.

tually exclusive properties. A complementary encoding augments a simple feature list description with a complementary list of "negative" features to explicitly represent those attributes an input does *not* have. This organization makes it possible to represent a disjunction of feature list properties as a conjunction of "negative" features.

More specifically, assume that we need to represent a nominal parameter V whose range is the finite, unordered set of mutually exclusive values $\{v_1, \ldots, v_n\}$. V could also be a tree-structured attribute, or even an ordinal attribute for which disjunctive abstractions are important. We can represent the values of V with a bit string of length $2n + 1$

$$v \longmapsto \varepsilon x_1 \ldots x_n y_1 \ldots y_n$$

where ε is a bit representing the property "at least one value from the range of V". Each x_i denotes the presence of value v_i as in a simple feature list encoding. Each $y_i \equiv \neg x_i$ is a negative property that denotes the *absence* of v_i. As with other conjunctive bit encodings, the properties shared by two values can be computed using a simple bitwise logical product. Such a comparison of two distinct values v and v' will reveal no common "positive" feature list properties, but it will extract ε and all of the common negative properties. This conjunction of negative properties represents the disjunction of the positive properties associated with each value.

For example, consider a nominal parameter C denoting color that takes on values from the set { red, orange, yellow, green, blue }. The bit string encodings for these values in the complementary manifold are as follows:

red	$\mapsto$	1 10000 01111	orange	$\mapsto$	1 01000 10111
yellow	$\mapsto$	1 00100 11011	green	$\mapsto$	1 00010 11101
blue	$\mapsto$	1 00001 11110			

Disjunctive generalizations emerge from the bitwise product of these encodings.

$$\begin{aligned} \text{red} \otimes \text{yellow} \otimes \text{blue} &= \texttt{1 00000 01010} \\ &\equiv \varepsilon \wedge \neg\text{orange} \wedge \neg\text{green} \\ &\equiv \text{red} \vee \text{yellow} \vee \text{blue} \end{aligned}$$

In this case ε, $\neg$orange and $\neg$green are the properties shared by red, yellow, and blue. The conjunction of these shared properties is equivalent to the desired disjunctive abstraction.

3.2 Manifold Encodings for Classifier Systems

The application of Hayes-Roth's ordinal and complementary manifolds to classifier systems is fairly straightforward. In this section we illustrate the many possibilities with a few specific examples.

3.2.1 Graded Comparisons and Intervals of Values

When classifier system messages are encoded using an ordinal manifold, classifiers represent intervals defined by combinations of receptive fields. Consider the example given above of an ordinal manifold in which the value 2.9 is represented by the string 00111000000 and the value 5.2 is represented by the string 00001110000. The most specific classifier string matching both values is 00##1##0000, a classifier that represents the interval $(2.5, 5.5)$. This is a reasonable generalization of the two inputs 2.9 and 5.2 if we assume that input values vary stochastically, within limits specified by r_i, from some underlying prototypical value. In this case, the prototypical value is hypothesized to be 4.0. The non-specific classifier ########### represents any set of values whose variability exceeds the specified limits.

Representing All Subintervals It is often useful to have a distinct classifier representation for every subinterval of the parameter range. One way to do this is illustrated by the following manifold which we will call the *complete ordinal manifold.* Assume without loss of generality that the range of the parameter X is the set of integers $\{1, \ldots, n\}$. We define receptive fields anchored at each of these integers with left radius $r_l = \lfloor (n-1)/2 \rfloor$ and right radius $r_r = \lceil (n-1)/2 \rceil$. Every value of X is represented in this manifold by a bit string $b_1 \ldots b_n$ where $b_j = 1$ if $x \in [j - r_l, j + r_r]$, $b_j = 0$ otherwise. For example, if $n = 5$ the complete ordinal manifold is specified by the receptive fields

$$\{[-1,3],[0,4],[1,5],[2,6],[3,7]\}$$

The integer parameter values are represented by the following bit strings:

$1 \mapsto$ 11100 $\quad 2 \mapsto$ 11110 $\quad 3 \mapsto$ 11111 $\quad 4 \mapsto$ 01111 $\quad 5 \mapsto$ 00111

Every subinterval $[x, y]$ of the parameter range is represented by the most specific classifier string that matches both x and y.

$[1,2] \mapsto$ 111#0 $\quad [1,3] \mapsto$ 111## $\quad [1,4] \mapsto$ #11## $\quad [1,5] \mapsto$ ##1##
$[2,3] \mapsto$ 1111# $\quad [2,4] \mapsto$ #111# $\quad [2,5] \mapsto$ ##11# $\quad [3,4] \mapsto$ #1111
$[3,5] \mapsto$ ##111 $\quad [4,5] \mapsto$ 0#111

Note that an integer belongs to an interval just in case its bit string representation matches the classifier representing that interval. Moreover, this encoding of the integers is a distance-preserving encoding. This is especially important for interpreting partial matches, because it gives them meaning in terms of the size of the semantic disparity.

Handling Large Parameter Ranges If n is large, we can achieve a more compact encoding by using a hierarchical combination of coarse and fine receptive fields, together with sets of classifiers. The basic organizing principle, which we will call a *hierarchical ordinal manifold,* is to represent each value at several levels of precision using a complete ordinal manifold at each level. The levels correspond to successively refined partitions of the parameter range. For example, if $n = 1000$ we can cover the entire parameter range $[1, 1000]$ with ten "coarse" receptive fields $\{[i - 499, i + 500] : i = 100, 200, \ldots, 1000\}$. We can use the centers of these fields to partition the range into ten subintervals $[i-99, i]$. Each partition element can be covered by ten "medium" sized receptive fields $\{[j-49, j+50] : j \in [i-99, i], j \equiv 0 \pmod{10}\}$. Repeating the procedure, we refine each partition element into ten smaller subintervals. Each element $[j - 9, j]$ of the refined partition is covered with ten "fine" grained receptive fields $\{[k-4, k+5] : k \in [j-9, j]\}$. This nested collection of partitions and receptive fields is derived in a convenient, straightforward way using divisors[3] of n. At each of the three levels of refinement, a parameter value belongs to exactly one partition element. The properties of a value in this manifold are determined by the ten receptive fields covering each partition element. Values are represented by bit strings of length 30 containing one bit for each property at each level. For instance,

```
                "coarse"     "medium"      "fine"
  50   ↦     1111100000   1111111110   0000111111
  376  ↦     1111111100   0011111111   1111111111

             [50,100]   ↦   1111100000   ####11111#   ##########
[50,376] ↦ { [101,300]  ↦   111111#000   ##########   ##########
             [301,370]  ↦   1111111100   #1111#####   ##########
             [371,376]  ↦   1111111100   0011111111   11111#####
```

In this manifold, every subinterval of $[1, 1000]$ can be represented using at most five classifiers.

Effects of Genetic Operators The representational power of ordinal manifolds is not really helpful in a classifier system unless the interpretations are preserved under manipulation by the genetic algorithm. It is clear that arbitrary classifier strings do not necessarily have a meaningful interpretation in an ordinal manifold. It follows that arbitrary crossovers and mutations of classifiers using this encoding can lead to uninterpretable strings. Note, however, that if we apply crossover to two classifiers representing intervals that share at least one value, we always get a legal resultant string. This means that an ordinal manifold is invariant under crossover whenever the strings involved have at least one message that they both match. Fortunately, this requirement is easily and routinely satisfied in many classifier system implementations[4]. It is always true in classifier systems that use a *restricted mating policy* to help discover specialized clusters of rules correlated with the various pattern categories in the environment (Booker, 1985). Consequently, if the population is initialized with legal strings and the restricted mating policy is in effect, the semantic disruption of the genetic algorithm is limited to the minimal effects of mutation.

This point is important and is worth discussing in more detail. In the examples given so far, subintervals of the parameter range are represented by the most specific classifier string that matches both endpoints of the interval. In fact, there are other more general classifier strings that also have an unambiguous interpretation as subintervals. This becomes clear if we define an equivalence relation over classifier strings based on the set of legal input strings they match. For example, given the complete ordinal manifold over the set of integers $\{1, 2, 3, 4, 5\}$, the strings `11100` and `###00` are equivalent because each one only matches the single input `11100`. Here is the complete set of equivalence classes for legal classifier strings in this example:

---

[3]The partitions do not have to be derived from divisors of n. Similar manifolds can be constructed using nested partitions based on other criteria such as remainder modulo word size.

[4]Note that this requirement is difficult to satisfy in a system like SAMUEL (Grefenstette, 1989) in which each string manipulated by the genetic algorithm is a complete *set* of rules

$$\begin{array}{lll}
[1,1] \mapsto \mathtt{\{1\#\}^3 0\{0\#\}^1} & [1,2] \mapsto \mathtt{\{1\#\}^3\#0} & [1,3] \mapsto \mathtt{1\{1\#\}^2\#\#} \\
[1,4] \mapsto \mathtt{\#1\{1\#\}^1\#\#} & [1,5] \mapsto \mathtt{\#\#\{1\#\}^1\#\#} & [2,2] \mapsto \mathtt{\{1\#\}^3 10} \\
[2,3] \mapsto \mathtt{1\{1\#\}^2 1\#} & [2,4] \mapsto \mathtt{\#1\{1\#\}^1 1\#} & [2,5] \mapsto \mathtt{\#\#\{1\#\}^1 1\#} \\
[3,3] \mapsto \mathtt{1\{1\#\}^3 1} & [3,4] \mapsto \mathtt{\#1\{1\#\}^2 1} & [3,5] \mapsto \mathtt{\#\#\{1\#\}^2 1} \\
[4,4] \mapsto \mathtt{01\{1\#\}^3} & [4,5] \mapsto \mathtt{0\#\{1\#\}^3} & [5,5] \mapsto \mathtt{\{0\#\}^1 0\{1\#\}^3}
\end{array}$$

We use the notation $\{s_1 \ldots s_m\}^k$ here to indicate a substring of length k in the alphabet $\{s_1, \ldots, s_m\}$.

The implications of this more general interpretation of classifier strings are revealed if we now look at the set of strings representing all subintervals containing a given input value:

$$\begin{array}{rcl}
1 & \in & [1,1],[1,2],[1,3],[1,4],[1,5] \\
 & \equiv & \mathtt{\{1\#\}^3\{0\#\}^2} \\
2 & \in & [1,2],[1,3],[1,4],[1,5],[2,2],[2,3],[2,4],[2,5] \\
 & \equiv & \mathtt{\{1\#\}^4\{0\#\}^1} \\
3 & \in & [1,3],[1,4],[1,5],[2,3],[2,4],[2,5],[3,3],[3,4],[3,5] \\
 & \equiv & \mathtt{\{1\#\}^5} \\
4 & \in & [1,4],[1,5],[2,4],[2,5],[3,4],[3,5],[4,4],[4,5] \\
 & \equiv & \mathtt{\{0\#\}^1\{1\#\}^4} \\
5 & \in & [1,5],[2,5],[3,5],[4,5],[5,5] \\
 & \equiv & \mathtt{\{0\#\}^2\{1\#\}^3}
\end{array}$$

For each input string $\vec{s}$, the set of all subintervals containing $\vec{s}$ is represented by the set of all possible classifier strings that match $\vec{s}$. This set happens to be the *closure* (Liepins & Vose, in review) of the set $\{\vec{s}, \vec{\#}\}$ under the family of all two point crossovers[5], where $\vec{\#}$ denotes the string of all #'s.

It is now straightforward to formally prove that ordinal manifolds are invariant under crossover when a restricted mating policy is in effect. The set of all possible classifier strings that match an input $\vec{s}$ is exactly the closure of $\{\vec{s}, \vec{\#}\}$ under crossover, and the restricted mating policy assures us that only strings belonging to this closure will be crossed. These statements can be made more precise using the formal characterization of crossover provided by Liepins and Vose (in review).

Regarding the effects of mutation, it is clear that small mutation rates will have minimal effects on the semantics of a population of classifier strings. However, these semantic disruptions can be eliminated entirely if we perform mutations at the "symbolic" level rather than at the bit level. For example, mutation could be structured to alter one of the bounds of a subinterval as in the SAMUEL (Grefenstette, 1989) approach to learning rules.

3.2.2 Disjunctive Generalizations

Hayes-Roth's formulation of complementary encoding uses two bits for each parameter value even though the bits carry equivalent information. This redundancy is necessary because he uses binary strings to represent both generalizations and

[5]The same is true for uniform crossover

instances. A binary generalization using a single bit per value cannot distinguish between values that are absent in all instances, and values that are present in some instances and absent in others. This distinction is easily made in the classifier language, however, by using the "#" symbol. Consequently, classifier system messages can be encoded using a complementary manifold that requires only one bit per parameter value. Given n parameter values, an input value is represented with a bit string of length $n + 1$

$$v \mapsto \varepsilon y_1 \ldots y_n$$

where ε is defined as before and y_i denotes the absence of value i. In the color parameter example cited earlier, classifier system messages for the five colors { red, orange, yellow, green, blue } would be encoded as follows:

red	$\mapsto$	1 01111	orange	$\mapsto$	1 10111
yellow	$\mapsto$	1 11011	green	$\mapsto$	1 11101
blue	$\mapsto$	1 11110			

Note that this encoding can also be used to denote ambiguous inputs such as

blue $\wedge$ green $\mapsto$ 1 11100

if we assume that ambiguities are indicated when two or more property detectors are "on" at the same time.

Classifiers that match messages encoded using this complementary manifold are generalizations that correspond to a conjunction of manifold properties. As with Hayes-Roth's binary generalizations, the classifiers represent disjunctive regularities in a straightforward manner

red $\vee$ yellow $\vee$ blue	$\mapsto$	1 #1#1#
red $\wedge$ (yellow $\vee$ blue)	$\mapsto$	1 01#1#
"any color"	$\mapsto$	1 #####

A disjunct of values is represented by the most specific classifier string that matches each of the values. The classifier concept descriptions use "#" for the properties that are part of a disjunctive generalization, "0" for positive properties that occur in all concept instances, and "1" for negative properties shared by all instances.

Note that the semantics of classifiers representing disjunctive regularities are easily maintained under genetic operators. The semantics of the special ε bit require that `01..11` be the only input string in which $\varepsilon = 0$. Accordingly, as long as no `0` is arbitrarily placed in that bit position, every classifier string can be interpreted as a meaningful disjunction.

4 Discussion

Feature manifold encodings provide a straightforward way to represent ordinal and nominal attributes in classifier systems. Generalizations in the classifier language based on these encodings are as expressive as those available in many conventional symbolic description languages for learning concepts in attribute-based instance spaces. This expressive power is achieved without compromising the simplicity, efficiency, and well-chosen building blocks characteristic of the classifier language syntax. These encodings directly refute characterizations of the classifier system concept description language as inherently "low level" (Carbonell, 1989; Grefenstette, 1989). The syntactic complexity of a "high level" language is not required to obtain significant descriptive power. We can rely, instead, on a language interpreter that remains "out of the loop" to translate an internal representation in the classifier language into an externally visible, syntactically familiar description. In this way, we achieve both expressive power and learning efficiency at the same time. The feature manifold encodings, and their corresponding decoding algorithms, make this possible in classifier systems. Moreover, many specialized heuristic rule learning operators designed to work with symbolic attributes (eg. (Grefenstette, 1991)) can be used just as effectively with appropriately encoded classifier system representations. This suggests that efforts to modify genetic learning systems to directly accomodate high level rule language constructs (Grefenstette, 1989; Schuurmans & Schaeffer, 1989) are not necessarily required or desirable in the classifier system approach.

The binary encodings described here are just a sample of what is possible using feature manifolds and receptive fields. Receptive fields can be used in many ways to extract a large variety of useful properties. The robustness of the genetic algorithm allows us to consider using several different encodings for a given parameter and have the system *learn* which is most informative. These possibilities have the potential to make the classifier concept description language substantially more expressive and powerful.

† Acknowledgements

Part of this work was completed while the author was employed at the Naval Research Laboratory.

References

Agre, P. E. (1988). *The dynamic structure of everyday life.* Doctoral dissertation, MIT Artificial Intelligence Lab (Technical Report No. 1085), Massachusetts Institute of Technology, Cambridge, MA.

Agre, P. E. & Chapman, D. (1988). Indexicality and the binding problem. *Proceedings of the AAAI Spring Symposium on Parallel Models of Intelligence* (pp. 1–9). Stanford University, March 22-24, 1988.

Belew, R. & Forrest, S. (1988). Learning and programming in classifier systems. *Machine Learning, 3*, 193–223.

Booker, L. B. (1982). *Intelligent behavior as an adaptation to the task environment.* Doctoral dissertation, Department of Computer and Communication Sciences, University of Michigan, Ann Arbor, MI.

Booker, L. B. (1985). Improving the performance of genetic algorithms in classifier systems. *Proceedings of the First International Conference on Genetic Algorithms and Their Applications* (pp. 80–92). Pittsburgh, PA: Lawrence Erlbaum.

Booker, L. B., Goldberg, D. E., & Holland, J. H. (1989). Classifier systems and genetic algorithms. *Artificial Intelligence, 40*, 235–282.

Carbonell, J. G. (1989). Introduction: Paradigms for machine learning. *Artificial Intelligence, 40*, 1–9.

Caruana, R. A., & Schaffer, J. D. (1988). Representation and hidden bias: Gray vs. binary coding for genetic algorithms. *Proceedings of the Fifth International Conference on Machine Learning* (pp. 153–161). Ann Arbor, MI: Morgan Kaufmann.

Dietterich, T. G. (1989). Limits of inductive learning. *Proceedings of the Sixth International Conference on Machine Learning* (pp. 124–128). Ithaca, NY: Morgan Kaufmann.

Forrest, S. (1985). A study of parallelism in the classifier system and its application to classification in KL-ONE semantic networks. Doctoral dissertation, Department of Computer and Communication Sciences, University of Michigan, Ann Arbor.

Goldberg, D. E. (1985). Dynamic system control using rule learning and genetic algorithms. *Proceedings of the Ninth International Joint Conference on Artificial Intelligence* (pp. 588–592). Los Angeles, CA: Morgan Kaufmann.

Grefenstette, J. J. (1989). A system for learning control strategies with genetic algorithms. *Proceedings of the Third International Conference on Genetic Algorithms* (pp. 183–190). Fairfax, VA: Morgan Kaufmann.

Grefenstette, J. J. (1991). A lamarkian approach to learning in adversarial environments. To appear in *Proceedings of the Fourth International Conference on Genetic Algorithms.* La Jolla, CA: Morgan Kaufmann.

Hayes-Roth, R. (1976). Patterns of induction and associated knowledge acquisition algorithms. In C. H. Chen (Ed.) *Pattern recognition and artificial intelligence.* New York: Academic Press.

Liepins, G. & Vose, M. (in review). Characterizing crossover in genetic algorithms. Submitted to *Annals of Mathematics and Artificial Intelligence.*

Mitchell, T. M. (1980). The need for biases in learning generalizations. Technical report CBM-TR-117, Department of Computer Science, Rutgers University, New Brunswick, NJ.

Schuurmans, D. & Schaeffer, J. (1989). Representational difficulties with classifier systems. *Proceedings of the Third International Conference on Genetic Algorithms* (pp. 328–333). Fairfax, VA: Morgan Kaufmann.

Whitehead, S. D. & Ballard, D. H. (1990). Active perception and reinforcement learning. *Proceedings of the Seventh International Conference on Machine Learning* (pp. 179–188). Austin, TX: Morgan Kaufmann.

Quasimorphisms or Queasymorphisms? Modeling Finite Automaton Environments

T. H. Westerdale
Department of Computer Science, Birkbeck College,
University of London, Malet Street, London WC1E 7HX,
England

Abstract

The paper examines models that are homomorphic images of the first component of a particular two component cascade decomposition of the environment. The bucket brigade is used to estimate model state values. The discussion is limited to finite automaton environments whose successive input symbols are selected by the system probabilistically, with independent probabilities, according to a probability distribution over the input symbols.

1 Introduction

In this paper we discuss models of the environments of adaptive systems, not models of the systems themselves. It is standard to regard a model as something like a homomorphic image of whatever is being modeled. (Zeigler, 1972) The homomorphism in many cases is only approximate. ((Zeigler, 1976) Ch. 13) Since an approximate homomorphism is not strictly a homomorphism, Holland calls an approximate homomorphism a quasi-homomorphism, or quasimorphism. (Holland et al., 1986) He regards a model as a quasimorphic image, and so he too thinks that a model is something like a homomorphic image.

Of course, like Humpty Dumpty, we can use a word to mean whatever we want it to mean, but I would suggest a different meaning of the word model, at least when, as in this paper, the model is of the system's environment. Rather than regarding a model as a quasimorphic image, I would regard it as what I shall call a queasymorphic image. A queasymorphism resembles a quasimorphism, but is from only part of the environment, the part over which the system has some control. We conceptually decompose the environment into the part over which the system has some control and the part over which it has no control. We then define our queasymorphism over the controllable part only. This gives us a model with a self correcting property.

This paper gives only a basic outline of the approach. It deals with only the simplest case, in which the system decides what action to take without regard to environment state, (what Holland called the payoff only case(Holland, 1969)). More complicated cases (e.g. where the system is a production system) have been partially analyzed, but that analysis is not reported here. That analysis is, however, built on the analysis reported here.

We will assume that the environment E is a finite automaton with a large number of states, with payoffs (real numbers) associated with the states. The action of the system is to supply a sequence of input symbols to the automaton E.

We call a finite automaton *strongly connected* if for every ordered pair of states there is an input string that takes the automaton from the first state to the second. We will assume the environment E is strongly connected.

In the case with which this paper deals, the system selects the input symbols in a probabilistic fashion. Each selected symbol is fed into the environment, producing a change in state. Every symbol in the input alphabet has attached to it a real number called its availability. At each time unit, the system selects an input symbol, selecting the various symbols with probabilities proportional to their availabilities. Thus the probability of any symbol being selected is independent of environment state and of which symbols were selected previously.

We assume that associated with each environment state is a fixed real number called the state's payoff. These state payoff numbers remain fixed and unalterable. They are a fixed part of the environment.

We think of the system as receiving payoff from the environment. The payoff received in a given time unit is the payoff attached to the state the environment moves to in that time unit. The system wants to maximize its payoff per time unit, and it does so by modifying the availability numbers. A given set of availabilities will result in a particular average payoff per time unit.

Notice that the language of the previous paragraph seems to imply that payoffs are non-negative. For simplicity, we use such language throughout this paper, even though payoffs can be negative.

While selecting symbols, the system can collect payoff statistics and use them to decide how to modify the availabilities. A model is the system's way of holding these statistics. The system holds a model of the environment and modifies the availabilities on the basis of what it sees in the model. The rule it follows in doing this is called its reward scheme. This paper does not examine or compare reward schemes. Nevertheless, we have to briefly mention an illustrative reward scheme in order to indicate what is required of a model.

This paper does not discuss production systems. In a production system, the symbol that the system inputs depends on the current environment state. The analysis given here has been extended to production systems via a simplified version of the "prescriptions" discussed in (Westerdale, 1986). The results so far seem to me encouraging. This paper does not discuss the extension of the analysis to production systems.

More importantly, this paper does not touch upon the really important question: How does the system learn the state diagram of the desired model? In adaptive production systems (e.g. classifier systems) the productions themselves can be thought of as representing model states or sets of model states. The hope is that the evolving population of productions will evolve towards an appropriate model. This paper suggests that an appropriate model would be a queasymorphic image.

The argument in this paper is laid out as follows. In the first part we discuss how models can be used. In the second part we describe models that are queasymorphic images and indicate why they can usefully be used in the ways discussed in the first part.

The first part of the paper comprises sections 3, 4, and 5. In this part we use a conceptually simple (but usually impractical) model that is simply a complete copy of the environment. In section 3 we define and discuss "state excess value" of environment states, and "input symbol excess value" of input symbols. In section 4 we show how the bucket brigade can be used on our conceptually simple model to estimate environment state excess values. Section 5 is a small section containing an important extension of the results of section 4, but section 5 does not form part of the main argument in the paper and can be omitted on first reading.

The second part of the paper comprises sections 6 through 12. In this part we develop the notion of a queasymorphic image model. We do this step-wise. We examine a sequence of models, beginning with the impractical model of sections 3–5, and ending with the queasymorphic image model. Each of these models is thought of as a finite automaton with payoffs attached to its states. It will be obvious that the bucket brigade can be used as in section 4 to estimate "state excess values" of model states. In section 6 we note that for a model to be useful it must give the same "input symbol excess values" as the environment gives. The first model does, since it is merely a copy of the environment. As we examine the sequence of models in the following sections, we show that each gives the same "input symbol excess values" as the previous model in the sequence. Section 7 gives us ammunition for showing this by demonstrating that if one model is the image of another under a payoff preserving homomorphism, then the two models give the same "input symbol excess values".

Section 8 defines the second model in the sequence, a certain two component cascade decomposition of the environment, and uses the result of section 7 to show that this model gives the same "input symbol excess values" as the first model. The first component of this model represents the part of the environment over which the system has some control, and the second part represents the part of the environment over which the system has no control.

The third model in the sequence is formed from the second model by throwing away its second component. This third model is the interesting one. In section 9 we show that it gives the same "input symbol excess values" as does the second model. In section 10, we show that this model can work even if no payoffs are attached to its states. In practice, models do not have such attached payoffs, and in the following two sections we will show that the fourth and fifth models don't need them either.

Sections 11 and 12 are merely the application of some well known ideas to our models. In section 11 we use an analogue of the well known equiresponse relation to produce a fourth model that is a homomorphic image of the third model. We then use the result of section 7. In section 12 we suggest replacing that homomorphism with something like Holland's quasimorphism, whose image is then our fifth and final model. We define queasymorphism. The definition uses our decomposition and our quasimorphism. Our fifth model is a queasymorphic image. Our development shows that this model is self correcting and that the bucket brigade of section 4 can be used on it. Given the third model, the development in sections 11 and 12 is rather trivial, and section 12 is rather informal. What is interesting is the third model. The probably more practical fifth model follows from it almost trivially.

The notation used in the formal arguments is standard, but I have introduced a few notational devices to avoid too many confusing subscripts and superscripts. Section 2 introduces the basic notation and states the well known mathematical facts that will be used in the rest of the paper. I have tried to be sure that all mathematical facts assumed are stated in section 2. In fact, I have even included in section 2 a few very short proofs of some of the facts stated there. This is chiefly to make my notation clear in a familiar context, but it has the convenient side effect of making the paper slightly more self contained.

We begin, then, by discussing the use of the conceptually simple model that is a complete copy of the environment. We begin with a closer look at the environment itself.

Let η_r be the availability of symbol r. We assume all availabilities are positive. Define $\bar{\eta}=\sum_r \eta_r$, and $\phi_r=\eta_r/\bar{\eta}$, so that ϕ_r is the probability of the system selecting input symbol r. Number the environment states 1 to N. Let $[r]$ be the $N\times N$ zero one matrix whose ij'th element is one if and only if symbol r carries the environment from state i to state j. Let $P=\sum_r \phi_r[r]$, so that P is the state transition probability matrix of the Markov chain formed by the system-environment complex.

We call this P *our transition probability matrix.* It is row stochastic (i.e. row sums 1). In the next section we briefly review some properties of row stochastic matrices.

2 Row Stochastic Matrices[1]

We can regard any complex $N\times N$ matrix as a linear transformation on an N dimensional complex vector space of row vectors. In this paper, we shall write vectors as row vectors. Their transposes will be column vectors. We write $\mathbf{e}$ for the vector all of whose entries are 1, and $\mathbf{e}_i$ for the vector whose i'th entry is 1 and whose other entries are 0. If we have a matrix A, we shall sometimes write simply A_{ij} to mean the ij'th entry in A. So for example, $P_{ij}=\mathbf{e}_i P \mathbf{e}_j^{\mathsf{T}}$. An $N\times N$ non-negative matrix A is called irreducible if for any integers i and j (between 1 and N), there is a power of A whose ij'th entry is positive. (This is not the usual definition,

[1] A (row) vector is called non-negative if all its entries are non-negative real numbers. A vector is called stochastic if it is non-negative and its entries sum to 1. A matrix is called non-negative if all its entries are non-negative real numbers. A matrix is called row stochastic if is is non-negative and all its row sums are 1. It is called bistochastic if it is non-negative, all its row sums are 1, and all its column sums are 1.

but it is equivalent and short.) Since the environment is strongly connected, our transition probability matrix is irreducible. We shall use the following well known theorem.

THEOREM (Frobenius-Perron). [2] If A is an $N \times N$ irreducible non-negative matrix then: There is a positive real eigenvalue q which is a simple root of the characteristic polynomial and which is the upper bound of the moduli of all eigenvalues. There is an eigenvector $\mathbf{z}$ that has all positive real entries and that has eigenvalue q. The set of all eigenvectors of eigenvalue q is a one dimensional subspace of the N dimensional complex vector space. If in addition A has no zero entries then q actually exceeds all other eigenvalues in modulus.

Notice that we can without loss of generality insist that the $\mathbf{z}$ is a stochastic vector. (Just multiply the $\mathbf{z}$ by the appropriate scalar.) Then $\mathbf{z}\mathbf{e}^{\mathsf{T}} = 1$. If in addition the A in the theorem happens to be row stochastic then $A\mathbf{e}^{\mathsf{T}} = \mathbf{e}^{\mathsf{T}}$. In this case, multiplying $\mathbf{z}A = q\mathbf{z}$ on the right by $\mathbf{e}^{\mathsf{T}}$ gives $1 = q$.

In this section, the letter P will stand for an arbitrary irreducible, row stochastic square matrix.

We see that P has a unique stochastic eigenvector with eigenvalue 1. The eigenvector has no zero entries. We write the eigenvector as $\tilde{\mathbf{p}}$. $\tilde{\mathbf{p}}P = \tilde{\mathbf{p}}$. The subspace of eigenvectors with eigenvalue 1 is one dimensional.

Now P^{T} and P have the same eigenvalues, so if the A in the theorem is P^{T}, then $q = 1$ again, and there is a unique stochastic eigenvector of P^{T} with eigenvalue 1. Clearly this eigenvector is $\frac{1}{N}\mathbf{e}$. Again the subspace of eigenvectors with eigenvalue 1 is one dimensional.

If P has no zero entries then the eigenvalue in the Frobenius-Perron theorem (1 in this case) actually exceeds all the other eigenvalues in modulus. So $\lim_{n\to\infty} P^n$ exists, as can easily be seen by writing P in Jordan normal form. We call such a limit P^{∞}.

We call a non-negative matrix *primitive* if it has a power with no zero entries.

FACT 1: If P is row stochastic and primitive, then P^{∞} exists and equals $\mathbf{e}^{\mathsf{T}}\tilde{\mathbf{p}}$.

PROOF: For some positive integer l, P^l has no zero entries. Then $(P^l)^{\infty}$ exists. It is obviously row stochastic. Now clearly, $(P^l)^{\infty}P^l = (P^l)^{\infty}$, so each row of $(P^l)^{\infty}$ is a stochastic eigenvector of P^l with eigenvalue 1. But P^l is irreducible and row stochastic, so there is only one such eigenvector. Clearly it is $\tilde{\mathbf{p}}$, since $\tilde{\mathbf{p}}P^l = \tilde{\mathbf{p}}$. So $(P^l)^{\infty} = \mathbf{e}^{\mathsf{T}}\tilde{\mathbf{p}}$. Then for each $k < l$ we have $\lim_{n\to\infty} P^{k+ln} = P^k(P^l)^{\infty} = P^k\mathbf{e}^{\mathsf{T}}\tilde{\mathbf{p}} = \mathbf{e}^{\mathsf{T}}\tilde{\mathbf{p}}$. Hence P^{∞} exists and equals $\mathbf{e}^{\mathsf{T}}\tilde{\mathbf{p}}$. Q.E.D.

It is the case that for any row stochastic matrix A, the limit $\lim_{n\to\infty} \frac{1}{n}\sum_{k=0}^{n-1} A^k$ exists. We write it as $\tilde{A}$. Showing that the limit exists is a bit of a fiddle (Gantmacher, 1959), even for an irreducible matrix like P. If P is also primitive, then it is obvious that the limit $\tilde{P}$ exists and is P^{∞}. In any case, from the definition of $\tilde{P}$, we

---

[2] Frobenius extended the theorem of Perron. ((Gantmacher, 1959) pp. 64 and 65) The wording given here includes all of Perron's theorem and some of the extensions of Frobenius.

see that $\tilde{P}P = \tilde{P}$, so the rows of $\tilde{P}$ are stochastic eigenvectors of P with eigenvalue 1. $\tilde{\mathbf{p}}$ is the only such eigenvector, so $\tilde{P} = \mathbf{e}^\mathsf{T}\tilde{\mathbf{p}}$. If $\mathbf{v}^{[0]}, \mathbf{v}^{[1]}, \mathbf{v}^{[2]}, \ldots$ is a sequence of stochastic vectors with $\mathbf{v}^{[n+1]} = \mathbf{v}^{[n]}P$, then $\lim_{n\to\infty} \frac{1}{n}\sum_{k=0}^{n-1} \mathbf{v}^{[k]} = \mathbf{v}^{[0]}\tilde{P} = \tilde{\mathbf{p}}$. So we see that if P is our transition probability matrix, then $\tilde{p}_i$ (the i'th entry of $\tilde{\mathbf{p}}$) is the proportion of time that the environment spends in the i'th state. In this case, $\tilde{\mathbf{p}}$ is called the *mean limiting absolute probability vector.*

The following facts are well known:

FACT 2: If B is a square matrix and $\lim_{k\to\infty} B^k = 0$, then $\sum_{k=0}^{\infty} B^k = (I-B)^{-1}$, the sum and inverse both existing.

PROOF: Suppose for an arbitrary vector $\mathbf{v}$, $\mathbf{v}(I-B)=0$. Then $\mathbf{v} = \mathbf{v}B$, $\mathbf{v} = \mathbf{v}B^n$, and letting $n \to \infty$ gives $\mathbf{v} = 0$. So $I - B$ is invertible. Now $(I-B)\sum_{k=0}^{n} B^k = I - B^{n+1}$. Multiply on the left by $(I-B)^{-1}$ and then let $n \to \infty$. Q.E.D.

FACT 3: If P is row stochastic and P^∞ exists, then $\sum_{k=0}^{\infty}(P^k - P^\infty) = (I - P + P^\infty)^{-1} - P^\infty$, the sum and inverse both existing.

PROOF: Let $B = P - P^\infty$. By induction on k, we can show $B^k = P^k - P^\infty$ for $k \geq 1$. So the sum in the Fact becomes $(\sum_{k=0}^{\infty} B^k) - P^\infty$, since the zero'th term becomes $B^0 - P^\infty$. Now we use Fact 2. Q.E.D.

For any row stochastic irreducible matrix P, let $\hat{P} = I - P + \tilde{P}$. By (Kemeny and Snell, 1960) Thm. 5.1.4, the Cesaro sum $\sum\!\!c_{k=0}^{\infty}(P^k - \tilde{P})$ exists. (I shall write Cesaro sums as $\sum\!\!c$ and ordinary sums as $\sum$.) Of course if P is primitive, then it is obvious by Fact 3 that even the ordinary sum exists.

FACT 4. $\sum\!\!c_{k=0}^{\infty}(P^k - \tilde{P}) = \hat{P}^{-1} - \tilde{P}$, the Cesaro sum and inverse both existing.

PROOF: We temporarily call the Cesaro sum C. We note that $\tilde{\mathbf{p}}(P^k - \tilde{P}) = 0$. Then since $\tilde{P} = \mathbf{e}^\mathsf{T}\tilde{\mathbf{p}}$, we have $\tilde{P}(P^k - \tilde{P}) = 0$ and $\tilde{P}C = 0$. So $\hat{P}C = (I-P)C$. But $P\tilde{P} = P\mathbf{e}^\mathsf{T}\tilde{\mathbf{p}} = \tilde{P}$, so from the definition of C we have $PC = C - (I - \tilde{P})$ or $(I-P)C = I - \tilde{P}$. Thus $\hat{P}C = I - \tilde{P}$. We also have $\hat{P}\tilde{P} = \tilde{P}$, so $\hat{P}(C + \tilde{P}) = \hat{P}C + \hat{P}\tilde{P} = I$. Thus $\hat{P}$ is invertible and $C + \tilde{P} = \hat{P}^{-1}$, or $C = \hat{P}^{-1} - \tilde{P}$. Q.E.D.

If P is primitive then this is merely Fact 3.

We note that $\tilde{\mathbf{p}}\hat{P} = \tilde{\mathbf{p}}$ and $\hat{P}\mathbf{e}^\mathsf{T} = \mathbf{e}^\mathsf{T}$, so

$$\tilde{\mathbf{p}} = \tilde{\mathbf{p}}\hat{P}^{-1} \quad \text{and} \quad \mathbf{e}^\mathsf{T} = \hat{P}^{-1}\mathbf{e}^\mathsf{T}. \tag{1}$$

3 State Values

We now return to our environment E. From now on, P will be our transition probability matrix. Then P is irreducible, and $\tilde{\mathbf{p}}$ is the mean limiting absolute probability vector.

Let $\mathbf{t}$ be the N dimensional vector whose i'th entry t_i is the payoff associated with the i'th environment state. Let $\bar{m} = \tilde{\mathbf{p}}\mathbf{t}^\mathsf{T}$, so $\bar{m}$ is the average payoff per unit time, the quantity the system is trying to make larger. Let $\mathbf{a} = \mathbf{t} - \bar{m}\mathbf{e}$. a_i, the i'th entry of $\mathbf{a}$, is the *excess payoff* of state i. Note that $\tilde{\mathbf{p}}\mathbf{a}^\mathsf{T} = 0$.

Using the definition of $\mathbf{a}$ and the equations $\tilde{P}=\mathbf{e}^{\mathsf{T}}\tilde{\mathbf{p}}$, $\tilde{\mathbf{p}}\mathbf{t}^{\mathsf{T}}=\bar{m}$, $P^k\mathbf{e}^{\mathsf{T}}=\mathbf{e}^{\mathsf{T}}$, and $\hat{P}^{-1}\mathbf{e}^{\mathsf{T}}=\mathbf{e}^{\mathsf{T}}$, we can easily verify $(P^k-\tilde{P})\mathbf{t}^{\mathsf{T}}=P^k\mathbf{a}^{\mathsf{T}}$ and $(\hat{P}^{-1}-\tilde{P})\mathbf{t}^{\mathsf{T}}=\hat{P}^{-1}\mathbf{a}^{\mathsf{T}}$. So multiplying the equation in Fact 4 on the right by $\mathbf{t}^{\mathsf{T}}$ gives $\sum\mathrm{c}_{k=0}^{\infty}(P^k\mathbf{a}^{\mathsf{T}})=\hat{P}^{-1}\mathbf{a}^{\mathsf{T}}$.

We define the *excess value* of the i'th state, c_i, to be how much more payoff than usual one would expect if one started out in state i. Now beginning in state i, the amount of payoff (on average) k time units later is $\mathbf{e}_i P^k\mathbf{t}^{\mathsf{T}}$. This is $\mathbf{e}_i P^k\mathbf{t}^{\mathsf{T}} - \bar{m}$ more than usual. From the definition of $\mathbf{a}$, we see that this excess is $\mathbf{e}_i P^k\mathbf{a}^{\mathsf{T}}$. The column vector $P^k\mathbf{a}^{\mathsf{T}}$ clearly has this excess as its i'th entry. If we let $\mathbf{c}$ be the vector whose i'th entry is c_i, we see immediately that $\mathbf{c}^{\mathsf{T}}$ is the Cesaro sum in the last paragraph. In fact, we shall regard this Cesaro sum as the *definition of* $\mathbf{c}$. $\mathbf{c}=(\sum\mathrm{c}_{k=0}^{\infty}(P^k\mathbf{a}^{\mathsf{T}}))^{\mathsf{T}}$. We cannot guarantee that the sum converges in the ordinary sense. But of course if P is primitive then it will. For reference, we re-state the definition and the equation at the end of the last paragraph.

$$\mathbf{c}^{\mathsf{T}}=\sum\mathrm{c}_{k=0}^{\infty}(P^k\mathbf{a}^{\mathsf{T}}) \tag{2}$$

$$\mathbf{c}^{\mathsf{T}}=\hat{P}^{-1}\mathbf{a}^{\mathsf{T}}. \tag{3}$$

By the Cesaro sum definition of $\mathbf{c}$, we see immediately that $\tilde{\mathbf{p}}\mathbf{c}^{\mathsf{T}}=0$ and

$$(I-P)\mathbf{c}^{\mathsf{T}}=\mathbf{a}^{\mathsf{T}}. \tag{4}$$

The *excess value of input symbol r* is how much more payoff than usual one would expect as a result of inputting r. It is easy to see that this is $\tilde{\mathbf{p}}[r]\mathbf{c}^{\mathsf{T}}$. The system will want to relatively increase the availabilities of symbols with high excess values. A method of adjusting availabilities is usually called a *reward scheme.* Suppose we write η'_r, $\bar{\eta}'$, and ϕ'_r for the rate of change of η_r, $\bar{\eta}$, and ϕ_r respectively. In (Westerdale, 1986) (Sec. IV, last paragraph) we discussed some advantages in making $\eta'_r=\kappa\eta_r\tilde{\mathbf{p}}[r]\mathbf{c}^{\mathsf{T}}$ for some constant κ. Then by differential calculus we have $\bar{\eta}'=0$ and $\phi'_r=\phi_r\kappa\tilde{\mathbf{p}}[r]\mathbf{c}^{\mathsf{T}}$. Thus ϕ'_r/ϕ_r is proportional to the excess value of r. In (Westerdale, 1986) we discussed a reward scheme that implicitly effects this ϕ'_r. The scheme is called profit sharing by Grefenstette (Grefenstette, 1988). In this article, by contrast, we want to concentrate on ways that the system could obtain estimates of state excess values. It could then, if it wished, use the excess values directly to make the availability adjustments. For example, in what we shall call our *direct reward scheme,* every time r moves the environment from state i to state j, η_r is adjusted by an amount proportional to c_j. This achieves the η'_r mentioned above and hence the ϕ'_r. In this paper we are not discussing reward schemes, so questions like whether we need to divide the adjustment by $\bar{\eta}$, etc., will not concern us.

Of course when the availabilities change, so do the other quantities we have been discussing. But at least these quantities are continuous functions of the availabilities. In fact we have

LEMMA 1: The entries in the objects, P, $\tilde{\mathbf{p}}$, $\bar{m}$, $\mathbf{a}$, $\tilde{P}$, $\hat{P}$, $\hat{P}^{-1}$, and $\mathbf{c}$ are rational functions of the availabilities.

PROOF: Obviously the entries in P are. Let us call an object in the above sequence "legitimate" if its entries are rational functions of entries in objects earlier in the sequence. If A is a square matrix, then we write $\underline{A}$ to mean $adj(A)$. [3] Then we know $\underline{A}A = |A|I$. Letting $A = \hat{P}$, we have $((1/|\hat{P}|)\underline{\hat{P}})\hat{P} = I$ and $\hat{P}^{-1} = (1/|\hat{P}|)\underline{\hat{P}}$. So $\hat{P}^{-1}$ is legitimate. Letting $A = P - I$, and noting that $P - I$ is singular ($\tilde{\mathbf{p}}(P - I) = 0$), we have $(\underline{P - I})(P - I) = 0$. Let $\mathbf{w}$ be the first row of $\underline{P - I}$. Then $\mathbf{w}(P - I) = 0$ and $\mathbf{w}P = \mathbf{w}$, so $\mathbf{w}$ is an eigenvector of P with eigenvalue 1. So is $\tilde{\mathbf{p}}$, and the subspace of such eigenvectors is one dimensional. So there is a scalar k such that $\mathbf{w} = k\tilde{\mathbf{p}}$. Then $\mathbf{w}\mathbf{e}^{\mathsf{T}} = k$, and $\tilde{\mathbf{p}} = (1/(\mathbf{w}\mathbf{e}^{\mathsf{T}}))\mathbf{w}$. So $\tilde{\mathbf{p}}$ is legitimate. $\tilde{P} = \mathbf{e}^{\mathsf{T}}\tilde{\mathbf{p}}$, so $\tilde{P}$ is legitimate. $\mathbf{c}$ is legitimate by (3), and the other objects are legitimate by their definitions. Q.E.D.

So the excess values of the input symbols are also rational functions of the availabilities.

4 The Bucket Brigade

The direct reward scheme requires that the system have available at all times an estimate of the excess value of the current environment state. It obtains this from statistics derived from the past record of inputs to the environment and payoffs from the environment. These statistics are typically summarized as a model of the environment, complete with estimates of excess values of states and a guess as to the current state. In a conceptually simple, if impractical, version of this method, the system holds a complete copy of the environment state diagram and uses this as a model, giving the model the same inputs as the real environment so that the model state always corresponds exactly to the environment state. It can then use a bucket brigade to estimate environment state excess values.

The idea behind the bucket brigade is that cash balances attached to the environment states are used to estimate excess values. We shall conveniently think of these cash balances as attached to environment states, though of course they would really have to be attached to the surrogate states, the states of the model, the states of the copy of the environment. Cash is passed from one state to another as the environment undergoes state transitions, and the various states' cash balances are supposed to approach the excess values of the states.

We discuss a simple bucket brigade. When the environment moves from state i to state j, a proportion ε of the cash of state j is passed to state i. $0 < \varepsilon < 1$. Of course on the previous time unit, state i lost a proportion ε of its cash to the previous state. In addition, when state i is visited, the payoff t_i is added to its cash balance. Thus if we write $v_i^{(n)}$ for the cash balance of state i at time n, we see that for a transition from i to j, the change in the cash balance of state i is the sum of these:

$\varepsilon v_j^{(n)}$, the cash passed back from state j;

$-\varepsilon v_i^{(n)}$, the cash that state i passed back to the previous state;

t_i, the payoff to state i.

---

[3] Terminology: The *cofactor* of the ij'th entry in A is formed by deleting the i'th row and j'th column of A, taking the determinant of what is left, and then changing the sign if $i + j$ is odd. $adj(A)$, sometimes called the *adjoint* of A (Gantmacher, 1959), is formed from A by first replacing each entry by its cofactor, and then transposing the resulting matrix.

If P_{ij} is the ij'th entry in matrix P, then the probability of the above transition is $\tilde{p}_i P_{ij}$. So on average the change per time unit of the cash of state i is given by the right hand side of equation (5). We pretend that this average change is the actual change of the state-i cash in each time unit, so we say

$$v_i^{(n+1)} - v_i^{(n)} = \sum_j \tilde{p}_i P_{ij}(\varepsilon v_j^{(n)} - \varepsilon v_i^{(n)} - t_i). \tag{5}$$

The reader will note that by averaging in this way, we are ignoring all unbiased sampling noise, even though the amount of such noise is different for different states.

Let $\mathbf{v}^{(n)}$ be the vector whose i'th entry is $v_i^{(n)}$, and let D be the diagonal matrix whose ii'th entry is $\tilde{p}_i$. Then by the last equation we see that we have

$$(\mathbf{v}^{(n+1)})^{\mathsf{T}} - (\mathbf{v}^{(n)})^{\mathsf{T}} = D(\varepsilon P(\mathbf{v}^{(n)})^{\mathsf{T}} - \varepsilon(\mathbf{v}^{(n)})^{\mathsf{T}} + \mathbf{t}^{\mathsf{T}}). \tag{6}$$

Then $\mathbf{e}((\mathbf{v}^{(n+1)})^{\mathsf{T}} - (\mathbf{v}^{(n)})^{\mathsf{T}}) = \bar{m}$, which means that on average, not surprisingly, $\bar{m}$ extra cash is added overall each time unit. Clearly, unless this is subtracted, the amount of cash will simply grow without bound. One way of subtracting this is to subtract $\bar{m}$ cash from a state every time the state is visited. Then the right hand side of (5) is changed by the subtraction of $\tilde{p}_i \bar{m}$, and the right hand side of (6) is changed by the subtraction of $\bar{m}D\mathbf{e}^{\mathsf{T}}$. The new equation (6) is then equivalent to $(\mathbf{v}^{(n+1)})^{\mathsf{T}} = D\mathbf{a}^{\mathsf{T}} + Q(\mathbf{v}^{(n)})^{\mathsf{T}}$, where $Q = \varepsilon DP - \varepsilon D + I$. Writing $\mathbf{v}^{(0)}$ as $\mathbf{v}$, we can now prove the following equation by induction on n.

$$(\mathbf{v}^{(n)})^{\mathsf{T}} = \left(\sum_{k=0}^{n-1} (Q^k D\mathbf{a}^{\mathsf{T}}) \right) + Q^n \mathbf{v}^{\mathsf{T}} \tag{7}$$

We now examine the matrix Q. The matrix $-\varepsilon D + I$ has entirely non negative entries. Therefore, so does Q. Q has a positive entry everywhere P does so Q is irreducible. $Q\mathbf{e}^{\mathsf{T}} = \mathbf{e}^{\mathsf{T}}$ and $\mathbf{e}Q = \mathbf{e}$ so Q is bi-stochastic. The matrix $-\varepsilon D + I$ is reflexive (has positive diagonal entries) and hence so is Q. Thus Q^N has no zero entries and Q is primitive. So Q^∞ exists and equals $\tilde{Q}$. So $\hat{Q} = I - Q + Q^\infty$ and from Fact 3 we have

$$\sum_{k=0}^{\infty} (Q^k - Q^\infty) = \hat{Q}^{-1} - Q^\infty, \tag{8}$$

the ordinary sum and inverse both existing. Now $Q^k\mathbf{e}^{\mathsf{T}} = \mathbf{e}^{\mathsf{T}}$, so $Q^\infty \mathbf{e}^{\mathsf{T}} = \mathbf{e}^{\mathsf{T}}$ and Q^∞ is row stochastic. $Q^\infty Q = Q^\infty$ so each row of Q^∞ is a stochastic eigenvector of Q with eigenvalue 1. But there is only one such eigenvector, and from $(\frac{1}{N}\mathbf{e})Q = \frac{1}{N}\mathbf{e}$ we see that it is $\frac{1}{N}\mathbf{e}$. Thus

$$Q^\infty = \tfrac{1}{N}\mathbf{e}^{\mathsf{T}}\mathbf{e}. \tag{9}$$

Now $\mathbf{e}D = \tilde{\mathbf{p}}$ and $\tilde{\mathbf{p}}\mathbf{a}^{\mathsf{T}} = 0$, so $\mathbf{e}D\mathbf{a}^{\mathsf{T}} = 0$. From this and (9) we have $Q^\infty D\mathbf{a}^{\mathsf{T}} = 0$. Thus if we multiply (8) on the right by $D\mathbf{a}^{\mathsf{T}}$, we obtain $\sum_{k=0}^{\infty}(Q^k D\mathbf{a}^{\mathsf{T}}) = \hat{Q}^{-1}D\mathbf{a}^{\mathsf{T}}$. From this and (7) we obtain $\lim_{n\to\infty}(\mathbf{v}^{(n)})^{\mathsf{T}} = \hat{Q}^{-1}D\mathbf{a}^{\mathsf{T}} + Q^\infty \mathbf{v}^{\mathsf{T}}$.

We define the vector $\mathbf{u}$ to be $\lim_{n\to\infty}\mathbf{v}^{(n)}$. Its i'th entry, u_i, is the limiting cash balance for state i. So from the last equation and (9) we have

$$\mathbf{u}^{\mathsf{T}} = \hat{Q}^{-1}D\mathbf{a}^{\mathsf{T}} + \tfrac{1}{N}\mathbf{e}^{\mathsf{T}}\mathbf{e}\mathbf{v}^{\mathsf{T}} \tag{10}$$

LEMMA 2: $(I-P)\mathbf{u}^\mathsf{T} = \frac{1}{\varepsilon}\mathbf{a}^\mathsf{T}$.

FIRST PROOF: In the limit, the change of cash at state i is zero. This change on average per visit to i is $a_i + (\sum_j P_{ij}\varepsilon u_j) - \varepsilon u_i$. The change per time unit is $\tilde{p}_i$ times this, but we are looking at the change per visit to i. So we have this equal to zero. Writing this as vectors gives $\mathbf{a}^\mathsf{T} + \varepsilon P\mathbf{u}^\mathsf{T} - \varepsilon\mathbf{u}^\mathsf{T} = 0$, from which the Lemma follows. Q.E.D.

DIRECT PROOF: We note that $\mathbf{e}\hat{Q} = \mathbf{e}$ so $\mathbf{e} = \mathbf{e}\hat{Q}^{-1}$. Thus $\mathbf{e}\hat{Q}^{-1}D\mathbf{a}^\mathsf{T} = \mathbf{e}D\mathbf{a}^\mathsf{T} = \tilde{\mathbf{p}}\mathbf{a}^\mathsf{T} = 0$. Now from $\hat{Q} = I - Q + Q^\infty$, (9), and $Q = \varepsilon DP - \varepsilon D + I$, we obtain $\hat{Q} = \varepsilon D(I-P) + \frac{1}{N}\mathbf{e}^\mathsf{T}\mathbf{e}$. We multiply this on the right by $\hat{Q}^{-1}D\mathbf{a}^\mathsf{T}$ and on the left by $\frac{1}{\varepsilon}D^{-1}$ and use $\mathbf{e}\hat{Q}^{-1}D\mathbf{a}^\mathsf{T} = 0$. We obtain $\frac{1}{\varepsilon}\mathbf{a}^\mathsf{T} = (I-P)\hat{Q}^{-1}D\mathbf{a}^\mathsf{T}$. We multiply (10) on the left by $I-P$ and use the last equation and $(I-P)\mathbf{e}^\mathsf{T} = 0$. The Lemma results. Q.E.D.

Let the vector $\mathbf{w} = \mathbf{u} - \frac{1}{\varepsilon}\mathbf{c}$. Then using (4) and Lemma 2, we obtain $(I-P)\mathbf{w}^\mathsf{T} = 0$, $P\mathbf{w}^\mathsf{T} = \mathbf{w}^\mathsf{T}$, and $\mathbf{w}P^\mathsf{T} = \mathbf{w}$, so $\mathbf{w}$ is an eigenvector of P^T with eigenvalue 1. But so is $\mathbf{e}$. And the Frobenius-Perron Theorem says that the space of such eigenvectors is one dimensional, so there is a scalar χ such that $\mathbf{w} = \chi\mathbf{e}$. From this and $\mathbf{u} = \frac{1}{\varepsilon}\mathbf{c} + \mathbf{w}$ we obtain

$$\mathbf{u} = \frac{1}{\varepsilon}\mathbf{c} + \chi\mathbf{e}. \tag{11}$$

We now determine the value of χ. Now $\mathbf{e}Q = \mathbf{e}$ and $\mathbf{e}D\mathbf{a}^\mathsf{T} = 0$, so multiplying (7) on the left by $\mathbf{e}$ gives $\mathbf{e}(\mathbf{v}^{(n)})^\mathsf{T} = \mathbf{e}\mathbf{v}^\mathsf{T}$. Letting $n \to \infty$ gives $\mathbf{e}\mathbf{u}^\mathsf{T} = \mathbf{e}\mathbf{v}^\mathsf{T}$. Now $\chi\mathbf{e}^\mathsf{T} = \mathbf{w}^\mathsf{T} = \mathbf{u}^\mathsf{T} - \frac{1}{\varepsilon}\mathbf{c}^\mathsf{T}$. Multiplying on the left by $\mathbf{e}$ and using $\mathbf{e}\mathbf{u}^\mathsf{T} = \mathbf{e}\mathbf{v}^\mathsf{T}$ and $\mathbf{e}\mathbf{e}^\mathsf{T} = N$ gives $N\chi = \mathbf{e}\mathbf{v}^\mathsf{T} - \frac{1}{\varepsilon}\mathbf{e}\mathbf{c}^\mathsf{T} = (\sum_j v_j) - \frac{1}{\varepsilon}(\sum_j c_j)$. From this and (11), we have

LEMMA 3: $u_i = \frac{1}{\varepsilon}c_i + \chi$, where $\chi = \frac{1}{N}((\sum_j v_j) - \frac{1}{\varepsilon}(\sum_j c_j))$.

Thus the limiting cash balances of the states (the u_i's) are proportional to the state values (the c_i's), except for a correction χ. The proportion is $\frac{1}{\varepsilon}$, so that if the bucket brigade passes back a hundredth of the cash ($\varepsilon = \frac{1}{100}$) then the limiting cash balances are a hundred times the values.

The amount of circulating cash is constant, so clearly this proportional relationship could not be exact unless the total initial cash happened to equal exactly $\frac{1}{\varepsilon}$ times the sum of the state values. In general it will differ. $N\chi$ is the amount by which it differs. $N\chi$ can be thought of as the unwanted extra circulating cash. (Of course it could be negative.) Lemma 3 says that, in the limit, this unwanted cash is shared equally between the states as an additive correction amount χ.

The additive correction causes less of a problem than might be expected, and the cash balance u_i is often a reasonable substitute for c_i. For example, if our direct reward scheme uses u_j instead of c_j, then $\eta'_r = \kappa\eta_r\tilde{\mathbf{p}}[r]\mathbf{u}^\mathsf{T}$. Then from (11) and $\tilde{\mathbf{p}}\mathbf{c}^\mathsf{T} = 0$ we have $\tilde{\mathbf{p}}\mathbf{u}^\mathsf{T} = \chi$, so $\bar{\eta}' = \kappa\bar{\eta}\chi$ and $\phi'_r = \phi_r\kappa\frac{1}{\varepsilon}\tilde{\mathbf{p}}[r]\mathbf{c}^\mathsf{T}$. The χ is normalized out.

5 Payoffs That Are Not Fixed

We have been assuming that the payoff of any given environment state is fixed. We will sometimes want to consider environments whose state does not completely determine the payoff. For example, being in state i may determine not a fixed payoff, but rather a probability distribution over payoffs. Or the payoff may depend not only on the current state, but also on the state transition or on the previous input symbol. In the last case, instead of having just payoff t_i for state i, we have a different payoff $t_i^{(r)}$ for every input symbol r. (This was the case assumed throughout (Westerdale, 1986)) But then we can define $t_i = \sum_r \phi_r t_i^{(r)}$ and continue with the analysis as before. We will find that we can treat the other cases similarly. We merely define t_i as the average payoff for state i. (The direct reward scheme then uses $\frac{1}{\varepsilon} t_i^{(r)} + u_j$ instead of merely u_j, so $\eta'_r = \kappa \eta_r (\frac{1}{\varepsilon} \tilde{\mathbf{p}} \mathbf{t}^{(r)^\mathsf{T}} + \tilde{\mathbf{p}}[r]\mathbf{u}^\mathsf{T})$,
$\bar{\eta}' = \bar{\eta} \kappa (\frac{1}{\varepsilon} \bar{m} + \chi)$, and $\phi'_r = \phi_r \kappa \frac{1}{\varepsilon} (\tilde{\mathbf{p}} \mathbf{t}^{(r)^\mathsf{T}} - \bar{m} + \tilde{\mathbf{p}}[r]\mathbf{c}^\mathsf{T})$.)

6 Model State Values

In our discussion of the bucket brigade, we assumed that the model we were using was a complete copy of the environment. We now want to investigate what happens if the model is an automaton that is different from the environment. It will, however, have the same input alphabet as the environment. As in all the models discussed in this paper, the system feeds the same selected input symbol to the model as it feeds to the environment. So the change in model state is determined by the selected input symbol, without the help of any output of the environment. It will be conceptually convenient to assume that, like the environment, the model too has fixed payoffs attached to its states. Then just looking at the model on its own, the model states have excess values. A reward scheme like the direct reward scheme could then use these model state excess values instead of environment state excess values. The hope is that this will give the same effect as if environment state excess values had been used. A glance at the formulae for ϕ'_r that we have looked at above indicates the following general rule: In most of the interesting reward schemes, the state excess values enter into the determination of ϕ'_r only through the input symbol excess values. Thus if our model gives the same input symbol excess values (using model payoffs) as does the environment (using environment payoffs), then the reward scheme can safely use model state excess values instead of environment state excess values.

We shall look at a couple of models that do give the same input symbol excess values as does the environment, and we shall show that they do. This will show that the reward scheme could use the model state excess values. We shall then turn our attention to how the system could estimate the model state excess values using the bucket brigade.

7 Some Homomorphic Images

We first examine what happens when the model $\bar{E}$ is a homomorphic image of the environment automaton E. $\bar{E}$, then, is a finite automaton with the same input alphabet as E. Let us use a bar to distinguish the various quantities associated with $\bar{E}$ (such as $\bar{N}$, $\overline{[r]}$, $\bar{P}$, $\bar{\tilde{\mathbf{p}}}$, $\bar{\mathbf{t}}$, $\bar{\mathbf{a}}$) from those associated with E. The states of $\bar{E}$ are numbered $0, 1, 2, ..., \bar{N}$, and the ij'th entry in $\overline{[r]}$ is 1 just if r carries state i to state j in $\bar{E}$. $\bar{\mathbf{e}}$ is the $\bar{N}$ dimensional vector consisting of all ones, and $\bar{\mathbf{e}}_i$ is the $\bar{N}$ dimensional vector whose i'th entry is 1 and whose other entries are 0.

If i is a state of E, then $\hat{\imath}$ is the state of $\bar{E}$ to which the homomorphism maps i. That is, we represent our homomorphism by a hat.

Let us be quite clear about what we mean when we say there is a homomorphism from E onto $\bar{E}$. The homomorphism is a function from the state set of E to the state set of $\bar{E}$. To say it is a homomorphism means merely that for any input symbol r and any states i and j of E, the following statement holds (where we represent the function by a hat): If r carries E from state i to state j, then it carries $\bar{E}$ from state $\hat{\imath}$ to state $\hat{\jmath}$.

Let H be the $N \times \bar{N}$ zero-one matrix defined as follows. Its rows correspond to states of E and its columns to states of $\bar{E}$. Its ij'th entry is 1 just if the homomorphism maps state i of E to state j of $\bar{E}$.

So $\mathbf{e}_i H = \bar{\mathbf{e}}_{\hat{\imath}}$. Suppose r carries E from state i to state j, so $\mathbf{e}_i[r] = \mathbf{e}_j$. Then r carries $\bar{E}$ from state $\hat{\imath}$ to state $\hat{\jmath}$, $\bar{\mathbf{e}}_{\hat{\imath}}\overline{[r]} = \bar{\mathbf{e}}_{\hat{\jmath}}$. Of course $\mathbf{e}_j H = \bar{\mathbf{e}}_{\hat{\jmath}}$. From the last four equations, we obtain $\mathbf{e}_i H\overline{[r]} = \mathbf{e}_i[r]H$. i is arbitrary, so $H\overline{[r]} = [r]H$. Multiplying by ϕ_r and summing over all r gives

$$H\bar{P} = PH. \tag{12}$$

Since the state diagram of E is strongly connected, so is the state diagram of $\bar{E}$. Therefore $\bar{P}$ is also irreducible.

$H\bar{\mathbf{e}}^\mathsf{T} = \mathbf{e}^\mathsf{T}$, so $(\tilde{\mathbf{p}}H)\bar{\mathbf{e}}^\mathsf{T} = 1$, and we see that $\tilde{\mathbf{p}}H$ is an $\bar{N}$ dimensional stochastic vector. Multiplying (12) on the left by $\tilde{\mathbf{p}}$ gives $(\tilde{\mathbf{p}}H)\bar{P} = (\tilde{\mathbf{p}}H)$, so $\tilde{\mathbf{p}}H$ is a stochastic eigenvector of $\bar{P}$ with eigenvalue 1. But we saw that for irreducible row stochastic matrices there is only one of these, in this case $\bar{\tilde{\mathbf{p}}}$, so $\bar{\tilde{\mathbf{p}}} = \tilde{\mathbf{p}}H$.

We now wish to discuss payoffs. We look at the special case in which the homomorphism preserves payoff. That is, $t_i = \bar{t}_{\hat{\imath}}$. So all the states of E that the homomorphism maps to $\hat{\imath}$ have the same payoff, and this is the same as the payoff that is attached to the model state $\hat{\imath}$. The last equation can be written $\mathbf{e}_i\mathbf{t}^\mathsf{T} = \bar{\mathbf{e}}_{\hat{\imath}}\bar{\mathbf{t}}^\mathsf{T}$ or $\mathbf{e}_i\mathbf{t}^\mathsf{T} = \mathbf{e}_i H\bar{\mathbf{t}}^\mathsf{T}$. This holds for all i so $\mathbf{t}^\mathsf{T} = H\bar{\mathbf{t}}^\mathsf{T}$. This equation is just another way of saying that the homomorphism preserves payoffs. So if it does, then we have $\bar{\tilde{\mathbf{p}}}\bar{\mathbf{t}}^\mathsf{T} = \tilde{\mathbf{p}}H\bar{\mathbf{t}}^\mathsf{T} = \tilde{\mathbf{p}}\mathbf{t}^\mathsf{T}$, and the average payoff per unit time is the same for both automata, so we can use $\bar{m}$ to represent it in either case.

$H\bar{\mathbf{a}}^\mathsf{T} = H(\bar{\mathbf{t}}^\mathsf{T} - \bar{m}\bar{\mathbf{e}}^\mathsf{T}) = \mathbf{t}^\mathsf{T} - \bar{m}\mathbf{e}^\mathsf{T} = \mathbf{a}^\mathsf{T}$, as expected. In addition $H\bar{P}^k\bar{\mathbf{a}}^\mathsf{T} = P^k H\bar{\mathbf{a}}^\mathsf{T} = P^k\mathbf{a}^\mathsf{T}$ and so by the Cesaro sum definitions of $\mathbf{c}$ and $\bar{\mathbf{c}}$ we have $H\bar{\mathbf{c}}^\mathsf{T} = \mathbf{c}^\mathsf{T}$. Thus $\bar{c}_{\hat{\imath}} = \bar{\mathbf{e}}_{\hat{\imath}}\bar{\mathbf{c}}^\mathsf{T} = \mathbf{e}_i H\bar{\mathbf{c}}^\mathsf{T} = \mathbf{e}_i\mathbf{c}^\mathsf{T} = c_i$, and the homomorphism preserves state excess values.

From $\bar{\tilde{\mathbf{p}}} = \tilde{\mathbf{p}}H$, $H\overline{[r]} = [r]H$, and $H\bar{\mathbf{c}}^\mathsf{T} = \mathbf{c}^\mathsf{T}$ we have $\bar{\tilde{\mathbf{p}}}\overline{[r]}\bar{\mathbf{c}}^\mathsf{T} = \tilde{\mathbf{p}}[r]\mathbf{c}^\mathsf{T}$. In other words, the excess value of r is the same in E and $\bar{E}$. Now the fact that one of our automata was our "environment" is irrelevant to the argument. What we have shown is

LEMMA 4: If we have two strongly connected finite automata with the same input alphabet and with payoffs attached to their states, and if there is a payoff preserving homomorphism from the first onto the second, then the excess values of the input symbols are the same in either automaton.

8 Decomposition

We now examine a model that is a particular two component cascade decomposition of the environment.

In any finite automaton, we call an input string σ a *synchronizer* if there is an automaton state s such that whatever the current state of the automaton, the state of the automaton after the input of σ is guaranteed to be s.

We can show how to decompose the finite automaton E into two automata, E_1 and E_2, where the input to E_1 is the same as the input of E, where the input to E_2 is a pair consisting of the input to E together with the state of E_1, where E_2 is a permutation automaton, and where E_1 has a synchronizer. The E_1E_2 complex will be strongly connected. Any strongly connected finite automaton can be decomposed in this way.

We call an input string a *semisynchronizer* of E if it is a synchronizer of E_1.

Let S, S_1, and S_2 be the state sets of E, E_1, and E_2 respectively. To say that the E_1E_2 complex is a decomposition of the strongly connected automaton E means of course that there is a homomorphism $h : S_1 \times S_2 \rightarrow S$ such that once the relation $s = h(s_1, s_2)$ holds (with s, s_1, and s_2 the current states of E, E_1, and E_2,) it will continue to hold as long as E and the E_1E_2 complex are fed the same input string. In this sense, knowing the state of E_1E_2 tells one the state of E. We say the state of E is *always given* (via h) by the state of E_1E_2.

We want to show that if E is the environment, then the system can use the decomposition E_1E_2 as a model.

The details of the decomposition are given in (Westerdale, 1988). The basic idea is simple. Let M be the function monoid of E. Partially order the idempotent elements of M by $x \leq y$ iff $xy = yx = x$. Let e be a minimal element in the ordering and let R be the range of the function e. S_1 is a collection of certain (possibly overlapping) subsets of S. A subset s_1 is a member of S_1 provided there is an element of M that maps the subset R onto the subset s_1. If $x \in M$ and $s_1 \in S_1$, then clearly the image of s_1 under x is also a member of S_1. The transition function of E_1 is thus given by the transition function of E in the obvious way. Any input string that induces the function e on S will be a synchronizer of E_1.

S_2 is R. Every subset in the collection S_1 is the same size as R. $h(s_1, s_2)$ is a member of the subset s_1. s_2 simply specifies which member. How it does this is not important, but for concreteness we suggest that $h(s_1, s_2)$ be the unique member of s_1 that is mapped to s_2 by e. The transition function for E_2 is the one that enforces this rule. It is clear then that E_2 is a permutation automaton.

Note that if we fix s_1 and regard $h(s_1, s_2)$ as a function of s_2 only, then this function is one to one.

The state diagram of the E_1E_2 complex is strongly connected (formal proof in (Westerdale, 1988)).

It would be nice if E_1 had fewer states than E, but this is not always the case. Sometimes it has more.

Suppose we attach payoffs to the states of E_1E_2 as follows. The payoff of $<s_1, s_2>$ is the same as the payoff of $h(s_1\ ,\ s_2)$ in E. Then h is a payoff preserving homomorphism and Lemma 4 applies in reverse. That is, it applies if the first automaton in the Lemma is E_1E_2 and the second automaton in the Lemma is the environment E. So the excess values of the input symbols are the same in E_1E_2 as in E. So with these attached payoffs, E_1E_2 can be used as a model, and the reward scheme can use the model state excess values.

9 The First Component as a Homomorphic Image

Thus the system can use E_1E_2 as a model. We now want to show that the system can use E_1 as a model on its own. We do this by showing that if appropriate payoffs are attached to the states of E_1, then the input symbol excess values are the same in E_1 as in E_1E_2, and hence the same as in E. E_1 is a homomorphic image of E_1E_2 in the obvious way. As above, we let S_1 be the state set of E_1 and S_2 the state set of E_2. In this section, we let S be the state set of E_1E_2 so $S = S_1 \times S_2$. The homomorphism is a function from S to S_1. We represent it as a hat. Consider $i \in S$. Then $i = <s_1, s_2>$, where $s_1 \in S_1$ and $s_2 \in S_2$. Our homomorphism carries i to s_1, so $s_1 = \hat{i}$. We write s_2 as $\ddot{i}$. So for any $i \in S$ we can say $i = <\hat{i}, \ddot{i}>$.

We shall use the notation of section 7. There, we had E and its homomorphic image $\bar{E}$. Here, we have E_1E_2 and its homomorphic image E_1. There, quantities without a bar, like P, were associated with E, and quantities with a bar, like $\bar{P}$, were associated with $\bar{E}$. Here, quantities without a bar, like P, are associated with E_1E_2, and those with a bar, like $\bar{P}$, are associated with E_1.

We examine a particular input symbol r and a particular $k \in S_1$ and $j \in S$. We prove two lemmas about them. We define $K = \{k\} \times S_2$, so $K \subseteq S$. We give E_1E_2 the input symbol r. We ask which members of K are carried to j by the input symbol r.

Take an arbitrary $i \in K$. Does r carry it to j? Well if it does, then when we give r to E_1 it must carry $\hat{i}$ to $\hat{j}$. But $\hat{i} = k$, so r must carry k to $\hat{j}$ in E_1. Thus we have a

CONDITION: If we give r to E_1, r carries state k to $\hat{j}$.

We have seen that if the condition is not met, then if we give r to E_1E_2, no member of K is carried to j. But what if the condition is met?

LEMMA 5: If the condition is met, then if we give r to E_1E_2, exactly one member of K is carried to j.

PROOF: Select $i \in K$ and ask whether i is carried to j. Now $i = <k, \ddot{i}>$ and $j = <\hat{j}, \ddot{j}>$ and our condition tells us that k is carried to $\hat{j}$ in E_1, so we have only to see whether $\ddot{i}$ is carried to $\ddot{j}$ in E_2. Now whichever member i of K we selected, the input to E_2 is the same, namely the input symbol r together with the state k

of E_1. Because E_2 is a permutation automaton, that input permutes the states of E_2 and so exactly one state s_2 of E_2 is carried to $\ddot{j}$ by this input. Thus i is carried to j if and only if $\ddot{i}$ is this s_2. By the definition of K we see that there is exactly one member i of K for which $\ddot{i} = s_2$. Q.E.D.

LEMMA 6: $\sum_{i \in K} [r]_{ij} = \overline{[r]}_{k\hat{j}}$.

PROOF: Suppose we give r to E_1E_2. Then the sum in the Lemma is the number of elements of K that r carries to j. By Lemma 5 this is 1 if the condition is met. We saw previously that it is 0 if the condition is not met. But another way of stating the condition is: "$\overline{[r]}_{k\hat{j}}$ is 1 rather than 0". Q.E.D.

We define H as before. So $\mathbf{e}_j H = \mathbf{e}_{\hat{j}}$ and so from this and the definition of K we have

$$H^{\mathsf{T}} \mathbf{e}_j^{\mathsf{T}} = \bar{\mathbf{e}}_{\hat{j}}^{\mathsf{T}} \quad \text{and} \quad \bar{\mathbf{e}}_k H^{\mathsf{T}} = \textstyle\sum_{i \in K} \mathbf{e}_i. \tag{13}$$

LEMMA 7: $H^{\mathsf{T}}[r] = \overline{[r]} H^{\mathsf{T}}$ and $H^{\mathsf{T}} P = \bar{P} H^{\mathsf{T}}$.

PROOF: Lemma 6 can be written $\sum_{i \in K} \mathbf{e}_i [r] \mathbf{e}_j^{\mathsf{T}} = \bar{\mathbf{e}}_k \overline{[r]} \bar{\mathbf{e}}_{\hat{j}}^{\mathsf{T}}$. By (13), this is $\bar{\mathbf{e}}_k H^{\mathsf{T}} [r] \mathbf{e}_j^{\mathsf{T}} = \bar{\mathbf{e}}_k \overline{[r]} H^{\mathsf{T}} \mathbf{e}_j^{\mathsf{T}}$. The choice of k and j was arbitrary. Consequently the last equation holds for all $k \in S_1$ and $j \in S$. So $H^{\mathsf{T}}[r] = \overline{[r]} H^{\mathsf{T}}$. The choice of r was arbitrary, so this holds for all r. Multiplying by ϕ_r and summing over all r gives $H^{\mathsf{T}} P = \bar{P} H^{\mathsf{T}}$. Q.E.D.

Let $\ddot{N}$ be the number of states of E_2. For each state in S_1, there are $\ddot{N}$ states of S that the homomorphism maps to it, so $\mathbf{e}H = \ddot{N} \bar{\mathbf{e}}$. For simplicity let $\delta = (1/\ddot{N})$. Then $\bar{\mathbf{e}}^{\mathsf{T}} = \delta H^{\mathsf{T}} \mathbf{e}^{\mathsf{T}}$. Furthermore, $(\delta \bar{\tilde{\mathbf{p}}} H^{\mathsf{T}}) \mathbf{e}^{\mathsf{T}} = \bar{\tilde{\mathbf{p}}} \bar{\mathbf{e}}^{\mathsf{T}} = 1$, so $\delta \bar{\tilde{\mathbf{p}}} H^{\mathsf{T}}$ is a stochastic vector. $(\delta \bar{\tilde{\mathbf{p}}} H^{\mathsf{T}}) P = \delta \bar{\tilde{\mathbf{p}}} \bar{P} H^{\mathsf{T}} = \delta \bar{\tilde{\mathbf{p}}} H^{\mathsf{T}}$, so $\delta \bar{\tilde{\mathbf{p}}} H^{\mathsf{T}}$ is a stochastic eigenvector of P with eigenvalue 1. By the Frobenius-Perron theorem, there is only one of these, namely $\tilde{\mathbf{p}}$, so

$$\tilde{\mathbf{p}} = \delta \bar{\tilde{\mathbf{p}}} H^{\mathsf{T}}. \tag{14}$$

Multiplying on the right by $\mathbf{e}_j^{\mathsf{T}}$ and using the first equation in (13) gives $\tilde{p}_j = \delta \bar{\tilde{p}}_{\hat{j}}$. Consider any states $k \in S_1$ and $s_2 \in S_2$. Write $<k, s_2>$ as j. Then the conditional probability that E_2 is in state s_2, given that E_1 is in state k, is clearly $\tilde{p}_j / \bar{\tilde{p}}_{\hat{j}}$. We see that this is δ. In other words, given that E_1 is in state k, the various states of E_2 are equally probable. Put another way. Given that E_1 is in state k, E_1E_2 is in one of the states of $K = \{\mathrm{k}\} \times S_2$, and these states of K occur with equal probability in E_1E_2.

We now need to attach payoffs to the states of E_1. To each state k of E_1, we attach payoff $\bar{t}_k$, where $\bar{t}_k$ is the (unweighted) average of the payoffs t_i for all states i of E_1E_2 such that $\hat{i} = k$. This makes sense. Given that E_1 is in state k, the states of E_2 are equiprobable. If we define K as before, it is the states $i \in K$ over which we are averaging t_i, and all the states of K are equiprobable. $\bar{t}_k = \delta \sum_{i \in K} t_i$. Letting $\bar{\mathbf{t}}$ be the $\bar{N}$ dimensional vector whose k'th entry is $\bar{t}_k$, we have $\bar{\mathbf{e}}_k \bar{\mathbf{t}}^{\mathsf{T}} = \delta \sum_{i \in K} \mathbf{e}_i \mathbf{t}^{\mathsf{T}}$. Using the second equation in (13) gives $\bar{\mathbf{e}}_k \bar{\mathbf{t}}^{\mathsf{T}} = \delta \bar{\mathbf{e}}_k H^{\mathsf{T}} \mathbf{t}^{\mathsf{T}}$. This holds for arbitrary k, so

$$\bar{\mathbf{t}}^{\mathsf{T}} = \delta H^{\mathsf{T}} \mathbf{t}^{\mathsf{T}}. \tag{15}$$

Multiplying on the left by $\bar{\tilde{\mathbf{p}}}$ and using (14) gives $\bar{\tilde{\mathbf{p}}}\bar{\mathbf{t}}^{\mathsf{T}}=\bar{m}$, and so $\bar{m}$, not surprisingly, can play the role of average payoff per unit time of E_1. The vector of excess payoffs of E_1 is $\bar{\mathbf{a}}=\bar{\mathbf{t}}-\bar{m}\bar{\mathbf{e}}$. Using the definitions of $\bar{\mathbf{a}}$ and $\mathbf{a}$, and also (15) and $\bar{\mathbf{e}}^{\mathsf{T}}=\delta H^{\mathsf{T}}\mathbf{e}^{\mathsf{T}}$, gives $\bar{\mathbf{a}}^{\mathsf{T}}=\delta H^{\mathsf{T}}\mathbf{a}^{\mathsf{T}}$. From this and Lemma 7 we obtain $\bar{P}^k\bar{\mathbf{a}}^{\mathsf{T}}=\delta H^{\mathsf{T}}P^k\mathbf{a}^{\mathsf{T}}$. Then from the Cesaro sum definitions of $\bar{\mathbf{c}}$ and $\mathbf{c}$ we have $\bar{\mathbf{c}}^{\mathsf{T}}=\delta H^{\mathsf{T}}\mathbf{c}^{\mathsf{T}}$. Again multiplying on the left by $\bar{\mathbf{e}}_k$, defining K as before, and using the second equation in (13) gives $\bar{c}_k=\delta\sum_{i\in K}c_i$. So the excess value of state k of E_1 is the unweighted average of the excess values of all states i of E_1E_2 for which $\hat{i}=k$, as one would expect.

From $\bar{\mathbf{c}}^{\mathsf{T}}=\delta H^{\mathsf{T}}\mathbf{c}^{\mathsf{T}}$, Lemma 7, and (14) we have $\bar{\tilde{\mathbf{p}}}\overline{[r]}\bar{\mathbf{c}}^{\mathsf{T}}=\tilde{\mathbf{p}}[r]\mathbf{c}^{\mathsf{T}}$. So the excess value of r is the same in E_1 as it is in E_1E_2. We saw earlier that it was the same in E_1E_2 as in E. Thus the system can use E_1 as a model, and the reward scheme can use the model state excess values. In the next section we ask how the system can estimate the excess values of the states of E_1.

E_1 is a particularly well behaved automaton, for we have:

LEMMA 8: The transition probability matrix $\bar{P}$ of E_1 is primitive.

PROOF: Let σ be a synchronizer of E_1 and let k be the state E_1 is in following the input of the synchronizer. Since the state diagram of E_1E_2 is strongly connected, there is, for each state $i\in S_1$, an input string ρ_i that carries state k to state i in E_1. (Notation comment: here i is a state of E_1, not of E_1E_2.) Let n be the length of σ and n' be the length of the longest of the ρ_i's. Now select any two states i' and i of E_1. Then for any input string τ, the string $\tau\sigma\rho_i$ carries state i' to state i. Thus there is a string of length $n+n'$ that carries state i' to state i. So we see that the matrix $\bar{P}^{n+n'}$ has all entries positive. Q.E.D.

So in particular we see that $\bar{P}^{\infty}$ exists and that the Cesaro sum in the definition of $\bar{\mathbf{c}}$ actually converges in the ordinary sense.

10 The First Component as a Model, Using Bucket Brigade

We now ask how the system can estimate the excess values of the states of E_1. Consider the following three automata: the environment E, its decomposition E_1E_2, and E_1 on its own. Each of these automata has payoffs attached to its states. E is a homomorphic image of E_1E_2. A homomorphism h maps each state of E_1E_2 to a state of E. We assume that, as above, each state of E_1E_2 has the same payoff attached as does the state of E to which the homomorphism h maps it. Now consider the payoffs attached to the states of E_1. Suppose k is a state of E_1. As above, we assume that its attached payoff is equal to the unweighted average of the payoffs attached to the E_1E_2 states in $\{\mathrm{k}\}\times S_2$.

Suppose we simultaneously run the three automata, giving them the same inputs. Suppose that to begin with, the state of E is the state to which the homomorphism maps the state of E_1E_2, and also that to begin with, E_1 is in the same state as the first component of E_1E_2. Then this will continue to be the case during the run. We see that during the run, E_1E_2 and E will always produce exactly the same payoffs.

Suppose that, at some time during the run, E_1 is in state k. Then E_1E_2 is in one of the states of $K = \{k\} \times S_2$. We have seen that the states of K are equally probable, given that the first component is in state k. So the average payoff from E_1E_2 (and hence from E), given that E_1 is in state k, is the unweighted average of the payoffs associated with the states in K. This is the E_1 payoff associated with state k. That is, when E_1 is in state k, the E payoff may differ from the E_1 payoff. Sometimes it's higher and sometimes lower, but on average it is the E_1 payoff.

Earlier we showed how the system could estimate the excess values of the states of E using the bucket brigade on E (or rather on a model that was a complete copy of E). The limiting cash balances could be used as state excess values by the reward scheme. In exactly the same way, the system can estimate E_1 state values by using the bucket brigade on E_1. For the same reason, the limiting cash balances can be used as state excess values by the reward scheme. Remember that when a state is visited, the bucket brigade adds that state's payoff to its cash balance. Thus one might think that as the run proceeds (running E_1, E_1E_2, and E), the E_1 bucket brigade needs to add to state k the current E_1 payoff each time k is visited. But in fact the E_1 bucket brigade can use the current E_1E_2 payoff instead. This is because, as we have seen, if we look only at times when E_1 is in state k (and E_1E_2 is in one of the states of K), the E_1E_2 payoff averages out to the E_1 payoff. But E payoff is the same as E_1E_2 payoff. So the E_1 bucket brigade can use E payoff instead.

Thus E_1 can be used as a model with no attached payoffs. The bucket brigade can be run on E_1, using the payoff from E, to obtain estimates of the E_1 state excess values.

It might seem as if this only works if we start E_1 in the right state, and never make a mistake in our E_1 state transition. But E_1 is a self correcting model. Every time a semisynchronizer is input, E_1 automatically goes into the right state. We can see this if we continue to pretend that E_1, E_1E_2, and E are running simultaneously, with E_1E_2 always correctly tracking E so that the state of E is always the state to which h maps the state of E_1E_2. When a semisynchronizer is input, this puts E_1 in the same state as the first component of E_1E_2, so now the state of E_1 bears the right relationship to the state of E.

The system need not know what the semisynchronizers are. It is enough that they occur with non zero probability. Then the system can use the E_1 bucket brigade and it will work provided it makes E_1 state transition errors infrequently enough that the self correction mechanism keeps E_1 in the right state most of the time.

So to summarize, the system can use E_1 (with no attached payoffs) as a model, attaching a cash balance to each state and using the bucket brigade to move the cash around, using in the bucket brigade the payoff from the environment. The reward scheme can then use the cash balances as if they were state excess values.

As a postscript, we comment on the relationship between E_1 payoffs and E payoffs. To each state k of E_1 corresponds a subset $K = \{k\} \times S_2$ of states of E_1E_2. E is a homomorphic image of E_1E_2. Call the homomorphism $h : S_1 \times S_2 \rightarrow S$. We have seen that if we restrict h to K, then this restriction is one to one. Therefore the image $h[K]$ of K is a set of size $\ddot{N}$. (In our construction of the decomposition, this is the set that formed the state k.) h preserves state payoffs and state values, so the E_1 payoff of k is the unweighted average of the E payoffs of the states in $h[K]$, and the E_1 excess value of k is the unweighted average of the E excess values of the states in $h[K]$. If the input probabilities change, the state values might well change, but the state payoffs (in E, in E_1E_2, and in E_1) remain just as they were. Incidentally, it is easy to see that if k is an E_1 state, then the excess value of k says how much more E payoff than usual one should expect eventually if E_1 is in state k now.

11 The Equipayoff Homomorphism

Of course E_1 has in general a large number of states, so a homomorphic image of E_1 might provide a model of a more manageable size. Obviously our model need only reflect those aspects of the environment that are relevant to payoff. We can use a fairly standard trick to construct a payoff preserving homomorphism.

We define an equivalence relation over the set of states of E_1 as follows. Two states i and j are equivalent if for any sequence ρ of inputs, the payoff sequence starting in i is identical to the payoff sequence starting in j. The payoffs are the ones we have attached to the states of E_1. We call the relation the *equipayoff relation.* The set of equivalence classes forms the set of states of our homomorphic image, and the homomorphism, called the *equipayoff homomorphism,* maps each state of E_1 to the equivalence class containing it.

Note that if two states are related, and if we input symbol r, then r carries these two states to two states that are also related (or possibly to the same state). If A is an equivalence class and r an input symbol, then r maps A to a set of states that is a subset of one of the equivalence classes. So we can form the homomorphic image by letting the equivalence classes be the states of the image. The state transition function of the image is the obvious one. (If A is an equivalence class, $s_1 \in A$, and r carries s_1 to s_1' in E_1, then in the image, r carries A to the equivalence class of which s_1' is a member.) The payoff associated with an image state, with a class, is the payoff of any member of the class (all the members have the same payoff.). The homomorphism maps each state of E_1 to the equivalence class of which it is a member. It is easy to see that this is a payoff preserving homomorphism.

The homomorphism we have just constructed is well known. Payoff is, of course, a kind of output. Our equipayoff relation is just what is usually called the equiresponse relation, and our homomorphism is the usual one based on that relation. We have merely used the term equipayoff instead of equiresponse because our output is payoff.

We have used the equipayoff relation to construct a homomorphic image of E_1. We could have used it to construct a homomorphic image of E. This is the more usual use. We prefer an image of E_1 as our model because an image of E_1 has a synchronizer, since E_1 does.

So our model M is a homomorphic image of E_1 under the equipayoff homomorphism. We can conceptually attach to each state of M the same payoff that is attached to any of the states of E_1 that map to it under the homomorphism. If we then just look at M as an automaton on its own, we can define the excess value of each state in the usual way. The homomorphism is payoff preserving so Lemma 4 applies and the excess values of input symbols are the same in M as in E_1 and hence the same as in E. Thus the system can use M as a model, and the reward scheme can use the model state excess values.

To estimate the excess values of the states of M, the system can use the usual bucket brigade. This clearly works if the bucket brigade uses the M payoffs. But now think of running M, E_1, E_1E_2, and E simultaneously, just as in the last section, but with M running too, and with the current state of M being always given by the equipayoff homomorphism from E_1. Then we see that the bucket brigade on M can just as well use E_1 payoffs since they are the same as the M payoffs. And it can just as well use E payoffs since E payoffs average out to E_1 payoffs in the way we described when we were trying to use E_1 as a model. Thus we can use M as a model in the same way we could use E_1. M too has a synchronizer and so it too is self correcting.

12 Queasymorphic Images

Holland suggests the term quasi-homomorphism or quasimorphism for something that is close to a homomorphism, and he suggests that what we want for a model of the environment E is a quasimorphic image of E. But such a model need not have a synchronizer. We have seen that a synchronizer is an advantage. I tentatively suggest the following terminology.

If E and $\bar{E}$ are strongly connected automata with the same input alphabet and with payoffs attached to their states, then a *sleazymorphism* from E to $\bar{E}$ is a payoff preserving homomorphism from E_1 to $\bar{E}$, where E_1 is the first component of our decomposition E_1E_2 of E, and where the payoffs associated with the states of E_1 are determined from those of E as above by taking the appropriate unweighted averages. Note that the sleazymorphism will be onto and that $\bar{E}$ will have a synchronizer. Our equipayoff homomorphism is such a sleazymorphism. I'll call it the equipayoff sleazymorphism.

Our model needs to be not the image under a quasi-homomorphism, but rather under a quasi-sleazymorphism, or *queasymorphism.* Holland is vague about the ways in which a quasimorphism can fail to be a genuine homomorphism. I too am vague about the ways in which a queasymorphism can fail to be a genuine sleazymorphism, but one obvious way suggests itself. If we change the payoffs on the states of E_1 to make some of the nearly equal payoffs actually equal, and then use not the original equipayoff homomorphism, but rather the analogous homomorphism derived from the changed payoffs, then we have what I would call a queasymorphism from E. A queasymorphic image might be a nice model with far fewer states than any sleazymorphic image. Note that states with even identical payoff will not be identified in the queasymorphic image if there is any input string that would map the two states to states with very different payoffs. How nearly equal must payoffs be before they are made equal? Various criteria suggest themselves. For example, we might insist they are more nearly equal if the states' probabilities are higher. [4]

[4] We do not here discuss whether sleazymorphisms are indeed morphisms. It may be wise to extend the notion of sleazymorphism to allow it to be from E_1 to the first component of our decomposition of $\bar{E}$, rather than to $\bar{E}$ itself. A model then would need to be a queasymorphic image that has a synchronizer.

13 Conclusion and Disclaimer

We have suggested that a model should be something like a homomorphic image of the first component of a particular two component cascade decomposition of the environment. This makes the model an image under what I call a queasymorphism.

The advantage of this approach is that such a model has a synchronizer, and hence is self correcting. Other approaches need a correction mechanism to be added to the model, but here the correction mechanism is an intrinsic part of model operation.

We have analyzed here only the basic case, where the system selects input symbols with independent probabilities, independent of environment state. Furthermore, the selected input symbols determine model state without the help of any output from the environment. Using a simplified version of the "prescriptions" in (Westerdale, 1986), the analysis has been partially extended to deal with production systems, but that extension is not discussed here.

More importantly, this paper does not touch upon the really important question: How does the system learn the state diagram of the desired model? In adaptive production systems (e.g. classifier systems) the productions themselves can be thought of as representing model states or sets of model states. The hope is that the evolving population of productions will evolve towards an appropriate model. Is it easier to evolve a self correcting queasymorphic image model or to evolve the more standard quasimorphic image model? This is not yet clear.

References

Gantmacher, F. R. (1959). *Applications of the Theory of Matrices.* Interscience Publishers Inc., New York, N.Y.

Grefenstette, J. J. (Oct. 1988). Credit assignment in rule discovery systems based on genetic algorithms. *Machine Learning*, 3(2/3):225–246.

Holland, J. H. (1969). Adaptive plans optimal for payoff-only environments.

Holland, J. H., Holyoak, K. J., Nisbett, R. E., and Thagard, P. R. (1986). *Induction: Processes of Inference, Learning and Discovery.* MIT Press, Cambridge, Mass.

Kemeny, J. G. and Snell, J. L. (1960). *Finite Markov Chains.* Van Nostrand, Princeton.

Westerdale, T. H. (June 1988). An automaton decomposition for learning system environments. *Information and Computation*, 77(3):179–191.

Westerdale, T. H. (May/June 1986). A reward scheme for production systems with overlapping conflict sets. *IEEE Trans. Syst., Man, Cybern.*, SMC-16(3):369–383.

Zeigler, B. P. (1976). *Theory of Modeling and Simulation.* John Wiley and Sons Inc., New York, N.Y.

Zeigler, B. P. (October 1972). Towards a formal theory of modeling and simulation: Structure preserving morphisms. *Journal of the Association for Computing Machinery*, 19(4):742–764.

Variable Default Hierarchy Separation in a Classifier System

Robert E. Smith
University of Alabama
Department of Engineering
Tuscaloosa, Alabama 35487

David E. Goldberg
University of Illinois
Department of General Engineering
Urbana, Illinois 61801

Abstract

A learning classifier system (LCS) is a machine learning system that incorporates a production-system framework and a genetic algorithm (GA) for rule discovery (Goldberg, 1989; Holland, 1975). A primary feature of LCSs is their potential to exploit overlapping sets of rules called *default hierarchies*. Default hierarchies increase rule set parsimony, enlarge the solution set, and lend themselves to graceful refinement by the GA (Holland, Holyoak, Nisbett, & Thagard, 1986). Traditionally, auction-based, specificity-biased credit allocation (CA) and conflict resolution (CR) schemes have been used to encourage default hierarchy formation in an LCS. Analyses presented in this paper suggest that these schemes cannot be expected to perform adequately in arbitrary LCS environments. This paper presents an alternate CA/CR that associates two measures with each classifier in place of the single, traditional *strength* measure. The first measure is a *payoff estimate*, which is tuned by the linear-update scheme usually used for strength. The second measure is a *priority factor* that is tuned to control the outcome of a *necessity auction*. In the necessity auction the winning classifier pays out the payoff estimate of its nearest competitor, rather than a fraction of its own payoff estimate. Results and analyses are presented that show that this CA/CR scheme can induce *variable bid separation* that responds to the demands of the LCS environment. Additional analyses show that this scheme allows an LCS to adequately exploit a broader class of default hierarchies than traditional schemes. Several avenues are suggested for further study.

Keywords: machine learning, classifier systems, default hierarchies, credit allocation.

1 Introduction

A learning classifier system (LCS) is a machine learning system that incorporates a production-system framework and a genetic algorithm (GA) for rule discovery (Goldberg, 1989; Holland, 1975; Holland et al., 1986). A primary feature of LCS is their potential to exploit overlapping sets of rules called *default hierarchies*. Default hierarchies increase rule-set parsimony, enlarge the solution set, and lend themselves to graceful refinement by the GA (Holland et al., 1986). Traditionally, auction-based, specificity-biased credit allocation (CA) and conflict resolution (CR) schemes have been used to encourage default hierarchy formation in an LCS. These CA/CR schemes have two major limitations. First, they encourage a fixed amount of *separation* (difference) between between the bids of competing classifiers in a default hierarchy. This fixed separation is pre-specified by LCS parameters and is not responsive to the demands of arbitrary environments. Second, these schemes assume that each classifier in a default hierarchy can be prioritized based on specificity. In general LCS problems and codings, specificity-based prioritization is only possible for a narrow class of default hierarchies. Note that these limitations can be expected to retard the GA's effectiveness as well. A narrower class of exploitable default hierarchies directly limits the density of solutions in the space the LCS-GA is attempting to explore. Also, an LCS can only provide the GA with effective fitness values for classifier sets that its CA/CR mechanism can adequately organize and evaluate. Therefore, limitations on the class of exploitable default hierarchies indirectly affects the GA by providing less reliable fitness information.

To cope with the limitations of the traditional LCS approach, this paper presents an alternate CA/CR scheme. This scheme associates two measures with each classifier (in place of the single, traditional *strength* measure). The first measure is a *payoff estimate*, which is tuned by the linear-update scheme usually used for strength. The second measure is a *priority factor* that is tuned to control the outcome of a *necessity auction*. A necessity auction requires that the auction's highest bidder only pay a slightly higher bid than the next-highest bidder. In the simulated necessity auction the winning classifier pays only the payoff estimate of its nearest competitor. The priority factor is used to bias conflict resolution. Results and analyses are presented that show that this CA/CR scheme can induce *variable bid separation* that responds to the demands of the LCS environment. Additional analyses show that this scheme allows an LCS to adequately exploit a broader class of default hierarchies than in traditional schemes.

To motivate these modifications to the LCS CA/CR scheme, one must consider the contribution of default hierarchy exploitation to an LCS's operation. The following section presents a detailed discussion of the role of hierarchical rule-set organization in the LCS paradigm.

2 The Importance of Default Hierarchies in LCSs

Default hierarchies are rule sets that allow the errors of imperfect *default* rules (rules that are correct for some situations, but incorrect for others) to be handled by *exception rules*. Exception rules act in the situations where a fully-matched default rule would be incorrect. Therefore, exceptions shield the defaults (and the entire system) from errors. An exception rule can be imperfect as well, and multiple layers of exceptions can be used to further refine performance. This sort of hierarchical rule-set structure allows for more parsimonious rule sets, expansion of the solution space, and graceful refinement of rule sets through the layered addition of exception rules (Holland et al., 1986). The ability to incrementally modify the rule set is significant in its relation to the GA, since the refinement of the rule set under genetic discovery can be expected to be gradual.

LCSs organize default hierarchies by favoring exception rules over defaults in their conflict resolution and credit allocation schemes. Several studies have demonstrated the emergence of default hierarchies in LCSs (Goldberg, 1983; Wilson, 1986; Riolo, 1988). However, doubts remain about the efficacy of traditional LCS CA/CR schemes for default hierarchy organization in arbitrary environments (Wilson & Goldberg, 1989).

In addition to the importance of default hierarchies to LCS mechanics, automatic organization of hierarchical rule sets is a feature that distinguishes LCSs from other automated learning methods. Consider a comparison between LCSs and *learning automata* (Narendra, 1986). A learning automaton generally consists of a stochastic automaton whose outputs are used as the control inputs to some environment. In the learning automata process, the probability of a transition from one state of the automaton to another state under a given input is adjusted based on the payoff gained from that transition. An LCS can be thought of as a stochastic automaton, where external messages represent the automaton's input, internal messages represent the automaton's state, the classifier list represents the automaton's next-state function, and effectors (along with the classifier list) represent the automaton's output function. The probability of firing a rule is adjusted based on the payoff associated with the state transitions that it suggests. The traditional LCS CA/CR process is similar to a number of learning automata probability update schemes; however, in learning automata the space of state transitions is partitioned such that probabilities can be adjusted directly. In an LCS, general classifiers represent transition templates that can overlap in a variety of ways. Also, the set of available transitions changes under the action of the GA. These differences preclude direct adjustment of transition probabilities. Because classifiers do not partition the state-transition space, the rigorous mathematical proofs of convergence of learning automata algorithms do not transfer directly to an LCS. Moreover, a partition of the state-transition space may be impractical in many computational systems. Overlap between classifiers affords LCSs the added parsimony and other advantages of default hierarchies. Thus, automatic hierarchical conflict resolution is an important research area in its own right.

To examine LCS schemes for hierarchical rule-set organization in detail, it is important to define default hierarchies and related structures more formally, a task undertaken in the next section.

3 Formal Definitions

This section defines default hierarchies and their components. Although these definitions are similar to those presented in Goldsmith (1989), they differ substantially in their intention and detail. Specifically, they are not based on details of classifier syntax such as specificity. Instead, they characterize default hierarchies in terms of the relationships between rules and the requirements of an LCS environment.

Any classifier can be characterized by its action and by the set of messages that fully satisfy its conditions. This study will refer to any set of messages on the LCS message list that will fully satisfy a given classifier as a matching *message-list configuration* for that classifier.

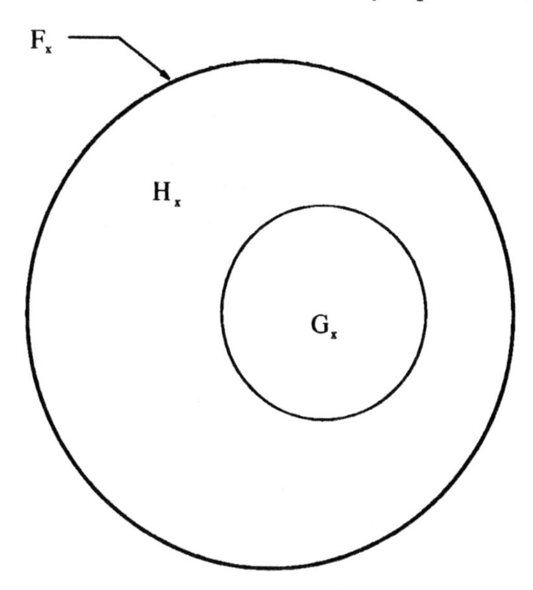

Figure 1: Message configurations that fully satisfy the conditions of classifier x. The set of matching message-list configurations for classifier x, F_x, is divided into two subsets: one for which the the classifier's action is correct, G_x, and one for which the classifier's action is incorrect, H_x.

Figure 1 shows a classifier x as a Venn diagram of its matching message-list configurations F_x. The message-list configurations are divided into a set where the action of classifier x is judged as correct, G_x, (relative to some given environment), and a set where the action of classifier x is judged as incorrect, H_x. In this discussion an incorrect action is loosely defined as an action that results in less payoff than an available alternative.

The competitive interactions of a pair of classifiers (d and e) in a given LCS environment can be characterized by two overlapping sets of matching message-list configurations and appropriate subsets (see Figure 2). The following notation describes these sets:

F_d - configurations that match d.

F_e - configurations that match e.

G_d - configurations for which d is matched and submits a correct action.

G_e - configurations for which e is matched and submits a correct action.

H_d - configurations for which d is matched and submits an incorrect action.

H_e - configurations for which e is matched and submits an incorrect action.

F_u - *undisputed configurations* where only one classifier from the pair is matched.

F_{ud} - configurations that are only matched by classifier d.

F_{ue} - configurations that are only matched by classifier e.

G_{ud} - undisputed configurations for which d is matched and submits a correct action.

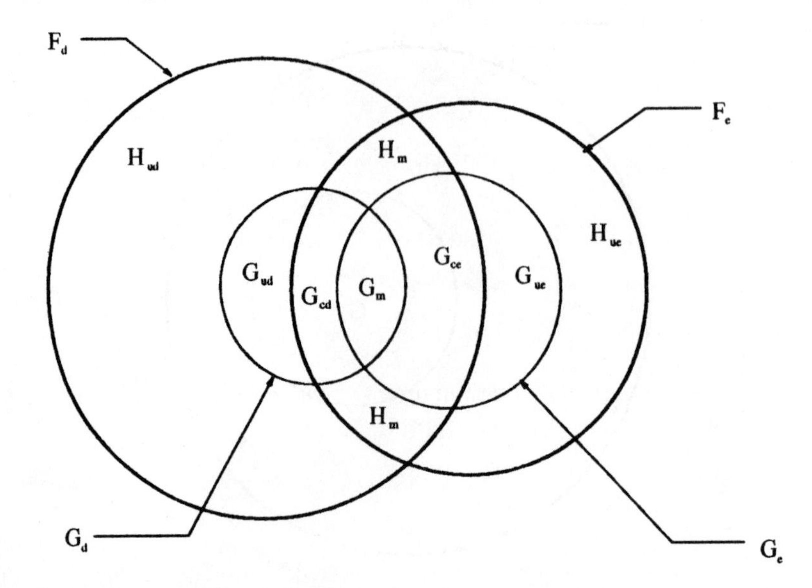

Figure 2: Message configurations matched by a pair of classifiers, e and d, and subsets that suggest their relationship to an LCS environment.

H_{ud} - undisputed configurations for which d is matched and submits an incorrect action.

G_{ue} - undisputed configurations for which e is matched and submits a correct action.

H_{ue} - undisputed configurations for which e is matched and submits an incorrect action.

$F_c = F_e \cap F_d$ - *conflict configurations* where both classifiers are matched.

G_{cd} - conflict configurations for which d submits a correct action and e submits an incorrect action.

G_{ce} - conflict configurations for which e submits a correct action and d submits an incorrect action.

G_{cm} - conflict configurations for which d and e mutually submit correct actions.

H_{cm} - conflict configurations for which d and e mutually submit incorrect actions.

Each set of configurations can be associated with a probability of occurrence for configurations in that set. For instance, $P(H_{cm})$ represents the occurrence probability of a conflict configuration for which both d and e are matched and submit incorrect actions.

Given these subsets and occurrence probabilities, the notion of a default-exception pair can be given a formal definition:

Definition 1 *A* default-exception pair *is a pair of classifiers* e *and* d *for which* $P(F_c) \neq 0$, $P(G_{ud}) \neq 0$, *and* $P(G_{ce}) > P(G_{cd})$. *Under this definition,* e *is the* exception *and* d *is the* default *(with respect to the pair* e, d*).*

A classifier can be a member of more than one default-exception pair, and can act as a default in one pair and as an exception in another.

The notion of an ideal default-exception pair can also be defined:

Definition 2 *A default-exception pair is defined to be* ideal *if $H_d \subset G_e$ and $P(H_e) = 0$.*

In other words, the exception in an ideal pair can correct all errors of the default, resulting in performance without error.

Definitions of default hierarchy and ideal default hierarchy follow naturally from the previous definitions:

Definition 3 *A* default hierarchy *is a set of rules where*

- *each rule in the set is either a default or an exception (under the previous definition of a default-exception pair) to at least one other rule in the set, and*
- *the rule set cannot be divided into subsets without separating at least one default-exception pair.*

The latter condition is only included to insure that the definition considers only a single, distinct default hierarchy, rather than groups of non-overlapping default hierarchies.

Definition 4 *A default hierarchy is defined to be* ideal

- *if, for each rule i that acts as a default to a set of exceptions 1, 2, 3, $\ldots n$, $H_i \subset \cup_{j=1}^{n} G_j$,*
- *and if, for each rule j that acts* only *as an exception, $P(H_j) = 0$.*

In an ideal hierarchy, layers of exceptions can correct all the errors of their associated defaults, and the rule set can perform without error.

To allow an exception to fire when both the default and exception match, the exception must receive priority in the conflict resolution process. If the correct set of priorities are maintained in a default hierarchy, exception rules consistently shield the default rules and prevent them from firing in the wrong situations. One can define ideal performance for an LCS conflict resolution scheme with respect to a given default-exception pair:

Definition 5 *Assume that on occurrence of a conflict configuration for a given default-exception pair, the CR scheme selects the default with probability P_d and the exception with probability $1 - P_d = P_e$.* Ideal performance *with respect to this pair occurs when the exception is always selected over the default by the CR mechanism ($P_e = 1$, $P_d = 0$).*

For an ideal default-exception pair, ideal CR performance results in performance without error.

Given these definitions, one can begin to characterize the performance of various CR schemes in terms of their ability to organize hierarchical rule-sets. The following section begins this characterization by examining the performance of traditional LCS CA/CR schemes on separating ideal default-exception pairs.

4 Ideal Default-Exception Pairs and Fixed Separation

The traditional LCS CA/CR scheme is based on an analogy to an auction. Each classifier's priority is based on its *bid*, B, which is, in turn, determined by the classifier's *strength*, S. Strength is a linearly-updated estimate of a classifier's contribution to LCS performance. In an attempt to allow the auction procedure to favor exceptions over defaults, each classifier's bid is biased based on the classifier's *specificity*, σ, where specificity is defined as the number of non-# positions over the total number of positions. Typically, a classifier's bid is determined as follows:

$$B = C\rho S;$$

where C is the bid constant, and the function $\rho = \rho(\sigma)$ is a monotonically increasing function of σ, where for $0 \leq \sigma \leq 1$, $0 \leq \rho(\sigma) \leq 1$. Note that each classifier calculates its bid as a function of its own strength and specificity alone.

Consider the strengths and bids of a single, ideal default-exception pair. Assuming ideal CR performance, and that each rule receives payoff R when it acts, the strength values are updated as follows:

$$S_d^{i+1} = S_d^i(1 - C\rho_d) + R,$$

$$S_e^{j+1} = S_e^j(1 - C\rho_e) + R,$$

where the subscript d denotes the default rule, the subscript e denotes the exception rule, and the superscripts denote the time step. By setting $i = i + 1$ and $j = j + 1$, the following steady-state strength values can be calculated:

$$S_d^{ss} = \frac{R}{C\rho_d}, \text{ and}$$

$$S_e^{ss} = \frac{R}{C\rho_e}.$$

This also yields the steady-state bids:

$$B_d^{ss} = B_e^{ss} = R.$$

This amounts to a proof by contradiction that the default-exception pair must make mistakes, since at the indicated steady state the LCS will not distinguish between the default and exception rules.

As pointed out in Goldberg (1989), one possible remedy for this problem is the addition of a classifier *tax*. Taxes are often used in LCSs to reduce the strengths of classifiers that serve no purpose, but whose strengths are not sufficiently lowered by the CA scheme to qualify them for deletion by the GA. They also serve the auxiliary purpose of providing steady-state default-exception separation. Consider a head tax, t_h, that is applied to each classifier during every system cycle. The strengths are updated as follows:

$$S_d^{i+1} = S_d^i(1 - C\rho_d - t_h) + R,$$

$$S_e^{j+1} = S_e^j(1 - C\rho_e - t_h) + R.$$

This results in the following steady states:

$$S_d^{ss} = \frac{R}{C\rho_d + t_h},$$

$$S_e^{ss} = \frac{R}{C\rho_e + t_h},$$

$$B_d^{ss} = \frac{C\rho_d R}{C\rho_d + t},$$

$$\text{and } B_e^{ss} = \frac{C\rho_e R}{C\rho_e + t_h}.$$

Let ΔB be the *bid separation* for this default-exception pair, which is defined as

$$\Delta B = B_e - B_d = \frac{RCt_h(\rho_e - \rho_d)}{(C\rho_e + t_h)(C\rho_d + t_h)}$$

Clearly, this is always a positive number for positive t_h, ρ_e, ρ_d, C, and R, since $\rho_e > \rho_d$. By specifying t_h and the $\rho(\sigma)$ function, an arbitrary amount of separation for a given ideal default-exception pair can be obtained. However, note that once the tax rate and $\rho(\sigma)$ are fixed, this separation is fixed for each default-exception pair with given values of σ.

Another method of inducing a fixed amount of bid separation is to bias the bid competition based on specificity while factoring some specificity bias out of the amount a classifier actually pays when fired (Riolo, 1988; Wilson, 1989). Let us assume that $\hat{\rho}(\sigma)$ is the specificity function used in payback, while $\rho(\sigma)$ remains the specificity function used in the bid competition. Further assume that $\hat{\rho}(\sigma)$ is a monotonically increasing function of σ, where for $0 \leq \sigma \leq 1, 0 \leq \hat{\rho}(\sigma) \leq 1$. In the absence of taxes, the resulting steady-state bids under this scheme are:

$$B_d^{ss} = \frac{\rho_d R}{\hat{\rho}_d}, \text{ and}$$

$$B_e^{ss} = \frac{\rho_e R}{\hat{\rho}_e}.$$

Note that ΔB is always positive under the following condition:

$$\frac{\rho_e}{\hat{\rho}_e} > \frac{\rho_d}{\hat{\rho}_d}.$$

This condition is satisfied in both Wilson's*Boole-2* system (Wilson, 1989) (where $\hat{\rho}(\sigma) = 1$ and $\rho(\sigma) = \sigma$) and in Riolo's LCS (Riolo, 1988) (where $\hat{\rho}(\sigma) = \sigma$ and $\rho(\sigma) = \sigma^n$, with $n > 1$). Note that although these schemes insure positive separation, the separation remains fixed for a given ratio of default and exception specificity.

At first glance, fixed separation may seem adequate for LCS hierarchical conflict resolution. Any positive separation seems to suggest that the LCS should favor the exception. However, several factors make fixed separation undesirable. First, one must take into account noise that may be present in an LCS. Noise is usually introduced into the LCS CR mechanism to allow new rules the possibility of activation and subsequent evaluation. Also, noise in an LCS can be the result of the system's interaction with an uncertain or noisy environment. Ideal CR performance can only occur if ΔB is greater than the maximum noise level. If the degree of separation of a default-exception pair is insufficient to overcome noise in the system, the system cannot adequately exploit the pair as a part of its rule-set structure. Since noise and the specificity ratios of useful default-exception pairs cannot be known beforehand, a correct form for the function $\rho(\sigma)$ and the values of LCS parameters cannot be specified for general utility in a fixed-separation CA/CR scheme. In addition to noise, the effects of taxes in arbitrary environments can make fixed separation undesirable.

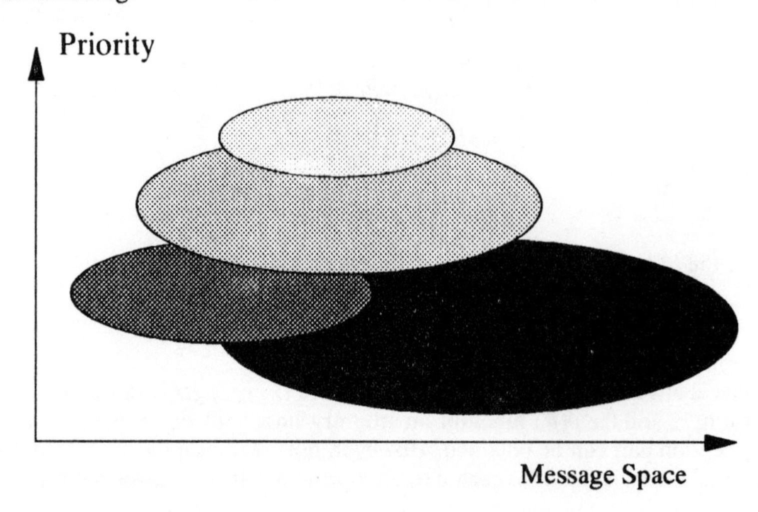

Figure 3: Default hierarchies can be formed where specificity does not correctly suggest priority.

A classifier can be taxed an arbitrary number of times between activations. The cumulative effects of repeated taxation can degrade fixed separation. Note that this applies to all taxes, not just to the head tax discussed above. Fixed separation can also be degraded by payoff delays that take place in the bucket brigade. Since the extent of the effects of noise, repeated taxes, and payoff delays cannot be known beforehand in arbitrary environments, the parameters that control fixed separation schemes cannot be specified generally.

Another important objection to specificity-based, fixed separation is that working default hierarchies can exist where specificity does not correctly prioritize the rule set. Figure 3 illustrates this possibility. Larger ovals represent more general (less specific) classifiers, an oval's horizontal position represents the range of configurations the classifier matches, and its vertical position represents its priority in the default hierarchy. The figure illustrates that classifiers can be placed into working default hierarchies in priority positions that are not necessarily related to their specificity. Moreover, the number of # positions in a classifier does not necessarily suggest the number of configurations that it matches in a given environment. Certain types of configurations may be presented more often than others. Only under the assumption of a uniform distribution of messages is specificity well correlated with the expected frequency of matches for a classifier. It is true that default hierarchies can be formed for any given environment such that specificity-based prioritization is possible; however, if the system is only able to exploit this restricted class of default hierarchies, expansion of the solution space through default hierarchies is limited. Additionally, if the system can only thoroughly evaluate rules that belong to specificity-based default hierarchies, rule fitness values are limited in their ability reflect actual rule utility, thus limiting the GA's ability to accurately select rules for deletion and reproduction.

As an alternative to specificity-biased, fixed separation schemes, the following sections introduce the *necessity auction* and the use of separate *priority factors* in the LCS CA/CR scheme as a combined method for improving the formation of default hierarchies. These additions allow classifiers to be prioritized into a default hierarchy based on their relative performance in bid competitions.

5 The Necessity Auction

The conventional LCS bid competition CA/CR scheme is a highly simplified model of auction dynamics. Each classifier calculates its bid as a fixed proportion of its own strength without regard to the bids of its competitors. Although this forms an easily implemented (linear) strength update scheme, it leads to steady-state strengths that are only related to individual classifier's performances and not to the performance of the system as a whole; this in turn leads to fixed bid separation. The *necessity auction*, a nonlinear scheme, is suggested to improve an LCS's simulation of auction dynamics and to induce bid separation that responds to the entire system's performance.

In many real auctions, each competitor calculates its bid based on the bids of its current competitors. Thus, bids have values that are related to the entire market. In a necessity auction, each bidder need only bid enough to out-bid its nearest competitor. The following procedure can be used to simulate this in an LCS:

1. Each classifier posts a portion of its strength as a *potential bid.*
2. Active classifiers are selected based on their potential bids.
3. Reward is distributed to active classifiers.
4. Instead of paying out the amount of its potential bid, a winning classifier pays out an *actual bid* that is equal to the potential bid of its nearest competitor. If there are no competitors, the classifier pays out a minimum bid.

Consider this scheme applied to a single, ideal default-exception pair. Assuming ideal CR performance, strengths are updated as follows:

$$S_d^{i+1} = S_d^i(1 - C\rho_d) + R \ ,$$

$$S_e^{j+1} = S_e^j - S_d^j C\rho_d + R \ .$$

At steady state, $S_d^{ss} = R/C\rho_d$, and $B_d^{ss} = R$. Under this condition, S_e remains unchanged; $S_e^{j+1} = S_e^j$. Therefore, any value of S_e which dictates sufficient bid separation is a valid steady state. This *variable separation* effect can be verified in simple LCS experiments.

Consider the simple, finite automaton environment shown in Figure 4. The symbol "s" indicates the automaton's state. The symbol "o" indicates the automaton's output, which is fed to the LCS as an environmental message. The symbol "i" indicates transitions that the automaton will take given i as the LCS output. The symbol "r" indicates the reward provided when the system takes the associated transition. Note that this automaton incorporates a single-step delay in reward. To obtain optimal reward, the LCS must emulate the logical OR function over the input bits. A single default-exception pair can be used to solve this problem:

0##&1##/111
000&1##/100

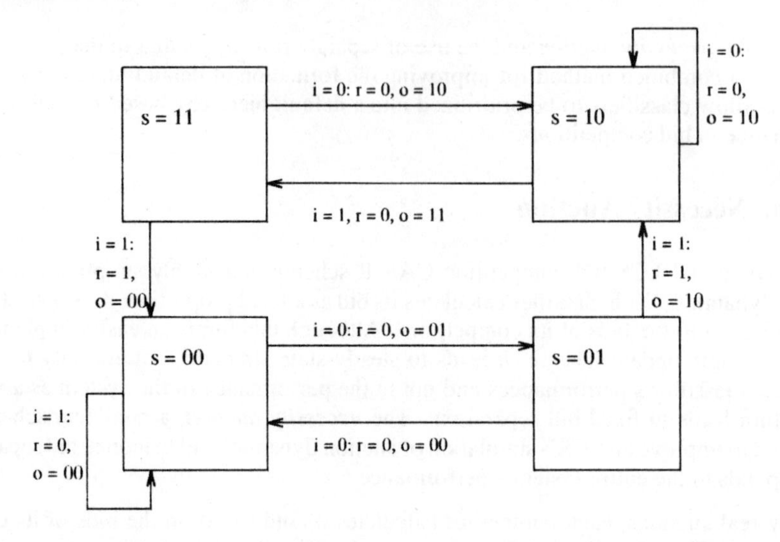

Figure 4: A simple, finite automaton environment with single-step reward delay.

where the convention of a leading 0 indicates an external message, and a leading 1 indicates an internal message. The effectors are assumed to copy the last internal message bit to the output, and the second (all–#) condition of each classifier couples the rules through the bucket brigade.

Figure 5 shows the results of an LCS applied to this environment with the conventional specificity-biased CA/CR scheme. In this experiment, $\rho = \sigma$, $C = 0.1$, $t_h = .001$, Gaussian noise with variance of 0.3 is added to the classifier bids, the GA is not active, the population is initialized with the correct default-exception pair, and the performance measure is the average reward over 16 cycles. Because of noise, repeated taxation, and reward delay, adequate, fixed separation becomes impossible and the system's performance remains erratic.

Figure 6 shows the results of a similar experiment with identical parameter settings and the necessity auction. Note that the bid separation and performance improve throughout the run. Experiments performed with other noise and tax parameters show similar results. These results show that the necessity auction can vary separation to suit the needs of the environment.

Although these preliminary experiments with this simple form of the necessity auction are encouraging, they are difficult to extend to multi-level default hierarchies. Consider a three-rule, two default-exception pair, ideal default hierarchy, where d is a default in one pair, ee is the exception in the other pair, and e acts as an exception to d and a default to ee. Assuming ideal CR performance for both pairs:

$$S_d^{i+1} = S_d^i(1 - C\rho_d) + R\,,$$

$$S_e^{i+1} = S_e^i - S_d^i C\rho_d + R\,,$$

$$S_{ee}^{i+1} = S_{ee}^i - S_e^i C\rho_e + R\,,$$

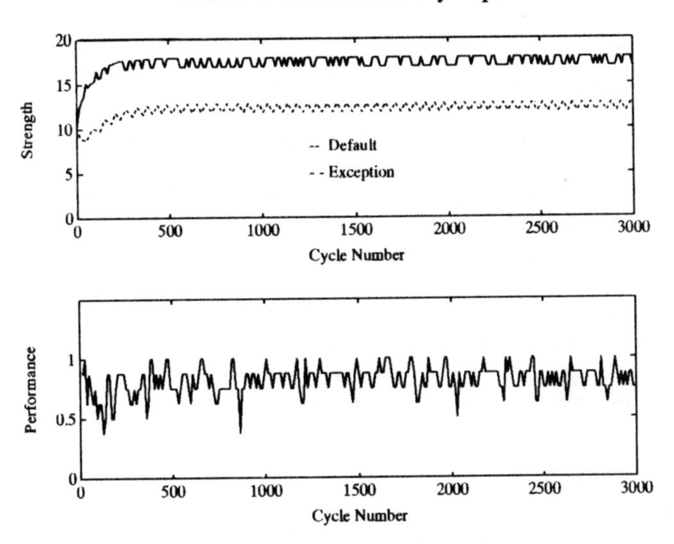

Figure 5: An experiment with an LCS using the conventional CA/CR scheme applied to the automaton of Figure 3.

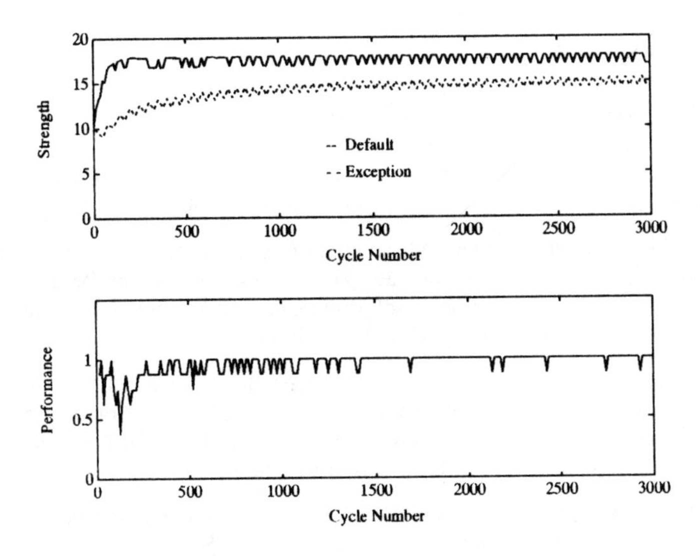

Figure 6: An experiment with an LCS using the necessity auction applied to the automaton of Figure 3.

Since $S^i_e C \rho_e > R$ for ideal CR performance,

$$S^{i+1}_{ee} = S^i_{ee} - \Gamma \ ,$$

where $\Gamma > 0$. Therefore, stable, positive separation between B_{ee} and B_e cannot exist, and ideal CR performance is proved impossible by contradiction. To clarify this deficiency, consider the reason for variable separation between B_e and B_d. The middle exception is able to separate above the default by using B_d as an estimate of the default's mean payoff. When this estimate reaches its maximum value, S_e becomes fixed. The second-level exception cannot use B_e in a similar fashion, because B_e is no longer an indicator of the middle exception's mean payoff. Instead, B_e plays the role of a *priority factor* that controls conflict resolution between the exception and the default. Following this line of reasoning, the next section examines the use of a separate priority factor for each classifier to extend variable-separation to multi-level default hierarchies.

6 Separate Priority Factors

Consider the maintenance of two separate factors for each classifier: one that estimates the classifier's expected payoff and another that suggests priority in the auction. The former will be called the *payoff estimate*, Π, and the later will be called the *priority factor*, Φ. For classifier x the payoff estimate is updated with the conventional, linear update procedure, while the priority factor uses the necessity auction, as follows:

$$\Pi^{i+1}_x = \Pi^i_x(1 - C) + CR \ ,$$

$$\Phi^{i+1}_x = \Phi^i_x - D\Pi^i_y + DR \ ,$$

where Π_y is the payoff estimate of classifier x's nearest competitor at time i, and D is the priority-update constant. The priority factor is not updated if no competitors exist. In the experiments presented here, the classifier's potential bid is determined by the product of its payoff estimate and its priority factor. The priority factor alone can also be used as the potential bid (Smith, 1991). Note that no specificity factor is used in resolving classifier conflicts. This scheme will be called the *priority tuning* CA/CR scheme.

Note that in addition to allowing variable separation for multi-level default hierarchies, the priority tuning scheme provides a separate payoff estimate that can be used as a fitness value for the GA. Since this value need not provide priority information to the CR scheme, it can serve as a clearer measure of a classifier's fitness.

By maintaining a separate priority factor, each classifier can use its nearest competitor's payoff estimate to update its own priority such that adequate separation occurs. In terms of the economic analogy, the priority factor can be thought of as a *cash reserve*, or as some other net-profit-based influence on the auction procedure.

Consider a four-state automaton as a test environment for the use of priority factors in an LCS. The four states (labeled A-D) are presented to an LCS in a cycle. Each state requires a different action, and each correct action provides a reward value of 1. Four classifiers can be formed into an ideal default hierarchy to perform the required task:

- A high-level **exception** classifier matches and provides the correct action in state A.
- A **middle**-level exception classifier matches in states A and B, but only provides the correct action in state B.

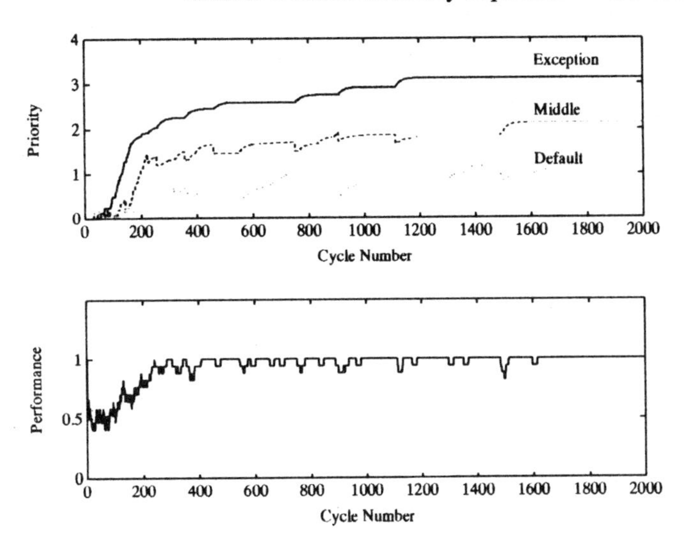

Figure 7: A simulation of priority tuning in an LCS for a four-level default hierarchy given an even start.

- A **default** classifier matches in states A, B, and C, but only provides the correct action in state C.
- A background default matches in all four states, but only provides the correct action in state D.

Experiments presented here focus on the exception, middle, and default classifiers. In the simulations presented, $C = D = 0.2$, and Gaussian noise with variance of 0.1 is added to each classifier's potential bid. The performance is defined as before.

Figure 7 shows the results of a simulation where initial values of $\Phi^0 = 0$ and $\Pi^0 = 0.5$ were used for all four classifiers. After the system gains information from repeated trials, the necessity auction correctly separates the 4-level default hierarchy. Note that late in the run, the priority factors of each default-exception pair are separated by about one. For the given amount of input noise, this separation should give an error rate of less than .001, as is reflected in the performance measure.

Figure 8 shows the results of a similar simulation, but with the priority values started from adverse conditions ($\Phi_d^0 = 4$, $\Phi_m^0 = 2$, $\Phi_e^0 = 0.5$, and $\Phi_b^0 = 0$). The default hierarchy reorders itself, since the default and middle rules lose priority when they fire in situations where they are not rewarded.

7 General Default-Exception Pairs

Up to this point, analyses and experimental results with the priority tuning scheme have only addressed ideal default hierarchies, where exception rules fully correct all the errors

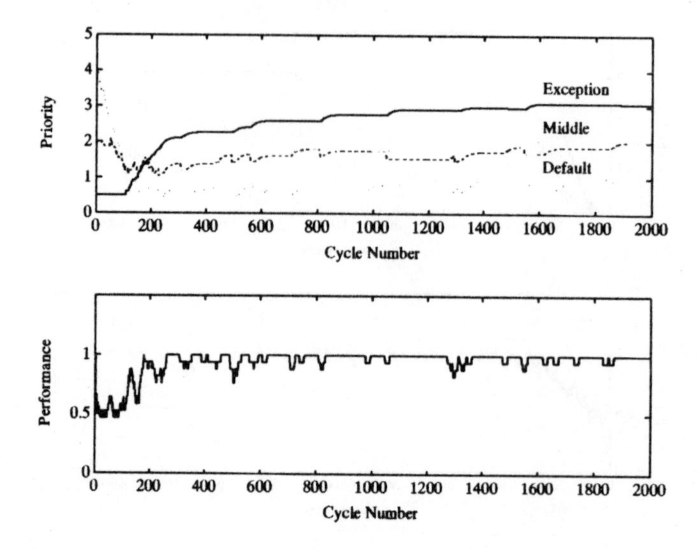

Figure 8: A simulation of priority tuning in an LCS for a four-level default hierarchy given an adverse start.

of their associated default rules. The results are encouraging, but their success is largely based on the assumption that each rule in the default hierarchy receives the maximum payoff level when the hierarchy functions perfectly. The bucket brigade may be able to insure this condition for ideal default hierarchies, even if the environment provides a variety of reward levels; however, one cannot expect that the ideal rule set will be available during much of the system's operation. In these situations, it is desirable for an LCS to organize general default hierarchies that improve system performance but do not correct all errors, to allow for graceful refinement of the rule set by the GA.

In a general default hierarchy, the difference between payoff estimates for competing classifiers may not tend to zero as adequate bid separation is obtained. As an example, an exception classifier could correct many of a default classifier's errors, but could itself cause errors in situations where it does not compete with the default. If the exception's payoff estimate remains lower than the default's payoff estimate when adequate bid separation exists, the separation will tend to degrade. In another example of a non-ideal default hierarchy, an exception classifier could only correct some, but not all, of a default classifier's errors. If the exception's payoff estimate remains higher than the default's payoff estimate after adequate bid separation is obtained, the necessity auction will cause the priority of the exception rule to rise without bound. Although this effect may be an accurate simulation of a net-profit biased, competitive auction, it is not desirable in an LCS. If a classifier's priority grows too large, newly introduced classifiers may not be evaluated sufficiently.

These observations bring up several important questions. What class of general default hierarchies can the priority tuning scheme be expected to exploit? Is this class broader than that of traditional schemes? Can the scheme be modified so that bid separation for imperfect default hierarchies remains bounded, and if so, how will this bound effect the scheme's ability to exploit general default hierarchies? To begin to address these questions, the following sections present an extended analysis of the priority tuning scheme, a modification to bound priorities, and experiments to evaluate the schemes' effectiveness on general default hierarchies.

7.1 Analysis of general default-exception pairs under priority tuning

Consider a general default-exception pair under the priority tuning scheme. The expected value of the payoff estimates for this pair can be explicitly stated in terms of the configuration occurrence probabilities:

$$E[\Pi_e^{t+1}] = E[\Pi_e^t](1-C) + CR\left[\frac{P(G_{ue}) + P_e\left(P(G_{ce}) + P(G_{cm})\right)}{P(F_{ue}) + P_e P(F_c)}\right];$$

$$E[\Pi_d^{t+1}] = E[\Pi_d^t](1-C) + CR\left[\frac{P(G_{ud}) + P_d\left(P(G_{cd}) + P(G_{cm})\right)}{P(F_{ud}) + P_d P(F_c)}\right].$$

Similarly, the expected priority factors are as follows:

$$E[\Phi_e^{t+1}] = E[\Phi_e^t] + DP_e\left[R(P(G_{ce}) + P(G_{cm})) - P(F_c)E[\Pi_d^t]\right];$$

$$E[\Phi_d^{t+1}] = E[\Phi_d^t] + DP_d\left[R(P(G_{cd}) + P(G_{cm})) - P(F_c)E[\Pi_e^t]\right].$$

Note that these values are only updated if a conflict occurs. The steady-state expected value of the payoff estimate can be calculated by setting $E[\Pi_e^{t+1}] = E[\Pi_e^t]$:

$$E[\Pi_e^{ss}] = R\left[\frac{P(G_{ue}) + P_e\left(P(G_{ce}) + P(G_{cm})\right)}{P(F_{ue}) + P_e P(F_c)}\right];$$

$$E[\Pi_d^{ss}] = R\left[\frac{P(G_{ud}) + P_d\left(P(G_{cd}) + P(G_{cm})\right)}{P(F_{ud}) + P_d P(F_c)}\right].$$

Previous arguments suggested that priorities can grow without bound in some situations. Therefore, instead of calculating a steady-state for these measures, one can examine conditions where the one-step, expected separation gain is non-negative $((E[\Phi_e^{t+1}] - E[\Phi_d^{t+1}]) - (E[\Phi_e^t] - E[\Phi_d^t]) \geq 0)$:

$$P_e\left[(P(G_ce) + P(G_cm)) - P(F_c)\frac{E[\Pi_d^t]}{R}\right] \geq P_d\left[(P(G_cd) + P(G_cm)) - P(F_c)\frac{E[\Pi_e^t]}{R}\right].$$

Assuming ideal CR performance, and that the steady-state expected payoff estimates can be used as the expected payoff estimates in this inequality[1], the condition for non-negative separation gain becomes:

$$\left[(P(G_{ce}) + P(G_{cm})) - P(F_c)\frac{P(G_{ud})}{P(F_{ud})}\right] \geq 0.$$

[1]Note that this assumption implies that expected payoff estimates are adjusted to their steady-states while priorities are held constant. Although this is not what really occurs in the priority tuning scheme, the assumption is valid for the arguments presented here, where ideal CR performance is also assumed. The assumption is also a reasonable approximation if changes in priorities are much smaller than those of payoff estimates.

Manipulation of the previous inequality yields:

$$\frac{P(G_{ce}) + P(G_{cm})}{P(F_c)} \geq \frac{P(G_{ud})}{P(F_{ud})}.$$

This condition defines the class of default-exception pairs for which ideal performance of the priority tuning CA/CR scheme can be maintained (where separation is adequate for ideal performance and separation gain is non-negative).[2] Qualitatively, the condition implies that ideal performance can be maintained if the percentage of the conflict region where the exception is correct is greater than or equal to the percentage of the undisputed region where the default is correct. Note that the class is independent of the rule specificities.

The indicated class of default-exception pairs is clearly more general than that of specificity-based schemes; however, the class does not include all general default-exception pairs, and it also can include pairs of classifiers that are not default-exception pairs. Although this is clearly a limitation on the priority tuning scheme, it may be a difficult limitation to overcome with any CA/CR scheme that only associates local performance measures with each classifier. The use of global information in the LCS is usually avoided, since no amount of such information is generally adequate, and since global information prevents efficient parallel implementation. Although the class of exploitable default-exception pairs under the priority tuning scheme is imperfect, it can be expected to enlarge the space of useful rule sets over that of traditional schemes. This scheme also frees the organization of default hierarchies from dependence on arbitrary, syntactic features of the classifiers (i.e. specificity) and allows an LCS to organize defaults hierarchies based on the demands of the environment alone.

7.2 Bounding bid separation

As indicated in previous arguments, the priority tuning CA/CR scheme could lead to undesirable, unbounded growth of bid separation for some general default hierarchies. One way to limit priority-factor growth is by limiting the effects of the necessity auction. Consider the following modification to the priority update scheme:

$$\Phi_x^{i+1} = \Phi_x^i + CR - \begin{cases} C\Pi_y & \text{if } \Phi_x - \Phi_y < M \\ C\Pi_x & \text{if } \Phi_x - \Phi_y \geq M \end{cases}$$

where M is a *margin* that limits separation. Limitation of the necessity auction's effects could also be obtained through a classifier tax. Although limiting the effects of the necessity auction is clearly a limit on the variable-bid-separation effect, priority continues to adapt to environmental conditions up to the margin. A bound on the system's maximum error rate can be derived for a given margin and a given input noise level. By expanding the margin sufficiently, any desired error rate should be achievable.

7.3 Experiments with general default hierarchies

To evaluate the effects of the priority-tuning scheme (with the margin modification) on an LCS applied to general default hierarchies, the environment used in previous section is modified to simulate an imperfect default hierarchy. Figure 9 shows the results of a simulation where the background default is only rewarded a maximum of 60% of the time.

[2] Note that this is a steady-state argument that does not address stability or convergence.

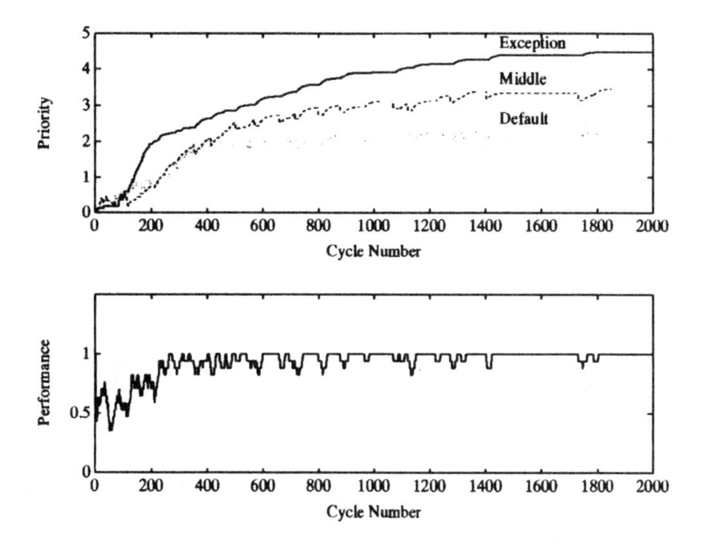

Figure 9: A simulation of priority tuning in an LCS for a four-level default hierarchy with a 60%-reliable default.

Note that the performance measure in this experiment considers all cases where the imperfect rule fires in its correct state as correct actions, whether reward is received or not. Therefore, this measure ranges from zero to one, and reflects the performance of the default hierarchy without regard to the default's imperfection A margin of 1.5 is applied. Despite the imperfection, the CA/CR scheme maintains bounded, adequate separation. Similar results were obtained with other reward probabilities for the default.

Figure 10 shows the results of a simulation where the middle-level exception is rewarded a maximum of 75% of the time. Note that the imperfect default-exception pair in this experiment does not fall into the class of priority-tuning exploitable pairs. The performance measure is defined as in the previous experiment. With the imperfect exception, the necessity auction has difficulty maintaining stable separation. This is because the middle-level exception is forced to pay out a bid that has a higher mean value than its own reward. The demands of online adaptation dictate that the system must lower the unreliable exception's priority factor when the exception fires and is not rewarded. Despite this limitation, the system maintains high performance (mean performance = .89 in the last 400 cycles) when compared to random choice (mean performance = .52), or when compared to the case where the default consistently beats the exception (mean performance = .75). Since the system can maintain some degree of separation when imperfect exceptions exist, rules can receive a more thorough fitness evaluation. Therefore, the GA can be expected to be more effective in improving the rule set.

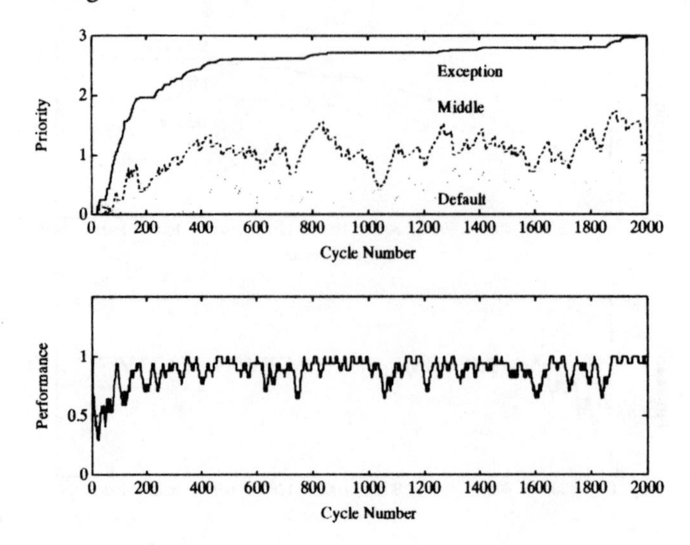

Figure 10: A simulation of priority tuning in an LCS for a four-level default hierarchy with a 75%-reliable middle-level exception.

8 Final Comments

Default hierarchies are an important aspect of the LCS paradigm. They improve rule-set parsimony, enlarge the space of adequate rule sets, and allow for graceful refinement of rule sets. These aspects of default hierarchies are particularly important in their relationship to the genetic rule discovery mechanism. In addition to the improvements default hierarchies make in LCS mechanics, automatic organization of hierarchical rule sets is in itself an interesting aspect of the LCS as a machine learning paradigm.

Previous LCS studies have used specificity-biased CA/CR schemes in attempts to organize default hierarchies. In effect, these studies assume that default hierarchies can be defined in terms of specificity. This paper uses an alternate definition of default hierarchy that does not include syntactic aspects of the classifiers (like specificity). Instead, default hierarchies are defined in terms of the occurrence probabilities of messages and the adequacy of classifier actions in response to these messages. This definition considers a more general class of rule sets that can be organized hierarchically for good LCS performance.

The primary conclusion of this study is that the global effect of default hierarchy organization can be obtained by the exploiting local information available from classifier competition. Experiments and analyses are presented that suggest that the priority-tuning scheme is a potentially useful technique for this purpose. The priority tuning scheme incorporates the use of separate priority factors and the necessity auction to determine correct organization of a broad class of general default hierarchies. The ability of this scheme to exploit a broader class of general default hierarchies should have a favorable effect on the genetic rule discovery process.

The preliminary results presented here are encouraging, but more analytical and experimental investigation is required. If automatic learning systems that store state-transition information in general form are to be effectively employed, hierarchical conflict resolution schemes must be investigated further. In particular, more research into the convergence and stability of hierarchical conflict resolution schemes is needed. Such investigations may suggest methods for further broadening the class of exploitable general default hierarchies in LCSs. Investigation of the effects of such schemes on the genetic algorithm are also necessary. The separation of the classifier fitness (payoff estimate) from its CR priority may aid in the GA's effectiveness, and in analysis of the GA's effect on rule-set structure in LCSs.

Acknowledgments

The authors gratefully acknowledge support provided by NASA under Grant NGT–50224 and support provided by the National Science Foundation under Grant CTS–8451610.

References

References

Goldberg, D. E. (1983). Computer–aided gas pipeline operation using genetic algorithms and rule learning (Doctoral dissertation, University of Michigan). *Dissertation Abstracts International, 44(10)*, 3174B. (University Microfilms No. 8402282)

Goldberg, D. E. (1989). *Genetic algorithms in search, optimization, and machine learning.* Reading, MA: Addison–Wesley.

Goldsmith, S. Y. (1989). *Steady state analysis of a simple classifier system.* Unpublished doctoral dissertation, University of New Mexico, Albuquerque.

Holland, J. H. (1975). *Adaptation in natural and artificial systems.* Ann Arbor: The University of Michigan Press.

Holland, J. H., Holyoak, K. J., Nisbett, R. E., & Thagard, P. R. (1986). *Induction: Processes of inference, learning, and discovery.* Cambridge: MIT Press.

Narendra, K. S. (1986). Recent developments in learning automata. In K. S. Narendra (Ed.), *Adaptive and learning systems: Theory and applications* (197–212). New York: Plenum Press.

Riolo, R. L. (1988). *Empirical studies of default hierarchies and sequences of rules in learning classifier systems.* Unpublished doctoral dissertation, University of Michigan, Ann Arbor.

Smith, R. E. (1991). *Default hierarchy formation and memory exploitation in learning classifier systems* (TCGA Report No. 91003). Tuscaloosa: The University of Alabama, The Clearinghouse for Genetic Algorithms.

Wilson, S. W. (1986). *Classifier system learning of a Boolean function* (Research Memo RIS–27r). Cambridge,MA: Rowland Institute for Science.

Wilson, S. W. (1989). Bid competition and specificity reconsidered. *Complex Systems, 2,* 705–723.

Wilson, S. W., & Goldberg, D. E. (1989). A critical review of classifier systems. *Proceedings of the Third International Conference on Genetic Algorithms*, 244–255.

PART 4

CODING AND REPRESENTATION

A Hierarchical Approach to Learning the Boolean Multiplexer Function

John R. Koza
Computer Science Department
Stanford University
Stanford, CA 94305 USA
Koza@Sunburn.Stanford.Edu
415-941-0336

ABSTRACT

This paper describes the recently developed genetic programming paradigm which genetically breeds populations of computer programs to solve problems. In genetic programming, the individuals in the population are hierarchical compositions of functions and arguments. Each of these individual computer programs is evaluated for its fitness in handling the problem environment. The size and shape of the computer program needed to solve the problem is not predetermined by the user, but instead emerges from the simulated evolutionary process driven by fitness. In this paper, the operation of the genetic programming paradigm is illustrated with the problem of learning the Boolean 11-multiplexer function.

1. Introduction and Overview

We start by reviewing previous work in the field of genetic algorithms and previous work in the field of induction of computer programs. We then describe the recently developed genetic programming paradigm and apply it to the problem of learning the Boolean 11-multiplexer problem.

2. Background

John Holland of the University of Michigan presented the pioneering formulation of genetic algorithms for fixed-length character strings in *Adaptation in Natural and Artificial Systems* (Holland 1975). The genetic algorithm is a highly parallel mathematical algorithm that transforms a population of individual mathematical objects (typically fixed-length binary character strings) into a new population using operations patterned after natural genetic operations such as sexual recombination (crossover) and fitness proportionate reproduction (Darwinian survival of the fittest). The genetic algorithm begins with an initial population of individuals (typically randomly generated). It then iteratively evaluates the individuals in the population for fitness with respect to the problem environment and produces a new population by performing genetic operations on individuals selected from the population with a probability proportional to fitness.

Representation is a key issue in genetic algorithm work because genetic algorithms directly manipulate the coded representation of the problem and because the representation scheme can severely limit the window by which the system observes the world. Fixed length character strings present difficulties in problems where the desired solution is hierarchical and where the size and shape of the solution is unknown in advance. The need for more powerful representations has been long recognized (De Jong 1988).

The structure of the individual mathematical objects that are manipulated by a genetic algorithm can be more complex than the fixed length character strings. Smith (1980) departed from the early fixed-length character strings by introducing variable length strings, including strings whose elements were if-then rules (rather than single characters).

Holland's introduction of the classifier system (1986) continued the trend towards increasing the complexity of the structures undergoing adaptation. The classifier system is a cognitive architecture into which the genetic algorithm is embedded so as to allow adaptive modification of a population of if-then rules (whose condition and action parts are fixed length binary strings) using a bucket brigade algorithm for allocation of credit.

Wilson (1987b) introduced hierarchical credit allocation into Holland's bucket brigade algorithm in to encourage the creation of hierarchies of rules in lieu of the exceedingly long sequences of rules that are otherwise characteristic of classifier systems. Wilson's efforts recognize the central importance of hierarchies in representing the tasks and subtasks (i.e. programs and subroutines) that are needed to solve complex problems.

Goldberg et. al (1989) introduced the messy genetic algorithm (mGA) which processes populations of variable length character strings. Messy genetic algorithms solve problems by combining relatively short, well-tested sub-strings that deal with part of a problem to form longer, more complex strings that deal with all aspects of the problem.

3. Background on Genetic Programming Paradigm

The recently developed genetic programming paradigm is a method of program induction which genetically breeds a population of computer programs to solve problems.

Work in program induction goes back to the 1950's. Friedberg's early work (1958, 1959) attempted to artificially generate entire computer programs to solve problems. Friedberg used a rather ineffective form of search to generate computer programs in a hypothetical assembly language for a hypothetical computer with a one-bit register.

Cramer (1985) applied genetic algorithms to program induction. Cramer used the genetic algorithm operating on fixed length character strings to generate computer programs with a fixed structure and reported on the difficult and highly epistatic nature of the problem.

Fujiki and Dickinson (1987) implemented analogs of the genetic operations from the conventional genetic algorithm to manipulate the individual if-then clauses of a LISP computer program consisting of a single conditional (COND) statement. The individual if-then clauses of Fujiki and Dickinsons' COND statement were parts of a strategy for playing the iterated prisoner's dilemma game.

In the recently developed genetic programming paradigm, the individuals in the population are compositions of functions and terminals appropriate to the particular problem domain. The set of functions used typically includes arithmetic operations, mathematical functions, conditional logical operations, and domain-specific functions. The set of functions must be closed in the sense that each function in the function set must be defined for any combination of elements from the range of every function that it may encounter and every terminal that it may encounter. The set of terminals used typically includes inputs (sensors) appropriate to the problem domain and constants. The search space is the hyperspace of all possible compositions of functions and terminals that can be recursively composed using the available functions and terminals.

The symbolic expressions (S-expressions) of the LISP programming language provide an especially convenient way to create and manipulate hierarchical compositions of functions and terminals. S-expressions in LISP correspond directly to the parse tree that is internally created by most compilers at the time of compilation. The parse tree is nothing more than a direct mapping of the given composition of functions (i.e. the given computer program). We need access to this parse tree to do crossover on the parts of computer programs. The LISP programming language gives us this convenient access to the parse tree, the ability to conveniently manipulate this program as if it were data, and the convenient ability to immediately execute a newly created parse tree.

The genetic programming paradigm genetically breeds computer programs to solve problems by executing the following two steps:

(1) Generate an initial population of random compositions (computer programs) of the functions and terminals of the problem.
(2) Iteratively perform the following until the termination criterion has been satisfied:
    (a) Execute each program in the population and assign it a fitness value according to how well it solves the problem.
    (b) Create a new population of computer programs by applying the following two primary operations:
        (i) Allow existing computer programs to survive with a Darwinian probability based on their fitness.
        (ii) Create new computer programs by genetically recombining randomly chosen parts of two existing programs (each chosen from the population with a Darwinian probability based on fitness).

The basic genetic operations for the genetic programming paradigm are fitness proportionate reproduction and crossover (recombination). Fitness proportionate reproduction is the basic engine of Darwinian reproduction and survival of the fittest. It operates here in the same way as it does for conventional genetic algorithm.

The crossover (recombination) operation for the genetic programming paradigm is a sexual operation that operates on two parental LISP S-expressions and produces two offspring S-expressions using parts of each parent. In particular, the crossover operation creates new offspring S-expressions by exchanging sub-trees (i.e. sub-lists) between the two parents. Because entire sub-trees (i.e. sub-lists) are swapped, this crossover operation always produces syntactically and semantically valid LISP S-expressions as offspring. For example, consider the two parental LISP S-expressions:

```
(OR (NOT D1) (AND D0 D1))

(OR (OR D1 (NOT D0)) (AND (NOT D0) (NOT D1)))
```

These two LISP S-expressions can be depicted graphically as rooted, point-labeled trees with ordered branches. The two parental LISP S-expressions are shown below:

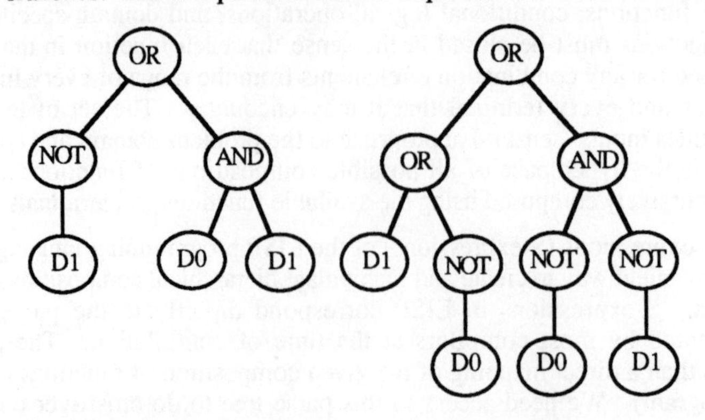

Assume that the points of both trees are numbered in a depth-first way starting at the left. Suppose that the second point (out of 6 points of the first parent) is randomly selected as the crossover point for the first parent and that the sixth point (out of 10 points of the second parent) is randomly selected as the crossover point of the second parent. The crossover points are therefore the NOT in the first parent and the AND in the second parent. The two crossover fragments are two sub-trees shown below:

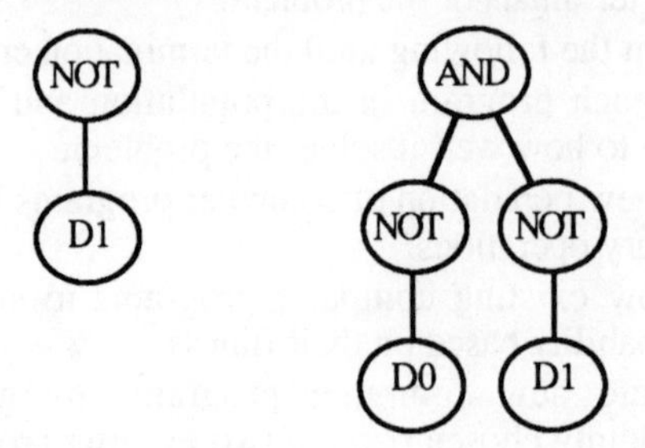

These two crossover fragments correspond to the bold, underlined sub-expressions (sub-lists) in the two parental LISP S-expressions shown above. The two offspring resulting from crossover are shown below.

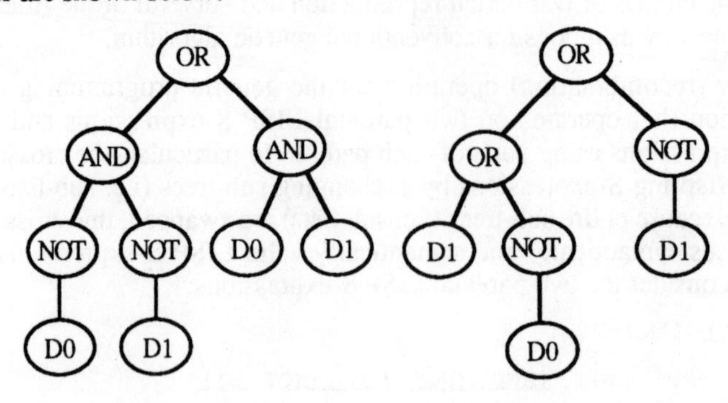

Note that the first offspring produced is an S-expression for the even parity function:

```
(OR (AND (NOT D0) (NOT D1)) (AND D0 D1)).
```

Although one might think that computer programs are so epistatic that they could only be genetically bred in a few especially congenial problem domains, we have shown that computer programs can be genetically bred to solve problems in a surprising variety of different areas of machine learning and artificial intelligence, including

- planning (e.g. navigating an artificial ant to find food along an irregular trail; developing a robotic action sequence that can stack blocks) [Koza 1990b],
- emergent behavior (e.g. discovering a computer program for locating food, carrying food to the nest, and dropping pheromones, which, when executed by all the ants in an ant colony, produces interesting higher level "emergent" behavior) [Koza 1991a],
- finding minimax strategies for games (e.g. differential pursuer-evader games; discrete games in extensive form) by both evolution and co-evolution [1991b],
- optimal control (e.g. centering a cart and balancing a broom in minimal time by applying a bang-bang force to the cart) (Koza and Keane 1990a, 1990b],
- discovering inverse kinematic equations (e.g. to move a robot arm to designated target points) [Koza 1991f],
- sequence induction (e.g. inducing a recursive procedure for generating sequences such as the Fibonacci and the Hofstadter sequences) [Koza 1989],
- symbolic "data to function" regression, integration, differentiation, and symbolic solution to general functional equations (including differential equations with initial conditions, integral equations, and inverse problems) [Koza 1990],
- empirical discovery (e.g. rediscovering Kepler's Third Law; rediscovering the well-known non-linear econometric "exchange equation" MV = PQ from actual, noisy time series data for the money supply, the velocity of money, the price level, and the gross national product of an economy) [Koza 1991d],
- automatic programming (e.g. solving pairs of linear equations, solving quadratic equations for complex roots, and discovering trigonometric identities),
- pattern recognition (e.g. translation-invariant recognition of a one-dimensional shape in a linear wrap-around retina),
- concept formation and decision tree induction [Koza 1991c],
- generation of random numbers (using entropy as fitness) [Koza 1991e], and
- simultaneous architectural design and training of neural nets [Koza and Rice 1991a].

A visualization of the application of the genetic programming paradigm to planning, emergent behavior, empirical discovery, inverse kinematics, game playing, and the Boolean 11-multiplexer problem discussed in this paper can be viewed in the *Artificial Life II Video Proceedings* videotape [Koza and Rice 1991b].

4. Boolean 11-Multiplexer Function

The problem of machine learning of a function requires developing a composition of functions that can return the correct value of the function after seeing specific examples of the value of the function associated with particular combinations of arguments.

In this paper, the problem is to learn the Boolean 11-multiplexer function. In general, the input to the Boolean multiplexer function consists of k "address" bits a_i and 2^k "data" bits d_i and is a string of length $k+2^k$ of the form $a_{k-1}...a_1a_0\ d_2k\text{-}1...d_1\ d_0$. The value of the multiplexer function is the value (0 or 1) of the particular data bit that is singled out by the k address bits of the multiplexer. For example, for the 11-multiplexer (where k = 3), if the three address bits $a_2a_1a_0$ are 110, then the multiplexer singles out data bit number 6 (i.e. d_6) to be its output. The Boolean multiplexer function with $k+2^k$ arguments is one of 2^{k+2^k} possible Boolean functions of $k+2^k$ arguments. Thus, the search space for the Boolean multiplexer is of size 2^{k+2^k}. When k=3, this search space is of size $2^{2^{11}} = 2^{2048}$, which is approximately 10^{616}.

The solution of the Boolean 11-multiplexer problem (involving a search space of size 10^{616}) will serve to show the interplay, in the genetic programming paradigm, of

- the genetic variation inevitably created in the initial random generation,
- the small improvements for some individuals in the population via localized hill-climbing from generation to generation,
- the way particular individuals become specialized so as to be able to correctly handle certain sub-cases of the problem (i. e. case-splitting),
- the creative role of crossover in recombining valuable parts of more fit parents to produce new individuals with new capabilities, and
- how the nurturing of a large population of alternative solutions to the problem (rather than a single point in the solution space) helps avoid false peaks in the search for the solution to the problem.

This problem will also serve to illustrate the importance of hierarchies in solving problems and making the ultimate solution understandable. Moreover, the progressively changing size and shape of the various individuals in the population in various generations shows the flexibility resulting from not determining the size and shape of ultimate solution or the intermediate results in advance

The five major steps in setting up the genetic programming paradigm require determining: (1) the set of terminals, (2) the set of functions, (3) the fitness function, (4) the parameters for the run, and (5) the criterion for designating a result and terminating a run.

The first step in applying the genetic programming paradigm to a problem is to select the set of terminals that will be available to the algorithm for constructing the computer programs (LISP S-expressions) that will try to solve the problem. For some problems (in particular, Boolean function learning problems), this choice is especially straight-forward and obvious. The set of terminals for this problem consists of the 11 inputs to the Boolean 11-multiplexer. The algorithm cannot distinguish as to whether the terminals are address lines or data lines. The terminal set for this problem is

T = {A0, A1, A2, D0, D1, ... , D7}.

The second step in applying the genetic programming paradigm to a problem is to select the set of functions. The set of functions for this problem is

$$F = \{AND, OR, NOT, IF\}.$$

This set of basic logical functions satisfies the closure property. This set is sufficient to realize any Boolean function. In addition, this set is a convenient set in that it often produces easily understood S-expressions.

The third step in applying the genetic programming paradigm to a problem is to determine the fitness function. Fitness is often evaluated over fitness (environmental) cases. The fitness cases here consist of the 2^{11} possible combinations of the 11 arguments along with the associated correct value of the 11-multiplexer function. For the 11-multiplexer (where k = 3), there are 2048 such combinations of arguments. In this particular paper, we use the entire set of 2048 combinations of arguments (i.e. we do not use sampling); however, sampling is an obvious option. The standardized fitness is the sum, taken over all 2048 fitness cases, of the Hamming distances between the Boolean value returned by the S-expression for a given combination of arguments and the correct Boolean value. This fitness measure is equivalent to the number of mismatches (i.e. sum of errors).

The fourth major step in using the genetic programming paradigm is selecting the values of certain parameters for running the algorithm. Population size is the most important parameter. It is 4000 here. Each new generation is created from the preceding generation by applying fitness proportionate reproduction to 10% of the population and by applying crossover to 90% of the population (with one parent selected proportionate to fitness). In selecting crossover points, 90% were internal (function) points of the tree and 10% were external (terminal) points of the tree. Mutation was not used. For the practical reason of conserving computer time, the depth of initial random S-expressions was limited to 5 and the depth of S-expressions created by crossover was limited to 20.

Finally, the fifth major step in using the genetic programming paradigm is the criterion for terminating a run and designating a result. We terminate a given run when either (1) the genetic programming paradigm produces a computer program whose standardized fitness attains the perfect score of zero, or (2) 51 generations (i.e. the initial random generation and 50 others) have been run.

Note that the first two of these five major steps in applying the genetic programming paradigm corresponds to the step (performed by the user) of determining the representation scheme in the conventional genetic algorithm operating on character strings (that is, determining the chromosome length, alphabet size, and the mapping between the problem and chromosomes). The last three of the these five steps apply to both methodologies.

In addition, note that the step (performed by the user) of determining the set of primitive functions in the genetic programming paradigm is equivalent to a similar step in other machine learning paradigms. For example, this same determination of primitive functions occurs in the induction of decision trees using ID3 (Quinlan 1986, Koza 1991c) when the user selects the functions that can appear at the internal points of the decision tree. Similarly, this same determination occurs in neural net problems when the user selects the external functions that are to be activated by the output of a neural network. The same user determination occurs in other machine learning paradigms (although the name given to this omnipresent determination varies and is often considered by the researcher to be implicit in the statement of his or her problem).

We illustrate the overall process by discussing one particular run of the Boolean 11-

multiplexer in detail. Later, we will present statistics involving multiple runs.

The process begins with the generation of the initial random population (i.e. generation 0). Predictably, the initial random population includes a variety of highly unfit individuals. Many individual S-expressions in this initial random population are merely constants, such as the contradictory `(AND A0 (NOT A0))`. Other individuals are passive and merely pass a single input through as the output, such as `(NOT (NOT A1))`. Other individuals inefficiently do the same, such as `(OR D7 D7)`. Some initial random individuals, such as `(IF D0 A0 A2)`, make a decision based on precisely the wrong argument (i.e. using a data bit, rather than address bits, to select the output). Many initial random individuals are partially blind in that they do not incorporate all 11 arguments that are necessary to solve the problem. Some initial random S-expressions are just nonsense, such as

```
(IF (IF (IF D2 D2 D2) D2 D2) D2 D2).
```

Nonetheless, even in this highly unfit initial random population, some individuals are somewhat more fit than others. For this particular run, the individuals in the initial random population had values of standardized fitness ranging from 768 mismatches (i.e. 1280 matches or hits) to 1280 mismatches (i.e. 768 matches). As it happens, a total of 23 individuals out of the 4000 in this initial random population tied with the best score of 1280 matches (i.e. 768 mismatches) on generation 0. One of these 23 best scoring individuals from generation 0 was the S-expression

```
(IF A0 D1 D2).
```

This individual has obvious shortcomings. Notably, this individual is partially blind in that it uses only three of the 11 necessary terminals of the problem. As a consequence of this fact alone, this individual cannot possibly be a correct solution to the problem. This individual nonetheless does some things right. For example, it uses one of the three address bits (A0) as the basis for its action. It could easily have done this incorrectly and used one of the eight data bits. In addition, this individual uses only data bits (D1 and D2) as its output. It could have used address bits. Moreover, if A0 (which is the low order binary bit of the 3-bit address) is T (True), this individual selects one of the three odd numbered data bits (D1) as it output. Moreover, if A0 is NIL, this individual selects one of the three even numbered data bits (D2) as its output. In other words, this individual correctly links the parity of the low order address bit A0 with the parity of the data bit it selects as its output. This individual is far from perfect, but it is far from being without merit. It is more fit than 3977 of the 4000 individuals in the population. In the valley of the blind, the one-eyed man is king.

In contrast, the worst individual in the population for the initial random generation had a standardized fitness of 1280 (i.e. only 768 matches). It is shown below:

```
(OR (NOT A1) (NOT (IF (AND A2 A0) D7 D3))).
```

The average standardized fitness for all 4000 individuals in the population for generation 0 is 985.4. This value of average standardized fitness for the initial random population forms the baseline and serves as a useful benchmark for monitoring later improvements in the average standardized fitness of the population as a whole.

The hits histogram is a useful monitoring tool based on the number of hits (i.e. matches). This histogram provides a way of viewing the population as a whole for a particular generation. The horizontal axis of the hits histogram is the number of hits (i.e.

matches) and the vertical axis is the number of individuals in the population scoring that number of hits. Fifty different levels of fitness are represented in the hits histogram for the population at generation 0 of this problem. In order to make this histogram legible for this problem, we have divided the horizontal axis into buckets of size 64. For example, 1553 individuals out of 4000 (i.e. about 39%) had between 1152 and 1215 matches (hits). The mode (i.e. highest point) of the distribution occurs at 1152 matches (hits). The figure below shows the hits histogram of the population for generation 0.

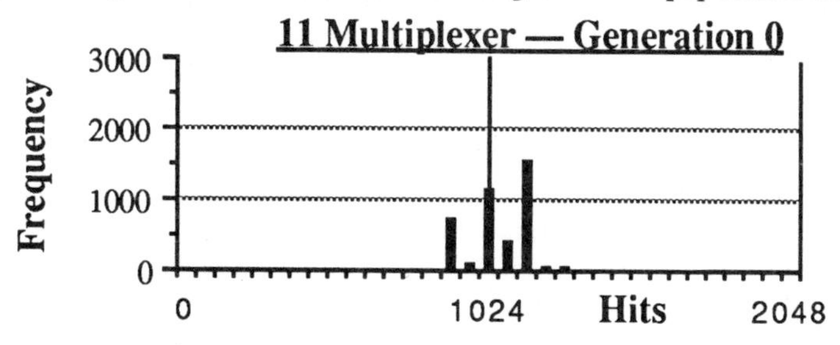

A new population is then created from the current population using the operations of Darwinian fitness proportionate reproduction and crossover. When these operations are completed, the new population (i.e. the new generation) replaces the old population. In going from the initial random generation (generation 0) to generation 1, the genetic programming paradigm works with the inevitable genetic variation existing in an initial random population. The initial random generation is an exercise in blind random search. The search is a parallel search of the search space over the 4000 individual points .

The figure below shows the standardized fitness (i.e. mismatches) for generations 0 through 9 of this run for the best single individual in the population, the worst single individual in the population, and the average for the population.

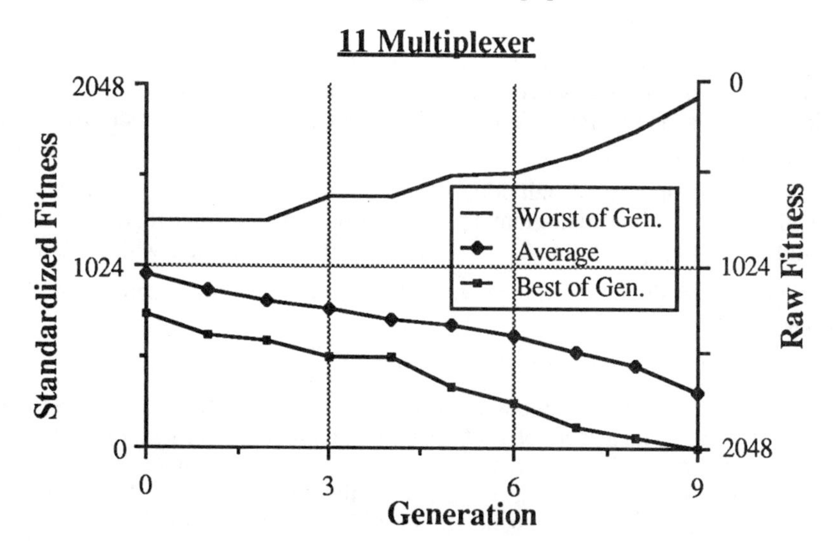

The average standardized fitness of the population immediately begins improving (i.e. decreasing) from the baseline value of 985.4 for generation 0 to about 891.9 for generation 1. As it happens, in this particular run, the average standardized fitness improves (i.e. decreases) monotonically between generation 2 and generation 9 and assumes values of 845, 823, 763, 731, 651, 558, 459, and 382, respectively. We usually see a generally improving trend in average standardized fitness from generation to generation, but not necessarily a monotonic improvement.

In addition, we similarly usually see a generally improving trend in the standardized fitness of the best single individual in the population from generation to generation. As it happens, in this particular run of this particular problem, the standardized fitness of the best single individual in the population improves (i.e. decreases) monotonically between generation 2 and generation 9. In particular, it assumes values of 640, 576, 384, 384, 256, 256, 128, and 0 (i.e. a perfect score), respectively.

On the other hand, the standardized fitness of the worst single individual in the population fluctuates considerably. For this particular run, the standardized fitness of the worst individual starts at 1280, fluctuates considerably between generations 1 and 9, and then deteriorates (increases) to 1792 by generation 9.

The standardized fitness for the best single individual in the population improved (i.e. dropped) to 640 mismatches (i.e. 1408 matches) for generations 1 and 2 of the run. Only one individual in the population attained this best score in generation 1, namely

```
(IF A0 (IF A2 D7 D3) D0).
```

Note that this individual performs better than the best individual from generation 0 for two reasons. First, this individual considers two of the three address bits (A0 and A2) in deciding which data bit to choose as output, whereas the best individual in generation 0 considered only one of the three address bits (A0). Second this best individual from generation 1 incorporates three of the eight data bits as its output, whereas the best individual in generation 0 incorporated only two of the eight potential data bits as output. Although still far from perfect, the best individual from generation 1 is less blind and more complex than the best individual of the previous generation.

Note that the size and shape of this best scoring individual from generation 1 differs from the size and shape of the best scoring individual from generation 0. The progressive change in size and shape of the individuals in the population is a characteristic of the genetic programming paradigm.

By generation 2, the number of individuals sharing this high score of 1408 hits rose to 21. The high point of the histogram for generation 2 has advanced from 1152 for generation 0 to 1280 for generation 2. There are now 1620 individuals with 1280 hits.

In generation 3, one individual in the population attained a new high score of 1472 hits. This individual is

```
(IF A2 (IF A0 D7 D4) (AND (IF (IF A2 (NOT D5) A0) D3 D2) D2)).
```

Generation 3 shows further advances in fitness for the population as a whole. The number of individuals with a score of 1280 hits (the high point for generation 2) has risen to 2158 for generation 3. Moreover, the center of gravity of the fitness histogram has shifted significantly from left to right. In particular, the number of individuals with 1280 hits or better has risen from 1679 in generation 2 to 2719 in generation 3.

In generations 4 and 5, the best single individual has 1664 hits. This score is attained by only one individual in generation 4, but is attained by 13 individuals in generation 5.

One of these 13 individuals is

```
(IF A0 (IF A2 D7 D3) (IF A2 D4 (IF A1 D2 (IF A2 D7 D0))))).
```

Note that this individual uses all three address bits (A2, A1, and A0) in deciding upon the output. It also uses five of the eight data bits. By generation 4, the high point of the histogram has moved to 1408 with 1559 individuals.

In generation 6, four individuals attain a score of 1792 hits. The high point of the histogram has moved to 1536 hits. In generation 7, 70 individuals score 1792 hits.

In generation 8, there are four best-of-generation individuals. They all attain a score of 1920 hits. The mode (high point) of the histogram has moved to 1664 and 1672 individuals share this value. Moreover, an additional 887 individuals score 1792.

In generation 9, one individual emerges with a 100% perfect score of 2048 hits. That individual is

```
(IF A0 (IF A2 (IF A1 D7 (IF A0 D5 D0))
              (IF A0 (IF A1 (IF A2 D7 D3) D1) D0))
       (IF A2 (IF A1 D6 D4) (IF A2 D4 (IF A1 D2 (IF A2 D7 D0)))))
```

This 100% correct individual is depicted graphically below:

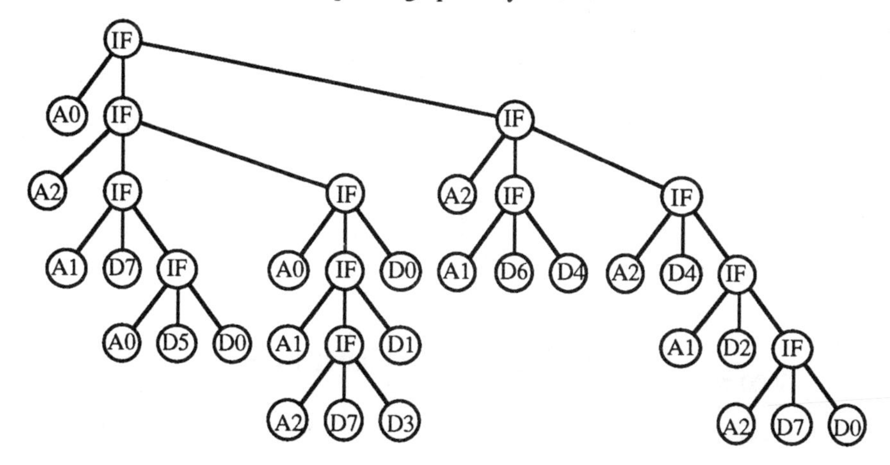

This 100% correct individual is a hierarchical structure consisting of 37 points (i.e. 12 functions and 25 terminals). Note that the size and shape of this solution emerged from the genetic programming paradigm. This particular size and this particular hierarchical structure was not specified in advance. Instead, it evolved as a result of reproduction, crossover, and the relentless pressure of fitness. In generation 0, the best single individual in the population had 12 points. The number of points in the best single individual in the population varied from generation to generation. It was 7, 16, 10, 10, 25, 48, 46, 54, and 60 for generations 1 through 9, respectively.

This 100% correct individual can be simplified to

```
(IF A0 (IF A2 (IF A1 D7 D5) (IF A1 D3 D1))
       (IF A2 (IF A1 D6 D4) (IF A1 D2 D0))).
```

When so rewritten, it can be seen that this individual correctly performs the 11-

multiplexer function by first examining address bits A0, A2, and A1 and then choosing the appropriate one of the eight possible data bits.

Table 1 shows, side by side, the hits histograms for the generations 1, 5, and 9 of this run. As one progresses from generation to generation, note the left-to-right "slinky" undulating movement of the center of mass of the histogram and the high point of the histogram. This movement reflects the improvement of the population as a whole as well as the best single individual in the population. There is a single 100% correct individual with 2048 hits is at generation 9; however, because of the scale of the vertical axis of this histogram, it is not visible in a population of size 4000.

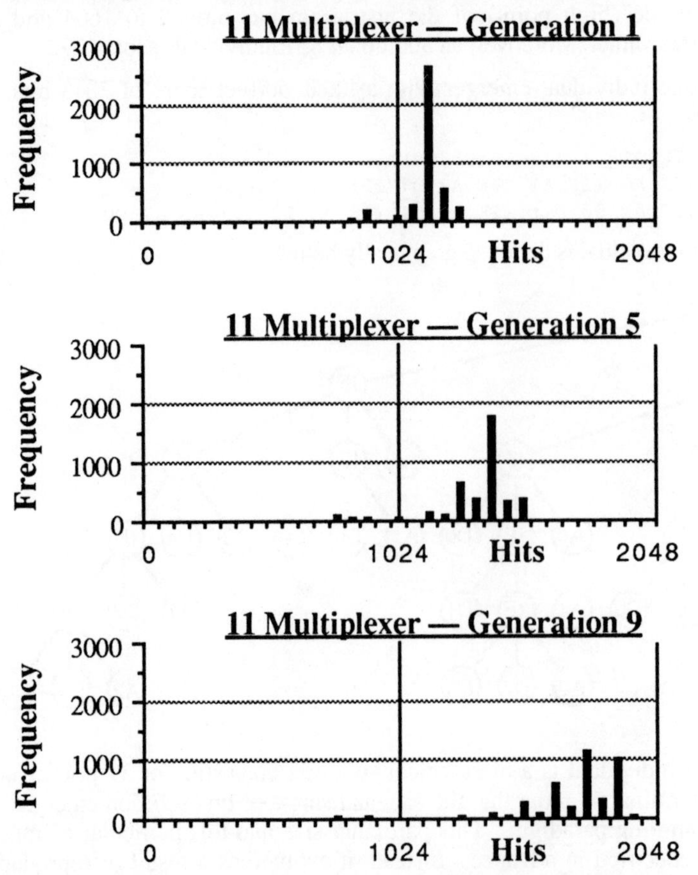

Further insight can be gained by studying the genealogical audit trail of the process. This audit trail consists of a complete record of the details of each operation that is performed. For the operations of fitness proportionate reproduction and crossover, the details consist of the individual(s) chosen for the operation and, for crossover, the particular points chosen within both participating individuals.

Construction of the audit trail starts with the individuals of the initial random generation (generation 0). Certain additional information such as the individual's rank location in the population (after sorting by normalized fitness) and its standardized fitness is also carried along as a convenience in interpreting the genealogy. Then, as each operation is performed to create a new individual for the next generation, a list is recursively formed consisting of the type of the operation performed, the individual(s) participating in the operation, the details of that operation, and, finally, a pointer to the audit trail(s) accumulated so far for the individual(s) participating in that operation.

An individual occurring at generation h has up to 2^{h+1} ancestors. The number of ancestors is less than 2^{h+1} to the extent that operations other than crossover are involved; however, crossover is, by far, the most frequent operation. For example, an individual occurring at generation 9 has up to 1024 ancestors. Note that a particular ancestor often appears more than once in this genealogy because all selections of individuals to participate in the basic genetic operations are skewed in proportion to fitness with reselection allowed. Moreover, even for a modest sized value of h, 2^{h+1} will typically be greater than the population size. This repetition, of course, does nothing to reduce the size of the genealogical tree.

Even with the use of pointers from descendants back to ancestors, construction of a complete genealogical audit trail is exponentially expensive in both computer time and memory space. Note that the audit trail must be constructed for each individual of each generation because the identity of the 100% correct individual(s) eventually solving the problem is not known in advance. Thus, there are 4000 audit trails. By generation 9, each of these 4000 audit trails recursively incorporates information about operations involving up to 1024 ancestors. The audit trail for the single 100% correct individual of interest in generation 9 alone occupies about 27 densely-printed pages.

The creative role of crossover and case-splitting is illustrated by an examination of the genealogical audit trail for the 100% correct individual emerging at generation 9. The 100% correct individual emerging at generation 9 is the child resulting from the most common genetic operation used in the process, namely crossover. The first parent from generation 8 had rank location of 58 (out of 4000, with a rank of 0 being the very best) in the population and scored 1792 hits (out of 2048). The second parent from generation 8 had rank location 1 and scored 1920 hits. Note that it is entirely typical that the individuals selected to participate in crossover have relatively high rank locations in the population since crossover is performed among individuals in a mating pool created proportional to fitness.

The first parent from generation 8 (scoring 1792) was

```
(IF A0 (IF A2 D7 D3)
       (IF A2 (IF A1 D6 D4) (IF A2 D4 (IF A1 D2 (IF A2 D7 D0)))))).
```

Note that this first parent starts by examining address bit A0. If A0 is T, the emboldened and underlined portion then examines address bit A2. It then, partially blindly, makes the output equal D7 or D3 without even considering address bit A1. Moreover, the underlined portion of this individual does not even contain data bits D1 and D5. On the other hand, when A0 is NIL, this first parent is 100% correct. In that event, it examines A2 and, if A2 is T, it then examines A1 and makes the output equal to D6 or D4 according to whether A1 is T or NIL. Moreover, if A2 is NIL, it redundantly retests A2 and then correctly makes the output equal to `(IF A1 D2 D0)`.

This first parent from generation is graphically depicted below:

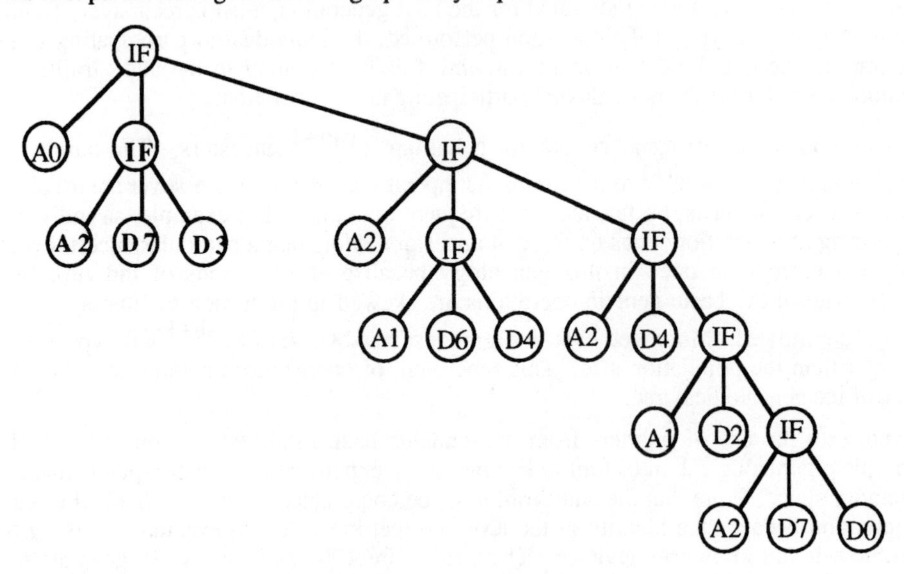

Note that the 100% correct portion of this first parent, namely, the sub-expression

```
(IF A2 (IF A1 D6 D4) (IF A2 D4 (IF A1 D2 (IF A2 D7 D0)))))
```

is itself a 6-multiplexer which tests A2 and A1 and correctly selects amongst D6, D4, D2, and D0. This becomes clear if we simplify this sub-expression to

```
(IF A2 (IF A1 D6 D4) (IF A1 D2 D0))
```

In other words, this imperfect first parent handles the even-numbered data bits correctly, but only partially correctly handles the odd-numbered data bits. The tree representing this first parent has 22 points. The crossover point chosen at random at the end of generation 8 was point 3 and corresponds to the second occurrence of the function IF. The crossover fragment consists of the emboldened and underlined sub-expression

(IF A2 D7 D3).

The second parent from generation 8 (scoring 1920 hits) was

```
(IF A0 (IF A0 (IF A2 (IF A1 D7 (IF A0 D5 D0))
                          (IF A0 (IF A1 (IF A2 D7 D3) D1) D0))
              (IF A1 D6 D4))
          (IF A2 D4 (IF A1 D2 (IF A0 D7 (IF A2 D4 D0))))))
```

This second parent has 40 points. The crossover point chosen at random for this second parent was point 5. The crossover fragment consists of the emboldened, underlined sub-expression. It correctly handles the case when A0 is T (i.e. the odd numbered addresses). It makes the output equal to D7 when the address bits are 111; it makes the output equal to D5 when the address bits are 101; it makes the output equal to D3 when the address bits are 011; and it makes the output equal to D1 when the address bits are 001.

Note that the 100% correct portion of this second parent, namely, the sub-expression

```
(IF A2 (IF A1 D7 (IF A0 D5 D0))
       (IF A0 (IF A1 (IF A2 D7 D3) D1) D0))
```

is itself a 6-multiplexer.

This second parent from generation 8 is graphically depicted below:

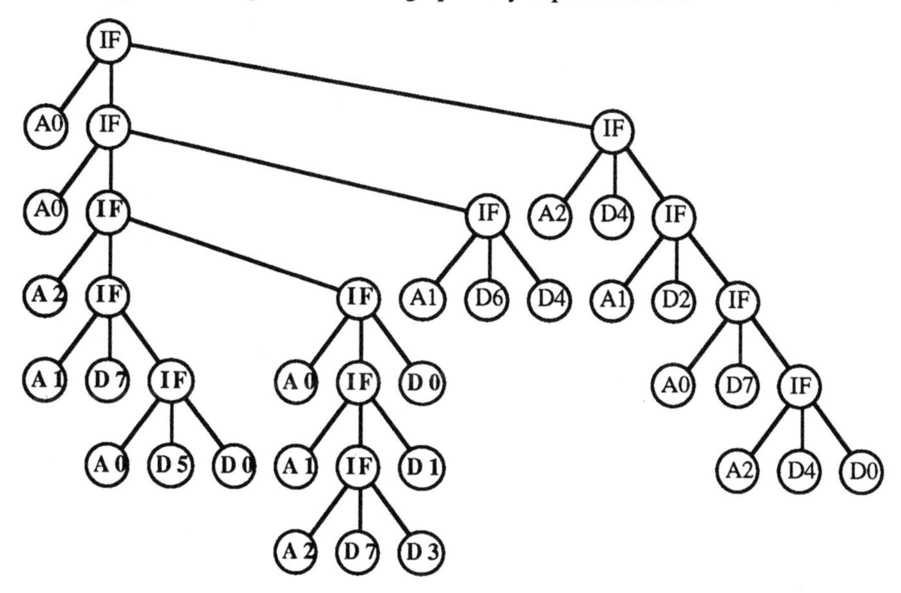

This embedded 6-multiplexer tests A2 and A1 and correctly selects amongst D7, D5, D3, and D1 (i.e. the odd numbered data bits). This fact becomes clearer if we simplify this sub-expression of this second parent to the following:

```
(IF A2 (IF A1 D7 D5) (IF A1 D3 D1)
```

This case splitting is graphically demonstrated by the following restatement of the 100% correct offspring into two 6-multiplexers:

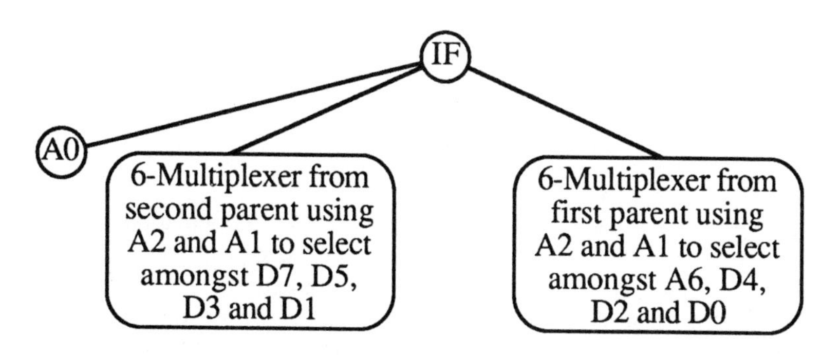

In other words, this imperfect second parent handles part of its environment correctly and part of its environment incorrectly. In particular, it handles the odd-numbered data bits correctly but only partially correctly handles the even-numbered data bits.

Even though neither parent is perfect, these two imperfect parents contain complementary, co-adapted portions which, when mated together, produce a 100% correct offspring individual. In effect, the creative effect of the crossover operation blends the two cases of the implicitly "case-split" environment into a single 100% correct solution.

Of course, not all crossovers between individuals are useful and productive. In fact, a large fraction of the individuals produced by the genetic operations are useless. But the existence of a population of alternative solutions to a problem provides the ingredients with which genetic recombination (crossover) can produce some improved individuals. The relentless pressure of natural selection based on fitness then causes these improved individuals to be preserved and to proliferate. Moreover, genetic variation and the existence of a population of alternative solutions to a problem makes it unlikely that the entire population will become trapped in local maxima.

Interestingly, the same crossover that produced the 100% correct individual also produced a "runt" scoring only 256 hits. In this particular crossover, the two crossover fragments not used in the 100% correct individual combined to produce an unusually unfit individual. This is one of the reasons why there is considerable variability from generation to generation in the worst single individual in the population.

As one traces the ancestry of the 100% correct individual created in generation 9 deeper back into the genealogical audit tree (i.e. towards earlier generations), one encounters parents scoring generally fewer and fewer hits. But if we look at the sequence of hits in the forward direction, we see localized hill-climbing in the search space occurring in parallel throughout the population as the creative operation of crossover recombines complementary, co-adapted portions of parents to produce improved offspring.

The genetic programming paradigm (as with genetic algorithms in general) contains probabilistic steps at several different points. As a result, we rarely obtain a solution to a problem in the precise way we anticipate and we rarely obtain the precise same solution twice. We can measure the number of individuals that need to be processed by a genetic algorithm to produce a desired result (i.e. 2048 matches) with a certain probability, say 99%. Suppose, for example, a particular run of a genetic algorithm produces the desired result with only a probability of success p_s after a specified choice (perhaps arbitrary and non-optimal) of number of generations N_{gen} and population of size N. Suppose also that we are seeking to achieve the desired result with a probability of, say, $z = 1 - \varepsilon = 99\%$. Then, the number K of independent runs required is

$$K = \frac{\log(1-z)}{\log(1-p_s)} = \frac{\log \varepsilon}{\log(1-p_s)}, \text{ where } \varepsilon = 1-z.$$

We ran 21 runs of the Boolean 11-multiplexer problem with a population size of 4,000 and 51 generations. We found that the probability of success p_s was 100% by generation 15 for all runs. Each run was aborted as soon as a perfect solution was found. A total of 980,000 individuals were processed. This is an average of 46,667 individuals per run. Since each run involves 4,000 individuals, a perfect solution is found in an average of 11.67 generations. The graph below shows the probability of success p_s of a run for various numbers of generations between 0 and 50 for a population size of 4000.

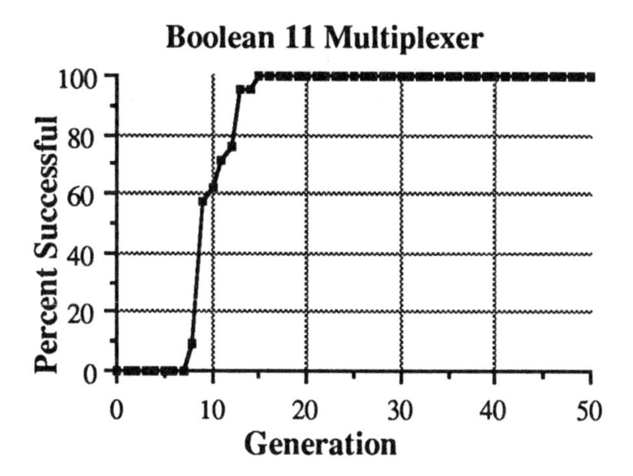

5. Hierarchies and Default Hierarchies

Note that the result of the genetic programming paradigm is always hierarchical. As we saw above, the solution to the 11-multiplexer problem was a hierarchy consisting of two 6-multiplexers. In one run where we applied the genetic programming paradigm to the simpler Boolean 6-multiplexer, we obtained the following 100% correct solution

```
(IF (AND A0 A1) D3 (IF A0 D1 (IF A1 D2 D0))).
```

This solution to the 6-multiplexer is also a hierarchy. It is a hierarchy that correctly handles the particular fitness cases where (AND A0 A1) is true and then correctly handles the remaining cases where (AND A0 A1) is false.

Default hierarchies often emerge from the genetic programming paradigm. A default hierarchy incorporates partially correct sub-rules into a perfect overall procedure by allowing the partially correct (default) sub-rules to handle the majority of the environment and by then dealing in a different way with certain specific exceptional cases in the environment. The S-expression above is also a default hierarchy in which the output defaults to

```
(IF A0 D1 (IF A1 D2 D0))
```

three quarters of the time. However, in the specific exceptional fitness case where both address bits (A0 and A1) are both T, the output is the data bit D3. Default hierarchies are considered desirable in induction problems (Holland 1986, Wilson 1987a) because they are often parsimonious and they are a human-like way of dealing with situations. Wilson's noteworthy BOOLE experiments (1987a) originally found a set of eight if-then classifier system rules for the Boolean 6-multiplexer that correctly (but tediously) handled each particular subcase of the problem. Subsequently, Wilson (1988) modified the credit allocation scheme and successfully produced a default hierarchy.

6. Non-Randomness of Results

The number of possible compositions using the set of available functions and the set of available terminals is very large. In particular, the number of possible trees representing such compositions increases rapidly as a function of the number of points in the tree. This is true because of the large number of ways of labeling the points of a tree with functions and terminals. The number of possible compositions of functions is, in particular, very large in relation to the 40,000 individuals processed in generations 0 through 9 in the particular run of the genetic programming paradigm described above.

There is a theoretic possibility that the probability of a solution to a given problem may be low in the original search space of the Boolean 11 Multiplexer problem (i.e. all Boolean functions of 11 arguments), but that the probability of randomly generating a composition of functions that solves the problem might be significantly higher in the space of randomly generated compositions of functions. The Boolean 11-multiplexer function is a unique function out of the $2^{2^{11}}$ (i.e. 2^{2048}) possible Boolean functions of 11 arguments and one output. The probability of randomly choosing zeroes and ones for the 2^{11} lines of a truth table so as to create this particular Boolean function is only 1 in $2^{2^{11}}$ (i.e. 2^{2048}). However, there is a theoretic possibility that the probability of randomly generating a composition of the functions AND, OR, NOT, and IF that performs the 11-multiplexer function might be better than 1 in 2^{2048}.

There is no *a priori* reason to believe that this is the case. That is, there is no *a priori* reason to believe that compositions of functions that solve the Boolean multiplexer problem are denser in the space of randomly generated compositions of functions than solutions to the problem in the original search space of the problem. Nonetheless, there is a possibility that this is the case, even though there is no *a priori* reason to think that it is the case.

To test against this possibility, we performed the following control experiment for the Boolean 11-multiplexer problem. We generated 1,000,000 random S-expressions to check if we could randomly generate a composition of functions that solved the problem. For this control experiment, we used the same algorithm and parameters used to generate the initial random population in the normal runs of the problem. No 100% correct individual was found in this random search. In fact, the high score in this random search was only 1408 hits (out of a possible 2048 hits) and the low score was 704 hits. Moreover, only 10 individuals out of 1,000,000 achieved this high score of 1408. The high point of the histogram among these 1,000,000 random individuals came at 1152 hits (with 183,820 individuals); the second highest point came at 896 hits (with 168,333 individuals); and the third highest point came at 1024 hits (with 135,379 individuals).

A similar control experiment was conducted for the Boolean 6-multiplexer problem (with a search space of 2^{2^6}, i.e. 2^{64}). Since the environment for the 6-mutliplexer problem had only 64 fitness cases (as compared with 2048 cases for the 11-multiplexer), it was practical to evaluate even more randomly generated individuals (i.e. 10,000,000) in this control experiment. As before, no 100% correct individual was found in this random search. In fact, no individual had more than 52 (of 64 possible) hits. As with the 11-multiplexer, the size of the search space (2^{64}) for the 6-mutliplexer is very large in relation to the number of individuals processed in a typical run solving the 6-mutliplexer.

We conclude that solutions to these problems in the space of randomly generated compositions of functions are not denser than solutions in the original search space of the problem. Therefore, *the results described in this paper are not the fruits of random search.*

As a matter of fact, we have evidence suggesting that the solutions are appreciably *sparser* in the space of randomly generated compositions of functions than solutions in the original search space of the problem. Consider, for example, the exclusive-or function. The exclusive-or function is the odd parity function with two Boolean arguments. The odd parity function of k Boolean arguments returns T (True) if the number of arguments equal to T is odd and returns NIL (False) otherwise. Whereas there are 2^{64} Boolean functions with six arguments and 2^{2048} Boolean functions with 11 arguments, there are only $2^{2^2} = 2^4 = 16$ Boolean functions with two Boolean arguments. That is, the exclusive-or function is one of only 16 possible Boolean functions with two Boolean arguments and one output. Thus, in the search space of Boolean functions, the probability of randomly choosing T's and NIL's for the 16 lines of a truth table that realizes this particular Boolean function is only 1 in 16.

We generated 100,000 random individuals using a function set consisting of the basic Boolean functions F = {AND, OR, NOT}. If randomly generated compositions of the basic Boolean functions that realize the exclusive-or function were as dense as solutions are in the original search space of the problem (i.e. the space of Boolean functions of 2 arguments), we would expect about about 6250 in 100,000 random compositions of functions (i.e. 1 in 16) to realize the exclusive-or function. Instead, we found that only 110 out of 100,000 randomly generated compositions that realized the exclusive-or function. This is a frequency of only 1 in 909. In other words, randomly generated compositions of functions realizing the exclusive-or function are about 57 times *sparser* than solutions in the original search space of Boolean functions.

Similarly, we generated an additional 100,000 random individuals using a function set consisting of the basic Boolean functions F = {AND, OR, NOT, IF}. We found that only 116 out of 100,000 randomly generated compositions realized the exclusive-or function (i.e. a frequency of 1 in 862). That is, with this new function set, randomly generated compositions of functions realizing the exclusive-or function are about 54 times *sparser* than solutions in the original search space of Boolean functions.

In addition, we performed similar experiments on two Boolean functions with three Boolean arguments and one output, namely, the 3-parity function and the 3-multiplexer function (i. e. the If-Then-Else function). There are only $2^{2^3} = 2^8 = 256$ Boolean functions with three Boolean arguments and one output. The probability of randomly choosing a particular combination of T's and NIL's for the 256 lines of a truth table is 1 in 256. If the probability of randomly generating a composition of functions realizing a particular Boolean function with three arguments equaled 1 in 256, we would expect about 39,063 random compositions per 10,000,000 to realize a particular Boolean function. However, after randomly generating 10,000,000 compositions of the functions AND, OR, and NOT, we found only 730 3-multiplexers and no 3-parity functions. That is, our randomly generated compositions of functions realizing the 3-multiplexer function are about 54 times *sparser* than solutions in the original search space of Boolean functions. The 3-parity function is presumably tens of thousands of times scarcer than one in 256..

These three results concerning the 3-parity function, the 3-multiplexer function, and the

2-parity (exclusive-or) function should not be too surprising since the parity and multiplexer functions have long been identified by researchers as functions that often pose difficulties for paradigms for machine learning, artificial intelligence, neural nets, and classifier systems (Wilson 1987a, Wilson 1988, Quinlan 1988, Barto et. al. 1985). In summary, as to these problems, compositions of functions solving the problem are substantially *less dense* than solutions are in the search space of the original problem.

The reader would do well to remember the origin of the concern that compositions of functions solving a problem might be denser than solutions to the problem are in the search space of original problem. In Lenat's work on discovering mathematical laws via heuristic search (1976) and other related work (Lenat 1983), the mathematical laws being sought were stated, in many cases, directly in terms of the list, i.e. *the* primitive data type of the LISP programming language. In addition, the lists in Lenat's artificial mathematician (AM) laws were manipulated by list manipulation functions that are unique or peculiar to LISP. Specifically, in many experiments in Lenat (1976), the mathematical laws sought were stated directly in terms of lists and list manipulation functions such as, CAR (which returns the first element of a list), CDR (which returns the tail of a list), etc. In Lenat's *mea culpa* article "Why AM and EURISKO appear to work" (Lenat and Brown 1984), Lenat recognized that LISP syntax may have overly facilitated discovery of his previously reported results, namely, mathematical laws stated in terms of LISP's list manipulation functions and LISP's primitive object (i.e. the list).

In contrast, the problem described in this paper is neither stated nor solved in terms of objects or operators unique or peculiar to LISP. The solution to the Boolean multiplexer function is expressed in terms of ordinary Boolean functions (such as OR, AND, NOT, and IF). Virtually any programming language can express solutions to this problem. The LISP programming language was chosen for use in the genetic programming paradigm primarily because of the many convenient features of LISP (most importantly, the fact that data and programs have the same form in LISP and that this common form corresponds to the parse tree of a computer program). The LISP programming language was *not* chosen because of the presence in LISP of the list as a primitive data type or because of LISP's particular functions for manipulating lists (e.g. CAR and CDR).

In summary, there is no *a priori* reason (nor any reason we have since discovered) to think that there is anything about the syntax of the programming language we chose to use here (i.e. LISP) that makes it easier to discover solutions to problems involving ordinary (i.e. non-list) objects and ordinary (i.e. non-list) functions. In addition, the control experiments verify that the results obtained herein are not the fruits of a random search.

7. Conclusions

We described the recently developed genetic programming paradigm and enumerated the five major steps for using it. We cited a variety of different problems which this paradigm has successfully solved. We described, in detail, one particular run in which the genetic programming paradigm learned the Boolean 11-multiplexer function and showed how the size and shape of the ultimate solution progressively evolved using genetic programming paradigm. We presented performance statistics for a number of runs that indicate the rapidity of the search technique and that the genetic programming paradigm performs far better than randomly.

References

Barto, A. G., Anandan, P., and Anderson, C. W. Cooperativity in networks of pattern recognizing stochastic learning automata. In Narendra,K.S. *Adaptive and Learning Systems*. New York: Plenum 1985.

Cramer, Nichael Lynn. A representation for the adaptive generation of simple sequential programs. In *Proceedings of an International Conference on Genetic Algorithms and Their Applications*. Hillsdale, NJ: Lawrence Erlbaum Associates 1985.

De Jong, Kenneth A. Learning with genetic algorithms: an overview. *Machine Learning*, 3(2), 121-138, 1988.

Fujiki, Cory and Dickinson, John. Using the genetic algorithm to generate LISP source code to solve the prisoner's dilemma. In Grefenstette, John J.(editor). *Genetic Algorithms and Their Applications: Proceedings of the Second International Conference on Genetic Algorithms*. Hillsdale, NJ: Lawrence Erlbaum Associates 1987.

Goldberg, David E., Korb, B., and Deb, K.. Messy genetic algorithms: Motivation, analysis, and first results. *Complex Systems*. Pages 493-530. 3(5) October 1989.

Holland, J. H. *Adaptation in Natural and Artificial Systems*. Ann Arbor, MI: University of Michigan Press 1975.

Holland, John H. Escaping brittleness: The possibilities of general-purpose learning algorithms applied to parallel rule-based systems. In Michalski, Ryszard S., Carbonell, Jaime G. and Mitchell, Tom M. *Machine Learning: An Artificial Intelligence Approach, Volume II*. P. 593-623. Los Altos, CA: Morgan Kaufmann 1986.

Koza, John R. Hierarchical genetic algorithms operating on populations of computer programs. In *Proceedings of the 11th International Joint Conference on Artificial Intelligence (IJCAI)*. San Mateo, CA Morgan Kaufmann 1989.

Koza, John R. *Genetic Programming: A Paradigm for Genetically Breeding Populations of Computer Programs to Solve Problems*. Stanford University Computer Science Department Technical Report STAN-CS-90-1314. June 1990. 1990a.

Koza, John R. Genetically breeding populations of computer programs to solve problems in artificial intelligence. In *Proceedings of the Second International Conference on Tools for AI*. Washington. November 6-9, 1990. 1990b

Koza, John R. Genetic evolution and co-evolution of computer programs. In Farmer, Doyne, Langton, Christopher, Rasmussen, S., and Taylor, C. (editors) *Artificial Life II, SFI Studies in the Sciences of Complexity*. Volume XI. Addison-Wesley, Redwood City CA 1991. 1991a.

Koza, John R. Evolution and co-evolution of computer programs to control independent-acting agents. In Meyer, Jean-Arcady and Wilson, Stewart W. *From Animals to Animats: Proceedings of the First International Conference on Simulation of Adaptive Behavior*. Paris. September 24-28, 1990. MIT Press, Cambridge, MA, 1991. 1991b.

Koza, John R. Concept formation and decision tree induction using the genetic programming paradigm. In Schwefel, Hans-Paul and Maenner, Reinhard (editors) *Parallel Problem Solving from Nature*. Springer-Verlag, Berlin, 1991. 1991c.

Koza, John R. A genetic approach to econometric modeling. In Bourgine, Paul and

Walliser, Bernard. *Proceedings of the 2nd International Conference on Economics and Artificial Intelligence*. Pergamon Press 1991. 1991d.

Koza, John R. Evolving a computer program to generate random numbers using the genetic programming paradigm. In Belew, Rik and Booker, Lashon (editors) *Proceedings of the Fourth International Conference on Genetic Algorithms*. San Mateo, Ca: Morgan Kaufmann Publishers Inc. 1991. 1991e.

Koza, John R. *Genetic Programming*. Cambridge, MA: MIT Press, 1991 (forthcoming). 1991f.

Koza, John R. and Keane, Martin A. Cart centering and broom balancing by genetically breeding populations of control strategy programs. In *Proceedings of International Joint Conference on Neural Networks, Washington, January, 1990*. Volume I. Hillsdale, NJ: Lawrence Erlbaum 1990. 1990a.

Koza, John R. and Keane, Martin A. Genetic breeding of non-linear optimal control strategies for broom balancing. In *Proceedings of the Ninth International Conference on Analysis and Optimization of Systems*. Berlin: Springer-Verlag, 1990. 1990b.

Koza, John R. and Rice, James P. Genetic generation of both the weights and architecture for a neural network. In *Proceedings of International Joint Conference on Neural Networks, Seattle, July 1991*. IEEE Press 1991. 1991a

Koza, John R. and Rice, James P. A genetic approach to artificial intelligence. In C. G. Langton *Artificial Life II Video Proceedings*. Addison-Wesley 1991. 1991b.

Lenat, Douglas B. AM: An Artificial Intelligence Approach to Discovery in Mathematics as Heuristic Search. PhD Dissertation. Computer Science Department. Stanford University. 1976.

Lenat, Douglas B. The role of heuristics in learning by discovery: Three case studies. In Michalski, Ryszard S., Carbonell, Jaime G. and Mitchell, Tom M. *Machine Learning: An Artificial Intelligence Approach, Volume I*. P. 243-306. Los Altos, CA: Morgan Kaufman 1983.

Lenat, Douglas B. and Brown, John Seely. Why AM and EURISKO appear to work. *Artificial Intelligence*. 23 (1984). 269-294.

Quinlan, J. R. Induction of decision trees. *Machine Learning* 1 (1), 81-106, 1986.

Quinlan, J. R. An empirical comparison of genetic and decision-tree classifiers. *Proceedings of the Fifth International Conference on Machine Learning*. San Mateo, CA: Morgan Kaufmann. 1988.

Smith, Steven F. *A Learning System Based on Genetic Adaptive Algorithms*. PhD dissertation. Pittsburgh: University of Pittsburgh 1980.

Wilson, Stewart. W. Classifier Systems and the animat problem. *Machine Learning*, 3(2), 199-228, 1987. 1987a

Wilson, S. W. Hierarchical credit allocation in a classifier system. *Proceedings of the Tenth International Joint Conference on Artificial Intelligence*, 217-220, 1987. 1987b.

Wilson, Stewart W. Bid competition and specificity reconsidered. *Journal of Complex Systems*. 2(6), 705-723, 1988.

A Grammar-Based Genetic Algorithm

Hendrik James Antonisse
AI Center W418
The MITRE Corporation
7525 Colshire Drive
McLean, VA 22102

ABSTRACT

High-level syntactically-based representations pose problems for applying the GA because it is hard to construct crossover operators that always result in legal offspring. This paper proposes a reformulation of the genetic algorithm that makes it appropriate to any representation that can be cast in a formal grammar. This reformulation is consistent with recent reinterpretations of GA foundations in set-theoretic terms, and concentrates on the modifications required to make the space of legal structures closed under the crossover operator. The analysis places no restriction on the form of the grammars.

Keywords: Inductive Bias, High-Level Representations, Crossover

1 Introduction

Until recently, workers in the area of computing applications who would like to apply the genetic algorithm (GA) to complex problems were stuck on the twin horns of a dilemma. If they wished to retain the guidance provided by GA theory they had to accept the feature-level, unstructured representation on which that theory had been based. If, on the other hand, they wished to use the more expressive relational representations afforded by, e.g., proposition-based inference systems, they had to forego the direct tie to GA theory. Vose has recast GA theory in set theoretic terms, where schemata are defined as predicates, or arbitrary subsets of the search space (Vose, 1991). This paper presents an approach to reformulating the genetic algorithm for the class of formal grammars and concentrates on the construction of crossover operators that appropriately reflect the search-space partitions, i.e., the semantics, that are syntactically expressed by the grammar.

The grammar-based approach to genetic algorithms may prove important for several reasons. One is that it results in a greatly increased level of control to programmers who wish to apply this algorithm to problems of interest (although see (Booker91) for a more traditional approach to GA programming in classifier systems). Secondly, it is expected to improve the performance of the GA for the class of problems whose solution spaces are expressible by grammars. This is a very large and interesting class of problems that contains, for instance, all well-specified computer programming languages. (Koza has been exploring this space with promising results, but without the formal basis being proposed here--see Section 5 (Koza, 1990)) Finally, the grammar-based approach to the GA brings into sharp focus the inductive bias at work in the traditional formulation of the algorithm. (Inductive bias is the bias imposed on solutions to a problem by virtue of the language in which solutions are cast. See, e.g., (Haussler, 1987) (Kelly, 1988), for investigations into this effect).

2 A Recapitulation of the Foundations of the GA

The genetic algorithm can be viewed as searching for a population of good solutions to an optimization problem. Given an objective function *f()* and a domain *D* that can be mapped into a language of strings of length *L*, the GA samples the language with *M* possible solution strings at a time. The objective is to carry a population *M* maximizing the total utility of the population: Maximize *sum i=1..M* $f(string_i)$. The algorithm approximates an interesting but computationally infeasible algorithm that I will call the *schema sampling algorithm (SSA)*. This algorithm, in turn, is an approximation of a more powerful algorithm I will call the *power-set sampling algorithm (PSSA)*. Each of these algorithms implements a search for good points in the solution space by exponentially biasing its search towards areas where good points have been observed so far.

2.1 Basis: The K-armed Bandit Analysis

GA's are often viewed as solutions to the K-armed bandit problem. A one-armed bandit is a gambling machine in which one puts a coin in the hope of achieving a payoff. Decision-making under uncertainty is sharply defined given the problem of maximizing one's return on a finite number of coins allotted to a number ("k") of one-armed bandits. If one does not know anything about the machines, there is no guaranteed optimal solution to this problem. However, a good solution exists if we are given the means and variances of the payoffs but not which machine has which mean/variance pairs (Holland, 1975):one should favor the best bandit so far but continue to sample the other bandits at exponentially decreasing rates.

2.2 The Power-Set Sampling Algorithm

How can this solution be exploited in an algorithm? Treating the evaluation of *f()* at a point in the solution space as a trial of a one-armed bandit, we may use multiple trials to build up statistics for that bandit. One may at the same time build up statistics of *sets* of bandits. In terms of searching the solution space, whenever we sample any point in the solution space we gather information on the expected payoff of *each set* of points that contains the sampled point. The solution points may be construed as 2^N bandits, where N is the number of points in the solution space. PSSA is then: Evaluate M solutions against the objective function f; update the estimated average payoffs of every possible set of solutions); generate a new set of M individuals so that we exponentially favor the odds of choosing new trials that are members of a set S_1 relative to those of S_2 whenever the expected value of S_1 is greater than S_2.; and repeat.

Assuming that the K-armed bandit analysis gives us a loss-minimizing strategy, the power-set algorithm imposes a well-motivated statistical bias on search. This statistical bias is at the heart of all work in the GA. The reason PSSA is presented is that it has the interesting property of embodying the beneficial statistical bias while avoiding *any* representational bias (modulo the encoding of *f()*)--it carries statistics on *all possible* solution sets. Unfortunately, it is hopelessly expensive to keep track of such statistics on any but the simplest problems.

2.3 The Schema Sampling Algorithm

The schema sampling algorithm addresses the problem that the power-set algorithm is infeasible by radically reducing the sets on which statistics are collected. SSA is based on partitions of the solution space into similarity templates called *schemata.* These may be denoted by introducing "wild card" positions in possible solution strings. For instance "110#" would denote the set of length-4 strings {1101, 1100} that share the common prefix "110". If we consider each position on the string to define a dimension of the solution space, the values of which are given by the alphabet defined on that position, it becomes clear that the schemata pick out hyperplanes of the space. Moreover, they do not pick out arbitrary, but specifically the *coordinate* hyperplanes of the space. Except for the restriction on the sets over which it collects statistics, SSA is the same as PSSA.

This algorithm can be effectively carried out, but it is infeasible for reasonably sized problems. The reason for this is most clearly apparent in step 3. The numbers of hyperplanes is exponential in the length of the strings encoding a problem under a given alphabet. Since statistics are to kept on all of these, the storage requirements for this algorithm are vast. Moreover, the updating procedure must update a sizable proportion of these at each iteration of the algorithm. For a problem encoded in a length 12 binary string using a population of 10 (still an extremely diminutive problem), the number of schemata is 3^{12}, or 531,441. The number of schemata updates per iteration is 10 X 2^{12}, or 40,960. In this respect the hyperplane algorithm represents a computationally naive algorithm.

2.4 The GA

The GA may be thought of as an approximation to SSA that approaches its behavior without direct statistics-keeping on schemata. Instead, it relies on an existing set of samples of the solution space and implements a tradeoff of resampling the same points in the space or exploring unsampled parts of the space using a set of genetic operators.. These operators include the crossover operator, in which one probabilistically chooses a pair of solution strings according to their measured goodness, randomly chooses a point along a the strings and cuts them there, and splices the first half of one string to the second half of the other, and the second half of the one to the first half of the other. E.g., for 11001 and 10111 crossing over at the third bit results in new strings as follows:

```
110     01        11011
    X         =
101     11        10101
```

The central result of GA theory is the schemata theorem (Holland, 1975), which gives an estimate of the disruption of the optimal number of samples of a hyperplane in time t+1 (where the "optimal number" is given by the schemata sampling algorithm) due to the effects of the genetic operators. This disruption imposes an upper limit on the GA's abilities to find complex, global relationships in its field of application. Thus the price paid for a feasible approximation to SSA is to depend on relatively short defining-length building blocks in the construction of problem solutions, i.e. it is to introduce a new, *proximity* bias into the search strategy.

To the degree that PSSA and SSA directly manipulate sets of solutions, they may be thought of as employing description languages that embody a *semantic ascent* over over the base-level description language of individual sampled solutions of the search space. (Semantic ascent occurs when one changes from a language in which one refers to a set of elements to a language in which one may refer to the same elements plus quantified sets of such elements, as in the move from sentential to predicate logic or from first-order to second-order logic (Quine, 1970)). Notice that the GA *does not* use semantic ascent.

This paper is about a generalization of the crossover operator for languages that embody semantic ascent, i.e. languages that are described through set-specifying grammars.

3 GA Limitations in Structured Domains

The last section described the three major biases inherited by and embodied in the GA. These are (1) the statistical bias deriving from the power-set sampling algorithm, (2) the similarity template bias derived from the schemata sampling algorithm, and (3) the bias towards relatively short building blocks induced by the crossover operator in the genetic algorithm itself. The focus of the remaining discussion in on the second of these biases.

The similarity template bias has the effect of a priori reducing the solutions considered in the SSA and GA. Other learning algorithms also reduce the set of possible solutions considered. In example-based learning, for instance, one usually limits the expressions considered for describing a set of examples to be at most K conjuncts of pure binary disjuncts of terms. Such a priori limits on the form of solutions considered represent the inductive biases (Haussler, 1987) of learning algorithms. One may see the GA as analogously embodying an inductive bias corresponding to the hypothesis language of similarity templates. Unfortunately this bias is hard-coded in the algorithm as it stands, and encodes a relatively weak hypothesis language that seems closely related to regular expressions. The only ammeliorating characteristic is the ability to extend "expressiveness" by increasing the population size samples by the algorithm. The rest of this paper presents a new approach that changes the role of inductive bias in the GA, allowing the inductive bias to be specified by the GA programmer in any hypothesis language expressed in a grammar.

3.1 Grammatical Expressions and Crossover

Ideally the domain of the GA is closed under the crossover operator. This fails to be achieved for many domains of interest. For instance, crossover is not closed if the problem domain includes structured expressions like production rules ("If the infection is primary-bacteria, the site one of the sterile sites, and suspected entry is gastro-intestinal, then the organism is probably bacteroides"). For purposes of illustration a domain of simple English sentences is introduced. The English fragment domain was chosen because it has a small, well-defined grammar in which the mechanical points regarding the old and new crossover operator may be clearly illustrated. However, the points are completely general and apply as well to grammar-defined inference systems. (In particular, well-defined grammars exist for many computationally interesting languages in the form of, e.g., BNF grammars. Both Inference Corporation in its shell "ART" and Gensym it its shell G2, for instance, use BNF grammars to define the syntax of their expert system applications. Such grammars now form the foundation of many expert system applications.)

Let the solution space of a problem (the areas where $f >= 0$) lie entirely within the subset of terms defined by the following simple grammar (from (O'Grady, et al, 1989)):

```
S      ::=> NP (M) (VP)
NP     ::=> (Det) (adjP) N (PP)
VP     ::=> V (NP) (PP)
PP     ::=> P NP
AdjP   ::=> (Spec) Adj
M      ::=> will | did
Det    ::=> a | the
N      ::=> mayor | town | crowd | Jack | Sally | Blairsburg
V      ::=> talk | meet
P      ::=> to | in | of
Adj    ::=> popular
```

Consider the sentences "Jack did meet Sally in Blairsburg" and "The popular mayor of the town will talk to the crowd." These are both sentence in the grammar. Let's say we chose a crossover point for the first of these:

Jack $_X$ did meet Sally in Blairsburg.

and ask what all the possible ways this might be crossed with the second. The possibilities are defined by the possible crossover points in the second sentence:

```
Jack The popular mayor of the town will talk to the crowd.
Jack popular mayor of the town will talk to the crowd.
Jack mayor of the town will talk to the crowd.
Jack of the town will talk to the crowd.
Jack the town will talk to the crowd.
Jack town will talk to the crowd.
Jack will talk to the crowd.     (***legal in the grammar***)
Jack talk to the crowd.
Jack to the crowd.
Jack the crowd.
Jack crowd.
```

Crossover is completely uninformed with respect to the grammatical structure of meaningful strings in the universe of discourse. Hence many of the strings that result from crossover are jibberish even in the extremely flexible context of natural language (especially if punctuation counts). In the formal language defined by the above grammar, only one of the possible crossovers is legal. The problem is how to pick this one out of the crowd.

3.2 GA's and Grammars

What are the implications of grammar-constrained solutions spaces for GA search? There are two of particular consequence: Firstly, grammar-like sets of string solutions often give the traditional GA search method insurmountable problems because vast stretches of the space of strings are GA-legal but grammar-illegal. The solution that is usually proposed, to just make the objective function zero at illegal points, has a direct negative effect on the ability of the GA to effectively search such spaces, according to the proportion of the zero areas. But consider a domain on which the GA is operating which *really is* determined by a grammar. What is the size of the language of the grammar compared to the set of all strings of the domain? It will usually be relatively small, and in the limit (for infinite languages) vanishingly so. Since the strings outside of the grammar give the GA no information (e.g., all evaluate to zero), but the algorithm continues to sample such portions of the solution space, the algorithm is often reduced to a virtual random walk in its search for meaningful solution points.

The second implication of a grammar-constrained space is that similarity templates are often simply the wrong partition of the solution space on which to "keep statistics." Now consider the space of sets of strings on which statistics are being collected by the schemata sampling algorithm. These sets are hard-coded and are defined irrespective of the problem at hand. There is no way, given the GA framework so far, of adding special provisions to bias the search in a more appropriate manner. Thus the GA is likely to follow false contours in the solution space that are artifacts of its similarity template bias.

4 A Grammar-Based GA

Notice that each nonterminal of a grammar corresponds to a subset of items in the domain. "NP" corresponds to the set of noun phrases, "M" to the set of modifiers, etc. Consider a new statistical algorithm constructed to collect statistics not on the similarity templates, but on just the sets describable in an arbitrary grammar, and in particular on those sets corresponding to the non-terminals of the grammar. The intent is that the expected value of every non-terminal in the grammar be computed from the expected values of its constituent productions.

The new algorithm works just like the power-set and schemata sampling algorithms, only the new loci for collecting statistics are the terms of the grammar. Clearly this can be done for the topmost non-terminal "S", since the expected value of that node is just the average value of all the sentences observed so far. To achieve this for all non-terminals, the credit for the evaluation of a string to a value under f must be distributed across the constituent *terminals* of the string as well. To achieve this a credit assignment function CA, must be invoked. For instance, let $f(sentence_i)$ result in v_1. Then $CA(sentence_i, f(sentence_i))$ results in an attribution of values to each term in $sentence_i$. (A straightforward approach is to let each term take the value v_i. For a fuller discussion of credit assignment within the GA framework, however, see (Holland, 1985) (Grefenstette, 1989)).

The grammar, *CA*, and the function *f* induce a statistics-gathering structure over the problem space. The grammars form near-arbitrary manifolds in the solution space. They define the subspaces of terminals that contain all possible derivation paths through the grammar. The each non-terminal corresponds to a manifold containing statistical estimates on which parts of the grammar have led to "good" sentences and which to "poor" sentences. A grammar manifold sampling algorithm analogous to the earlier schema sampling algorithm may now be constructed that organizes statistics-gathering activity according to the partitions of a domain defined by the grammar. Call this algorithm GMSA.

4.1 Grammar-based Crossover

This section introduces a grammar-based crossover operator that can be implemented for any grammar. Any sentence resulting from the operator is guaranteed to be a member of the grammar. The grammar-based algorithm stands in the same relation to the GMSA algorithm as the GA stands to SSA. Grammar-based crossover is implemented as follows:

0. Choose an arbitrary crossover point in a sentence. Notice that the crossover point will always correspond to a location in the derivation tree as well as a location in the sentence.
1. Generate two sentence fragments by splitting the sentence at the crossover point.
2. Mark the trailing edge of the leading sentence fragment and the leading edge of the trailing fragment with a tag that signifies the crossover location in the derivation tree. The tag is composed of the nearest production in the derivation whose left-hand side derives both the last term of the leading sentence fragment and the first of the trailing fragment. It includes previous productions up to the start node.
3. Fragments may be crossed-over if their they are leader-trailer pairs and if their tags unify, i.e., if the tags are equal.

The key to the new crossover is that any crossover point corresponds to a split in the derivation tree. Let the crossover points in the example sentences and the corresponding derivation-tree crossover points be:

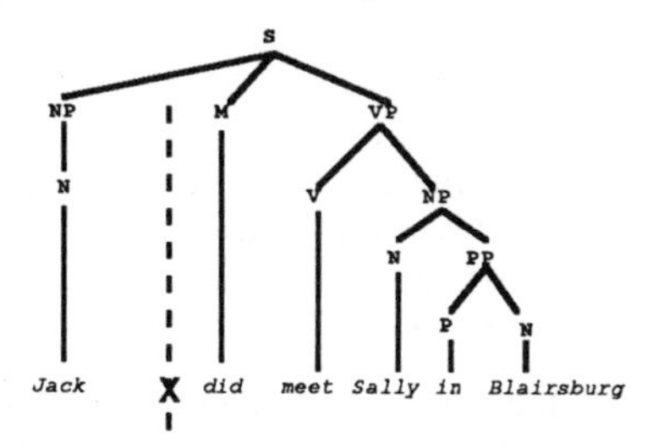

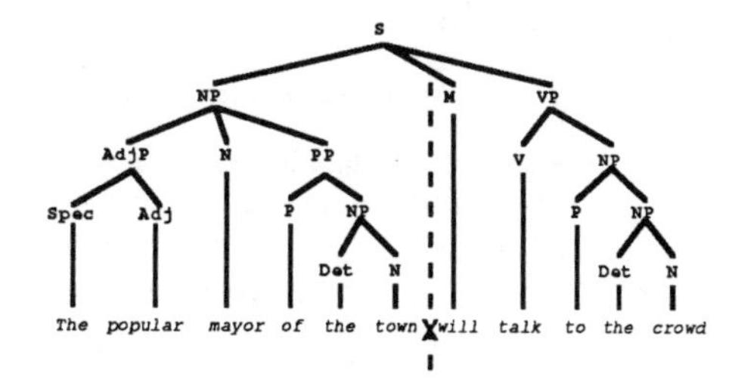

The resulting sentence fragments arc:

Jack [S -> NP * M VP]
[S -> NP * M VP] **did meet Sally in Blairsburg.**

The popular mayor of the town [S -> NP*M VP]
[S -> NP*M VP] **will talk to the crowd.**

Now crossover can be applied in this (miniscule) pool of fragments. Let the leading fragment of the first sentence attempt to crossover with the trailing fragment of the second sentence. Since their respective tags are equal, the crossover is allowed. The result is the following sentence:

Jack [S -> NP * M VP]
[S -> NP * M VP] **will talk to the crowd.**

Jack will talk to the crowd.

Likewise for the other sentence fragments:

The popular mayor of the town [S -> NP*M VP]
[S -> NP*M VP] **did meet Sally in Blairsburg.**

The popular mayor of the town did meet Sally in Blairsburg.

The converse to the above case also holds. Tags that do not unify indicate sentence fragments that, when spliced together, do not form member sentences of the language defined by the grammar. Had the crossover point of the second sentence been elsewhere the sentences may or may not have been capable of being crossed. For instance:

Jack did [S -> NP M * VP]
[S -> NP M * VP] **meet Sally in Blairsburg.**
The popular mayor of the town [S -> NP * M VP]
[S -> NP * M VP] **will talk to the crowd.**

do not crossover, while:

The popular mayor [S -> NP M VP] [NP -> AdjP N * PP]
[S -> NP M VP] [NP -> Adj N * PP] **of the town did meet...**

The popular mayor of the town did meet Sally in Blairsburg.

do crossover..Notice that the preceding productions had to be included in the second case--the embedding productions are carried along to distinguish between, e.g., noun phrases that are followed, in one case, by a verb phase, and in another by a terminator.

The above operator is analogous to single-point crossover in the traditional GA. A possible extension of the grammar-based crossover operator is to only carry single-production tags and operate on partial sentences, as in:

... the popular mayor [NP -> AdjP N * PP]
[NP -> AdjP N * PP] **of the town ...**

... the popular mayor of the town ...

or to use the tags to search for the crossover point, by sliding one tag structure across the other until segments of the tag unify. This should not be computationally expensive because of the highly constrained structure of the tags. An approach using the latter form of crossover seems to be analogous to multiple-point crossover in the unstructured GA.

Notice that no restrictions have been made on the form of the grammars in the discussions above. The impact of ambiguous grammars, however, is that crossover pairs which would form legal offspring, but whose respective fragments are derived from different productions, will not be recognized as legal without further processing. A grammar-based GA will, therefore, not search a space as efficiently with an ambiguous grammar as an unambiguous grammar defining a manifold of interest in that space.

4.2 Parity Space: An Example of a Grammar Manifold

Considerable work has been done in the GA community to characterize the problems that lead the algorithm astray. In particular, classes of GA-hard and GA-deceptive problems have been studied and the features that make them difficult for the GA described. Bethke, following a suggestion from Barto, provided the ground-breaking work in this with his Walsh-transform analysis of the algorithm (Bethke, 1981). Additional insight and extensions have been provided by (Goldberg, 1987) and (Holland, 1987). This section presents a class of problems that is difficult for the traditional GA but for which the difficulty is eliminated in the grammar-based genetic algorithm.

Consider a function *f() = b() g()*, and let *b()* correspond to the parity function. That is, *b(i)* on some binary input string *i* is 1 if the number of bits in *i* is even, 0 if it is odd. Let *g()* be a function that, on its own, is amenable to the GA (such as, for instance, x^2+x). Notice that the solution space of *f()* is defined by the (GA-nice) function *g()*, except that it is shot half-full of zero points at even-parity *i*. Parity is a generalization of the XOR function which,it was pointed out (Benke, 1989), would also give the GA problems. To see why this is so, notice that (1) there are no "building blocks" for the parity function--the defining length of each good schema is the length of the encoding string, and (2) a simple, traditional GA with replacement takes a population of good solutions to the parity problem (*f(i)* = 1 for even-numbers of 1's in *i*) and, over time, produces a random solution.

It might be thought that the root of the inability of the GA here is with the crossover operator. This is not so. If we return to the analysis of the schemata sampling algorithm we see that the problem is that for the parity function, every *coordinate* hyperplane in the solution space has a value of 0.5--half of its members will be even parity, half will be odd. No differential sampling will occur because the subsets of strings on which statistics are being kept do not correspond to any meaningful partitions of the string space relative to this function. The problem lies not with the inductive bias in the SSA that the GA inherits. It is precisely this bias that the grammar-based GA is designed to control--we need only find an appropriate grammar. In this case it is derived directly from the finite string automaton that implements (an even string length) parity function:

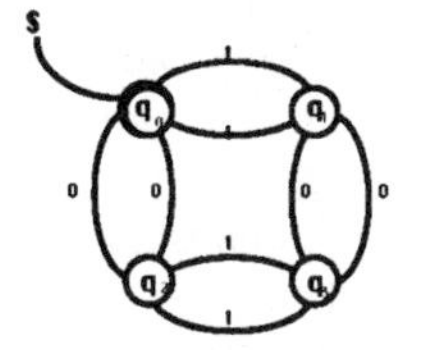

This FSA is quivalent to the grammar:

$$
\begin{aligned}
S &::> q_0 \\
q_0 &::> 1q_1 \mid 0q_2 \mid e \\
q_1 &::> 1q_0 \mid 0q_3 \\
q_2 &::> 1q_3 \mid 0q_0 \\
q_3 &::> 1q_2 \mid 0q_1
\end{aligned}
$$

We may now fold this grammar into the crossover operator in a way similar to that described above. In this case, only state information need be retained to guarantee legal strings. Only even parity strings are produced. For example:

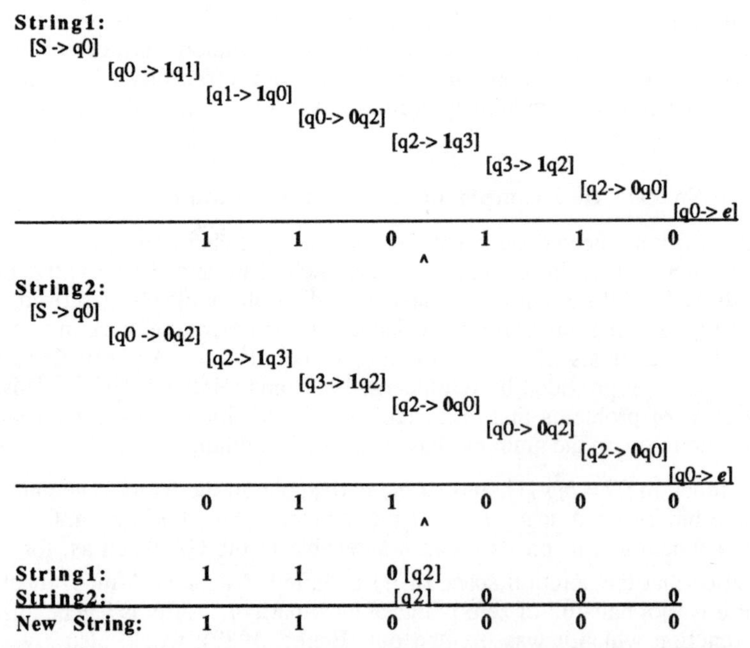

The result is to factor out of the genetic search exactly that half of the solution space which has odd parity. In effect, the grammar-based crossover implements a transformation of the space such that it has only legal solution points. The genetic search may now proceed to find solutions in the manifold of contiguous solutions defined by the transformation. (It remains, of course, problem-dependent whether genetic search will be effective in searching this space.)

5 Previous Approaches, Future Directions

The new crossover is designed to operate directly on a generative "blueprint" of the structures that are the target of the search. The leverage afforded by "genotypic" search for "phenotypic" characteristics is alluded to but seldom explicitly implemented in GA programs (although see (Harp et al, 1989) for genetic search of blueprints of neural network architectures, and (Antonisse & Keller, 1987) and (Koza, 1989) for other partially-grammatic approaches). Both the earlier mapping approaches (best exemplified by (Forrest, 1985)) and the modified crossover approaches (such as (Fujika, 1987), (Whitley, et al, 1989), and (Davis, 1989)) accept the similarity template foundations of the GA. The recent work of Koza ((Koza, 1989)) is intuitively along the same lines as the approach described here. It depends on a simple tree grammar to guarantee legal S-expressions that define programs. It is less general than the approach in this paper, being geared specifically for the special tree representation, and suffers from a dependence on variable naming conventions to bind variables across the crossover point. Variable binding is a non-local mechanism that has traditionally presented particular problems for GA search (Antonisse & Keller, 1987). In the approach outlined in this paper, variable binding may be expressed as a context sensitive aspect of a language.

The future directions for this work fall into two categories, empirical investigations and theoretical work. The theoretical work involves recasting the coordinate hyperplane analysis in the original proof of the schemata theorem as a set-theoretic analysis based on grammar subsets. This would, hopefully, bear fruit in the form of a generalized schemata theorem. Of more immediate and practical import, however, is to develop a base of experience with high-level GA's. For instance, we need to develop a clear picture of the overhead incurred by the crossover tags. Although it should be possible to make the test for tag unification inexpensive and to keep the length of the tags manageable, the process of unification incurs a cost that must be traded against the expected leverage the grammar-based representation gives on the search space. In general, we need to manage language fragments efficiently in the crossover phase. This can probably be handled most effectively using an explicit derivation tree. These do not seem to be insurmountable problems. Their solutions lead directly to much a more flexible, more widely applicable genetic algorithm.

References

(Antonisse & Keller, 1987) "Genetic Operators for High-level Knowledge Representations", H. J. Antonisse and K. S. Keller, Proc. 2nd Int. GA Conf., pp. 69-76, 1987.

(Antonisse, 1989) "A New Interpretation of Schema Notation that Overturns the Binary Encoding Constraint", H. J. Antonisse, Proc. 3rd Int. GA Conf., 1989.

(Benke, 1989) G. Benke, Personal communication, the MITRE Corporation, August, 1989.

(Bethke, 1981) "Genetic Algorithms as Function Optimizers", A. D. Bethke, Doctoral dissertation, University of Michigan, Dissertation Abstracts International, 41(9), 3503B. (University Microfilms No. 8106101).

(Booker91) "Representing Attribute-Based Concepts in a Classifier System", L. Booker, these Proceedings.

(Davis, 1989) "Adapting Operator Probabilities in Genetic Algorithms", L. Davis, Proc. 3nd Int. GA Conf., pp. 61-69, 1989.

(Forrest, 1985) "Implementing Semantic Network Structures Using the Classifier System", S. Forrest, Proc. Int. GA Conf., pp. 24-44, 1985.

(Fujiko, 1987) "Using the genetic algorithm to generate LISP code to solve the prisoner's dilemma", C. Fujiko and J. Dickinson, Proc. 2nd Int. GA Conf., pp. 236-240, 1987.

(Goldberg, 1987) "Simple genetic algorithms and the minimal deceptive problem", D. E. Goldberg, in L. Davis (Ed.), *Genetic Algorithms and Simulated Annealing* (pp 74-88), London, Pitman, 1987.

(Goldberg, 1989) *Genetic Algorithms in Search, Optimization, and Machine Learning*, D.E. Goldberg, Addison-Wesley, Reading PA, 1989.

(Grefenstette, 1988) "Credit Assignment in Rule Discovery Systems", J. J. Grefenstette, *Special Issue on Genetic Algorithms*, Machine Learning, Volume 3, Number 2/3, October, 1988.

(Harp et al, 1989) "Towards the Genetic Synthesis of Neural Networks," S. A. Harp, T. Samad, and A. Guha, Proc. 3nd Int. GA Conf., pp. 360-369, 1989.

(Haussler, 1987) "Bias, Version Spaces and Valiant's Learning Framework", D. Haussler, Proceedings of the Fourth International Conference on Machine Learning, University of California, Irvine, CA, June 22-25 , pp. 324-337, 1987.

(Holland, 1975) *Adaptation in Natural and Artificial Systems*, J. H. Holland, University of Michigan Press, Ann Arbor, MI, 1975.

(Holland, 1985) "Properties of the Bucket Brigade", J. H. Holland, Proc. Int. GA Conf., pp. 1-8, 1985.

(Holland, 1987) "Genetic algorithms and Classifier Systems, Foundations and Future Directions", J. H. Holland, Proc. 2nd Int. GA Conf., pp. 82-89, 1987.

(Kelly, 1988) "Theory Discovery and the Hypothesis Language", K. T. Kelly, Proceedings of the Fifth International Conference on Machine Learning, University of Michigan, Ann Arbor, June 12-14 , pp. 325-339, 1988.

(Koza, 1989) "Hierarchical Genetic Algorithms the Operate on Populations of Computer Programs", J. R. Koza, Proceedings of the Eleventh International Joint Conference on Artificial Intelligence, pp. 768-780, 1989.

(O'Grady, et al, 1989) *Contemporary Linguistics, An Introduction*, W. O'Grady, M. Dobrovolsky, and M. Aronoff, St. Martins, NY, 1989.

(Quine, 1970) Philosophy of Logic, W. V. Quine, Foundations of Philosophy Series, Prentice-Hall, Englewood Cliffs, NJ, 1970.

(Vose, 1991) "Generalizing the Notion of Schema in Genetic Algorithms", M. D. Vose, to appear in Artificial Intelligence, 1991.

(Whitley et al, 1989) "Scheduling Problems and Traveling Salesmen: The Genetic Edge Recombination," D. Whitley, T. Starkweather, and D'A. Fuquay, Proc. 3nd Int. GA Conf., pp. 133-141, 1989.

Genetic Algorithms for Real Parameter Optimization

Alden H. Wright
Department of Computer Science
University of Montana
Missoula, Montana 59812

Abstract

This paper is concerned with the application of genetic algorithms to optimization problems over several real parameters. It is shown that k-point crossover (for k small relative to the number of parameters) can be viewed as a crossover operation on the vector of parameters plus perturbations of some of the parameters. Mutation can also be considered as a perturbation of some of the parameters. This suggests a genetic algorithm that uses real parameter vectors as chromosomes, real parameters as genes, and real numbers as alleles. Such an algorithm is proposed with two possible crossover methods. Schemata are defined for this algorithm, and it is shown that Holland's Schema theorem holds for one of these crossover methods. Experimental results are given that indicate that this algorithm with a mixture of the two crossover methods outperformed the binary-coded genetic algorithm on 7 of 9 test problems.

Keywords: optimization, genetic algorithm, evolution

1 Introduction

In this paper, we are primarily concerned with the following optimization problem:

Maximize $f(x_1, x_2, \ldots, x_m)$

where each x_i is a real parameter subject to $a_i \leq x_i \leq b_i$

for some constants a_i and b_i.

Such problems have widespread application. Applications include optimizing simulation models, fitting nonlinear curves to data, solving systems of nonlinear equations, engineering design and control problems, and setting weights on neural networks.

2 Background

Genetic algorithms have been fairly successful at solving problems of this type that are too ill-behaved (such as multimodal and/or non-differentiable) for more conventional hill-climbing and derivative based techniques.

The usual method of applying genetic algorithms to real-parameter problems is to encode each parameter as a bit string using either a standard binary coding or a Gray coding. The bit strings for the parameters are concatenated together to give a single bit string (or "chromosome") which represents the entire vector of parameters. In biological terminology, each bit position corresponds to a gene of the chromosome, and each bit value corresponds to an allele. If $\overline{x}$ is a parameter vector, we will denote the corresponding bit string by the corresponding uppercase letter **X**. Thus, the problem is translated into a combinatorial problem where the points of the search space are the corners of a high-dimensional cube.

For example, if there are two parameters x_1 and x_2 with ranges $0 \le x_1 < 1$ and $-2 \le x_2 < 2$, and four bits are used to represent each parameter, then the point $(x_1, x_2) = (\frac{3}{16}, 1)$ would be represented by the bit string 0011 1100 using binary coding and 0010 1010 using Gray coding.

In this paper we will consider genetic algorithms where a chromosome corresponds to a vector of real parameters, a gene corresponds to a real number, and an allele corresponds to a real value.

Such algorithms have been suggested for particular applications [Lucasius and Kateman, 1989] for a chemometrics problem (finding the configuration of a DNA hairpin which will fit NMR data), and [Davis, 1989] in the area of using meta-operators for setting operator probabilities in a standard genetic algorithm. In [Antonisse, 1989] an argument is presented for the use of non-binary discrete alphabets in genetic algorithms. Antonisse argues that with a different interpretation of schemata, there are many more schemata using non-binary alphabets than with binary alphabets. In a sense, this paper extends his argument to real alphabets.

3. Binary Coding and Gray Coding

If a single parameter x_i has lower and upper bounds a_i and b_i respectively, then the standard way of binary coding x_i using n bits is to let real values between $a_i + k\frac{b_i - a_i}{2^n}$ and $a_i + (k+1)\frac{b_i - a_i}{2^n}$ correspond to the standard binary code for the integer k for $0 \le k < 2^n$. For example, if $a_i = 0$ and $b_i = 4$ and $n = 5$, then the real values between $\frac{3}{8}$ and $\frac{1}{2}$ would correspond to the binary code 00011.

To avoid talking about intervals, we will refer to the binary code for the integer k above as corresponding to the left end of the interval, namely $a_i + k\frac{b_i - a_i}{2^n}$. Thus, in the above example, we would refer to the binary code 00011 as corresponding to the real number $\frac{3}{8}$.

Gray coding is another way of coding parameters into bits which has the property that an increase of one step in the parameter value corresponds to a change of a single bit in the code. The conversion formula from binary coding to Gray coding is:

$$\gamma_k = \begin{cases} \beta_1 & \text{if } k = 1 \\ \beta_{k+1} \oplus \beta_k & \text{if } k > 1 \end{cases}$$

where γ_k is the kth Gray code bit, β_k is the kth binary code bit, bits are numbered from 1 to n starting on the left, and $\oplus$ denotes addition mod 2. The conversion from Gray coding to binary coding is:

$$\beta_k = \sum_{i=1}^{k} \gamma_i$$

where the summation is done mod 2. For example, the binary code 1101011 corresponds to the Gray code of 1011110.

4 Crossover

Crossover is a reproduction technique that takes two parent chromosomes and produces two child chromosomes. A commonly used method for crossover is called one-point crossover. In this method, both parent chromosomes are split into left and a right subchromosomes, where the left subchromosomes of each parent are the same length, and the right subchromosomes of each parent are the same length. Then each child gets the left subchromosome of one parent and the right subchromosome of the other parent. The split position (between two successive genes) is called the crossover point. For example, if the parent chromosomes are 011 10010 and 100 11110 and the crossover point is between bits 3 and 4 (where bits are numbered from left to right starting at 1), then the children are 011 11110 and 100 10010. We will call crossover applied at the bit level to bit strings <u>binary crossover</u>, and crossover applied at the real parameter level <u>real crossover</u>.

In considering what binary crossover does in a real parameter space, we first consider the special case where the crossover point falls between the codes for two parameters. In this case, one child gets some of its parameters from one parent, and some of its parameters from the other parent. For example, if the parents as bit strings are **X** and **Y**, corresponding to parameter vectors $\bar{x} = (x_1, x_2, \ldots, x_m)$ and $\bar{y} = (y_1, y_2, \ldots, y_m)$, and the split point is between component x_i and x_{i+1}, then one child corresponds to the parameter vector $(x_1, x_2, \ldots, x_i, y_{i+1}, \ldots, y_m)$ and the other corresponds to $(y_1, y_2, \ldots, y_i, x_{i+1}, \ldots, x_m)$. Thus, in this case, binary crossover is exactly the same as real crossover. This can be seen pictorially most easily in the case where $m = 2$: see Figure 1.

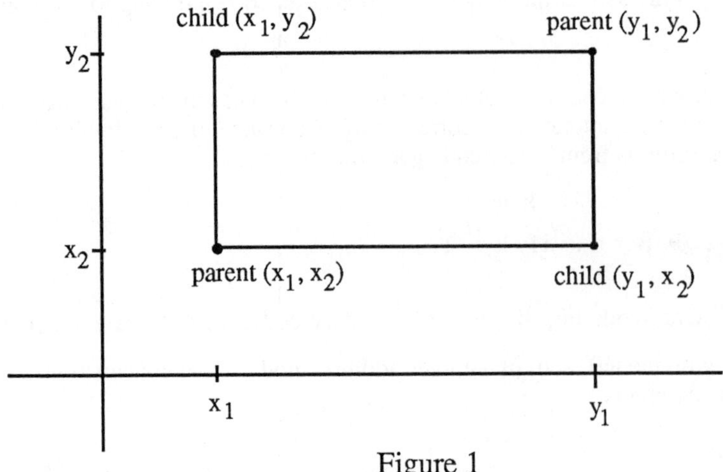

Figure 1

In general, if the split is between parameter i and parameter $i+1$, each child lies in the intersection of an i dimensional hyperplane determined by one parent, and an $m-i$ dimensional hyperplane determined by the other.

Next, suppose that a crossover point is chosen within the code for a parameter. We first consider the case where binary coding is used, and we limit ourselves to consideration of the one parameter in which the crossover point occurs. Then the part of the binary code of the parameter to the left of the crossover point will correspond to the more significant bits, and the part to the left of the split will correspond to the less significant bits. Thus, a child gets the more significant part of the parameter of one parent and the less significant part of the other parent. One can view the child as being a "perturbation" of the first parent, where the size of the perturbation is determined by the difference in the less significant bits of the parents. In fact, if the crossover point is between bit k and bit $k+1$, then the perturbation corresponds to changing some of bits $k+1$ to n of one of the parents. If $R_i = b_i - a_i$ is the size of the range for the parameter, then the maximum size of the perturbation is $R_i 2^{-k}$.

For example, suppose that the range of the parameter goes from 0 to 1, the parameter is coded by 5 bits, and the parameter values of the two parents are: $\frac{5}{32}$ and $\frac{27}{32}$. Then the corresponding binary codes are 001 01 and 110 11. Suppose that the crossover point is between bits 3 and 4, so that the codes of the children are 001 11 and 110 01 which correspond to parameter values of $\frac{7}{32}$ and $\frac{25}{32}$. The less significant bits of the parents are 01 and 11 respectively, and the difference between them corresponds to $\pm \frac{2}{32}$ in parameter space. Note that the first child's parameter value of $\frac{7}{32}$ is the first parent's parameter value perturbed by $\frac{2}{32}$, and the second child's parameter value of $\frac{25}{32}$ is the second parent's parameter value of $\frac{27}{32}$ perturbed by $-\frac{2}{32}$.

Also, note that if the more significant parts of the two parents are the same (as would frequently happen in a nearly converged population), then one child gets the parameter value of one of the parents, and the other child gets the parameter value of the other parent.

Now suppose that Gray coding is used, and that a crossover point occurs within the code for a parameter. From the formulas given for converting between binary coding and Gray coding, it can be seen than the most significant k bits of the binary code determine the most significant k bits of the Gray code and vice versa. Thus, again each child can be viewed as a perturbation of the parent from which it received its more significant bits. Since changing the $(k+1)^{st}$ Gray bit code affects only the $(k+1)^{st}$, $(k+2)^{nd}$ etc. binary code bits, the maximum size of the perturbation is the same as in the binary code case.

We can summarize the above in the following theorem:

Theorem 1. Let $\mathbf{X}$ and $\mathbf{Y}$ be the bit strings corresponding to real parameter vectors $\bar{x}$ and $\bar{y}$. Let $\mathbf{Z}$ be obtained from $\mathbf{X}$ and $\mathbf{Y}$ by one-point binary crossover where the crossover point lies between bits k and $k+1$ of parameter x_i and y_i. We assume that $\mathbf{Z}$ gets the bits to the left of the crossover point from $\mathbf{X}$, and those to the right of the crossover point from $\mathbf{Y}$. Then the real parameter vector $\bar{z}$ corresponding to $\mathbf{Z}$ can also be obtained from $\bar{x}$ and $\bar{y}$ by a real one-point crossover, where the crossover point is between x_i and x_{i+1}, followed by a perturbation of parameter x_i of size at most $R_i 2^{-k}$.

In the case of two-point crossover, $\bar{z}$ can be obtained from $\bar{x}$ and $\bar{y}$ by a two-point real crossover followed by perturbations of two parameters.

5 Mutation

Mutation is a common reproduction operator used for finding new points in the search space to evaluate. When a chromosome is chosen for mutation, a random choice is made of some of the genes of the chromosome, and these genes are modified.

In the case of a binary-coded genetic algorithm, the corresponding bits are "flipped" from 0 to 1 or from 1 to 0. Normally the probability that any given bit will be chosen is low, so that it would be unlikely for the code for a parameter to have more than one bit mutated.

When a bit of a parameter code is mutated, we can think of the corresponding real parameter as being perturbed. The size of the perturbation depends on the bit or bits chosen to be mutated. Let $R_i = b_i - a_i$ be the size of the range for the parameter x_i. If binary coding is used, then mutating the kth bit corresponds to a perturbation of $R_i 2^{-k}$. If Gray coding is used, changing the kth bit of the Gray code can affect all bits of the corresponding binary code from the kth bit to the nth bit. Thus, the magnitude of the corresponding perturbation can be up to $R_i 2^{-k+1}$, and perturbations of all sizes from $R_i 2^{-n}$ up to this maximum are possible.

The direction of the perturbation is determined by the value of the bit that is mutated. Under binary coding, changing a 0 to a 1 will always produce a perturbation in the positive direction, and changing a 1 to a 0 will always produce a perturbation in the negative direction. Under Gray coding, changing a 0 to a 1 may perturb the parameter in either direction.

6 Schemata

In the theory of genetic algorithms ([Holland, 1975] or [Goldberg, 1989]), a schema is a "similarity template" which describes a subset of the space of chromosomes. In a schema, some of the genes are unrestricted, and the other genes contain a fixed allele. In the binary case, a schema is described by a string over the alphabet {0, 1, *}, where a * means that the corresponding position in the string is unrestricted and can be either a 0 or a 1. For example, the string 01*1*0 describes the schema: {010100, 010110, 011100, 011110}.

To see the meaning of binary-coded schemata in terms of real parameters, let us restrict ourselves to a single parameter. Again, suppose that the size of the range of the parameter is R, and that the parameter is coded using n bits. Then the schemata all of whose * symbols are contiguous at the right end of the string correspond to connected intervals of real numbers. For example, if the range of the parameter is from 0 to 1, and if $n = 5$, then the schema 01*** corresponds to the parameter interval from $\frac{1}{4}$ to $\frac{1}{2}$ in either binary or Gray coding, and 011** corresponds to the interval from $\frac{3}{8}$ to $\frac{1}{2}$ in binary coding and the interval from $\frac{1}{4}$ to $\frac{3}{8}$ in Gray coding. In general, such a contiguous schema containing k *'s corresponds to a parameter interval of length $R\,2^{k-n}$ in both binary and Gray coding. Any single parameter schema whose *'s are not all contiguous at the right end corresponds to a disconnected union of intervals.

Going back to multi-parameter shemata, those schemata which correspond to a connected interval for each parameter correspond naturally to rectangular neighborhoods in parameter space.

Non-connected schemata can be used with binary coding to take advantages of periodicities that are a power of two relative to the corresponding parameter interval. It is unclear what the interpretation of non-connected Gray coding schemata is. It appears to the author that for "most" objective functions, the connected schemata are the most meaningful in that they capture locality information about the function.

Connected schemata, and the real crossover operation, would be especially relevant for a separable objective function, namely a function f such that $f(x_1, x_2, \ldots, x_m) = f_1(x_1) + f_2(x_2) + \ldots + f_m(x_m)$.

We call those schemata that correspond to connected sets of parameter space connected schemata.

For a single parameter which is coded with n bits, there are 2^n connected schemata with no *'s, 2^{n-1} connected schemata with one *, 2^{n-2} schemata with two *'s, etc., for a total of $\sum_{k=0}^{n} 2^{n-k} = 2^{n+1} - 1$ connected schemata. For all m parameters, there are $(2^{n+1} - 1)^m$ connected binary schemata. This contrasts with 3^{mn} unrestricted binary schemata.

7 A Real-Coded Genetic Algorithm

The above analysis of binary-coded genetic algorithms applied to real parameter spaces can be used to help design a genetic algorithm whose chromosomes are vectors of floating point numbers and whose alleles are real numbers.

To design a standard genetic optimization algorithm, the following things are needed:

1. A method for choosing the initial population.
2. A "scaling" function that converts the objective function into a nonnegative fitness function.
3. A selection function the computes the "target sampling rate" for each individual. The target sampling rate of an individual is the desired expected number of children for that individual.
4. A sampling algorithm that uses the target sampling rates for the individuals to choose which individuals are allowed to reproduce.
5. Reproduction operators that produce new individuals from old individuals.
6. A method for choosing which reproduction operator to apply to each individual that is to be reproduced.

The basic algorithm does step 1 to choose an initial population. Steps 2 to 6 are done to go from one population (or generation) to the next. These steps are repeated until the a convergence criterion is satisfied or for a predetermined number of generations.

In a standard binary-coded genetic algorithm, only steps 5 and 6 use the bit string representation of a chromosome. Thus, to design a real genetic algorithm, we can take the methods for doing steps 1 to 4 from the corresponding methods for standard genetic algorithms.

In our suggested algorithm, each population member is represented by a chromosome which is the parameter vector $\bar{x} = (x_1, x_2, \ldots, x_m) \in R^m$, and genes are the real parameters. Three reproduction operators are suggested: crossover, mutation, and linear crossover (another form of combination of two parents explained below).

Real crossover was defined above in the section on crossover.

If one starts with a finite population, then the crossover operation only allows one to reach a finite number of points in parameter space, namely those whose parameter components are selected from the corresponding parameter components of population members. Intuitively, one can reach those points which are intersections of the hyperplanes that go through the initial population points and that are parallel to the coordinate hyperplanes. In fact, if the initial population size is p, one can reach at most

p^m parameter vectors using crossover. One purpose of mutation should be to allow arbitrary points of parameter space to be reached.

In designing a real mutation operator, there is first the question of whether the point in R^m (parameter vector) should be mutated, or whether individual parameters should be mutated. [Schwefel, 1981] and [Matyas, 1965] and others base methods for global real-parameter optimization on using mutations of points in R^m. One can envisage functions where mutation in any of the coordinate directions would be unlikely to improve fitness, whereas motion in other directions could substantially improve fitness. For these functions, mutation in R^m would be much better than mutation in the coordinate directions with a lower mutation rate. However, we will see below that it is more difficult to make mutation in R^m compatible with the schema theorem, so we have chosen to mutate individual parameters.

There is both the problem of choosing the distribution of the sizes of mutations, and the problem of how to keep the mutated point within range. One method that we have tried is as follows: Choose a mutation rate and a maximum mutation size. The probability that a given parameter will be mutated is the mutation rate. Once a parameter is chosen for mutation, the direction of the mutation is chosen randomly with probability $\frac{1}{2}$. If the parameter is farther from the edge of the range than the maximum mutation size in the direction of the mutation, then the mutated parameter is chosen using a uniform distribution over the interval between the parameter and the parameter plus or minus the maximum mutation size. If the parameter is closer to the edge of the range than the maximum mutation size, then the new value of the parameter is chosen using a uniform distribution over the interval between the parameter and the edge of the range. To make this explicit, suppose that the original parameter value is x, the range is $[a, b]$, the maximum mutation size is M, and that the mutation has been chosen to be in the positive direction. Then the mutated parameter is chosen from the range $[x, \min(M,b)]$ using a uniform probability distribution. Experiments indicated that this mutation scheme, when applied repeatedly to a point in an interval without selection pressure, gave an almost uniform distribution of points over a large number of mutations.

The above mutation method does not correspond very closely to the mutation in a binary coded algorithm, in that the binary-coded mutation more heavily favors mutations with a small magnitude. This is especially true if a large number of bits are used to encode each real parameter.

A problem with real crossover is illustrated in Figure 2. The ellipses in the figure represent contour lines of the objective function. A local minimum is at the center of the inner ellipse. Points 1 and 2 are both relatively good points in that their function value is not too much above the local minimum. However, any point generated using crossover from these points will be much worse than either point.

To get around this situation, we propose another form of reproduction operator that we call linear crossover. From the two parent points p_1 and p_2 three new points are generated, namely $\frac{1}{2}p_1 + \frac{1}{2}p_2$, $\frac{3}{2}p_1 - \frac{1}{2}p_2$, and $-\frac{1}{2}p_1 + \frac{3}{2}p_2$. The point $\frac{1}{2}p_1 + \frac{1}{2}p_2$ is the midpoint of p_1 and p_2, while $\frac{3}{2}p_1 - \frac{1}{2}p_2$ and $-\frac{1}{2}p_1 + \frac{3}{2}p_2$

lie on the line determined by p_1 and p_2. In this paper, the best two of the three points are selected.

Linear crossover has the disadvantage that it would be highly disruptive of schemata and is not compatible with the version of the schema theorem given below.

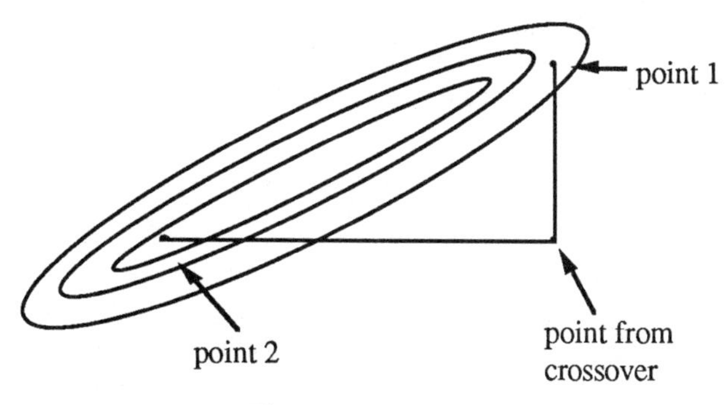

Figure 2

8 Schemata Analysis for Real-Allele Genetic Algorithms

As explained in Section 6, we feel that, in a binary-coded genetic algorithm for a real-parameter problem, the most meaningful schemata for functions without periodicities of power 2 are those that restrict some parameters to a subinterval of their possible range. Thus, the logical way to define a schema for a real-allele genetic algorithm is to restrict some or all of the parameters to subintervals of their possible ranges. We have already assumed that the range of parameter x_i is $a_i \le x_i \le b_i$.

Thus, parameter space is $\mathcal{P} = \prod_{i=1}^{m} [a_i, b_i]$, and a schema $\mathcal{S}$ is specified by choosing real values α_i and β_i such that $a_i \le \alpha_i \le \beta_i \le b_i$. Then the corresponding subset of parameter space is $\prod_{i=1}^{m} [\alpha_i, \beta_i]$. We will use the symbol $\mathcal{S}$ to refer either to the sequence of pairs $((\alpha_1, \beta_1), \ldots, (\alpha_m, \beta_m))$ or the corresponding subset of parameter space $\prod_{i=1}^{m} [\alpha_i, \beta_i]$.

We can do the same kind of analysis that is done for traditional genetic algorithms. For this analysis, the important thing about a schema is which parameters it restricts. To each schema $\mathcal{S}$ we associate a template as follows. If a parameter is unrestricted, i. e., $\alpha_i = a_i$ and $\beta_i = b_i$, then we associate a * with that parameter. If a parameter is restricted, i. e. $\alpha_i > a_i$ or $\beta_i < b_i$, then let I_i denote its interval $[a_i, b_i]$. Then the template is the sequence of I_i's and *'s corresponding to the parameters. For example, if parameter space is [-1,1] × [-1,1] × [-1,1], and if the schema is [-1,1] × [0,1] × [-1,0], then the corresponding m-tuple would be $*I_2I_3$.

Let the average fitness of the population members which are in schema $\mathcal{S}$ be $f(\mathcal{S})$, and let the average fitness of all population members be $\bar{f}$. Suppose that we use one-point crossover with a randomly selected crossover point. If selection and sampling are done so that the probability of an individual being selected for reproduction is the ratio of its fitness to the average fitness of the population, then the expected proportion of individuals of schema $\mathcal{S}$ that are selected for reproduction is:

$$p(\mathcal{S}, t+1) \geq p(\mathcal{S}, t)\,\frac{f(\mathcal{S})}{\bar{f}}\left(1 - C\,\frac{\delta(\mathcal{S})}{\ell - 1}(1 - p(\mathcal{S}, t))\right)(1 - M)^F$$

In this equation, ℓ is the chromosome length, F is the number of restricted (fixed) genes, $\delta(\mathcal{S})$ is the defining length (the distance between its outermost restricted genes), C is the probability of crossover, and M is the probability that a gene will be mutated. The above is Holland's Schema Theorem (Corollary 6.4.1 of [Holland, 1975]), and is stated for genetic algorithms with any discrete set of alleles for each gene.

The proof of the Schema theorem depends neither on the form of the objective function nor on how the restricted genes are restricted. Thus, it also applies to real-coded genetic algorithms using real mutation and real crossover as reproduction operators. (In the real case, the chromosome length ℓ is the number of real parameters m.)

9 Experimental Results

The above algorithm was implemented by modifying the Genesis program of John Grefenstette. The five problems of De Jong (F1 to F5) [De Jong, 1975] plus problems F6 and F7 of [Schaffer, Caruana, Eshelman, and Das, 1989] were included, along with two additional problems. Problem F5R is De Jong's problem 5 (Shekel's Foxholes) rotated 30 degrees in the plane before evaluating. Thus, the formula for F5R is:

$$f_{5r}(x_1, x_2) = f_5\left(\frac{\sqrt{3}}{2}x_1 + \frac{1}{2}x_2, -\frac{1}{2}x_1 + \frac{\sqrt{3}}{2}x_2\right)$$

where f_{5r} is De Jong's f_5 (Shekel's Foxholes). The last function, F8, corresponds to the solution of the following system of 5 equations in 5 unknowns:

$$x_k^2 + \sum_{i=1}^{5} x_i - 4 = 0 \qquad \text{for } k = 1, 2, \ldots, 5.$$

This system was converted to the objective function for an optimization problem by summing the absolute values of the left sides of the above equations. The parameter ranges for all variables was $-7 \leq x_k \leq 0$. In this region, there is a single multiplicity 1 solution (global minimum): (-4, -4, -4, -4, -4). There are a number of local minima on the edges of the region.

The global minimum for all problems had an objective function value of zero.

All problems were run with two point crossover, the elitist strategy, Baker's selection procedure, a scaling window of 5 generations, a population size of 20, and a crossover rate of 0.8. For the binary-coded algorithm, Gray coding was used. Both spinning and convergence thresholds were turned off. For all functions except F8, each experiment consisted of running the genetic algorithm for 1000 trials (function evaluations). For function F8, 5000 trials were used. Two results were saved for each experiment: the best performance, and the offline performance. The best performance is the smallest value of the objective function obtained over all function evaluations. The offline performance is the average over function evaluations of the best value obtained up to that function evaluation.

In order to select good parameter settings, multiple runs were done for both the binary and the real-coded algorithms with different settings for mutation rates (real and binary) and mutation sizes (real). What is presented below are the best results over all of these settings of the mutation parameters.

For the binary-coded algorithm, 1000 experiments were done at each mutation rate from 0.005 to 0.05 in steps of 0.005. Thus, for each function, 10 mutation rates were tried. For the real-coded algorithm using real crossover only, 1000 experiments were done at each combination of a mutation size and a mutation rate. The mutation sizes ranged from from 0.1 to 0.3 in steps of 0.1, and mutation rates ranged from 0.05 to 0.3 in steps of 0.05. Thus, for each function, 18 combinations of mutation size and rate were tried. The same was done using 50% real crossover and 50% linear crossover.

For each function and each type of algorithm (binary, real with all real crossover, and real with both real and linear crossover), the mutation settings that gave the optimal performance measure is shown below. The standard deviation is computed over the 1000 experiments at that mutation setting.

Table 1: Best Values of Best Performance

Type of Algorithm	Function	Mutation Rate	Mutation Size	Best Performance	Standard Deviation
Binary	F1	0.035		0.0062	0.0111
Real	F1	0.300	0.2	0.00058	0.00102
Real/linear	F1	0.050	0.1	0.00004	0.00004
Binary	F2	0.050		0.0192	0.0326
Real	F2	0.300	0.3	0.0119	0.0237
Real/linear	F2	0.300	0.1	0.0107	0.0200
Binary	F3	0.030		1.4740	0.9012
Real	F3	0.300	0.3	1.0490	1.0908
Real/linear	F3	0.300	0.3	0.5730	0.9506
Binary	F4	0.015		9.7311	4.0645
Real	F4	0.300	0.2	3.5671	1.9498
Real/linear	F4	0.200	0.1	-1.0906	0.7828
Binary	F5	0.050		1.0116	0.1795
Real	F5	0.300	0.3	10.5626	7.5863
Real/linear	F5	0.1	0.2	9.7598	6.5275
Binary	F5R	0.045		8.6699	5.7736
Real	F5R	0.200	0.3	11.7435	6.5181
Real/linear	F5R	0.150	0.2	11.4997	6.2822

Binary	F6	0.050		0.0310	0.0414
Real	F6	0.100	0.3	0.2071	0.1261
Real/linear	F6	0.050	0.1	0.0200	0.0266
Binary	F7	0.050		0.5888	0.3751
Real	F7	0.250	0.3	2.9721	1.3622
Real/linear	F7	0.050	0.1	0.0584	0.0256
Binary	F8	0.040		0.36223	0.9677
Real	F8	0.150	0.3	0.8290	3.5439
Real/linear	F8	0.050	0.3	0.1783	1.5616

Table 2: Best Values of Offline Performance

Type of Algorithm	Function	Mutation Rate	Mutation Size	Offline Performance	Standard Deviation
Binary	F1	0.045		0.5094	0.2553
Real	F1	0.300	0.3	0.4871	0.2831
Real/linear	F1	0.050	0.2	0.3604	0.1483
Binary	F2	0.030		1.2453	1.0705
Real	F2	0.300	0.1	1.2018	1.0537
Real/linear	F2	0.150	0.1	1.1673	1.0104
Binary	F3	0.030		4.4235	0.9691
Real	F3	0.300	0.3	4.2511	1.3463
Real/linear	F3	0.50	0.3	3.6280	1.2226
Binary	F4	0.015		36.3860	7.6296
Real	F4	0.300	0.2	32.0367	6.9149
Real/linear	F4	0.050	0.1	6.8157	1.9028
Binary	F5	0.045		14.1197	12.1777
Real	F5	0.300	0.3	34.1292	38.5147
Real/linear	F5	0.300	0.3	23.3482	14.2119
Binary	F5R	0.045		19.9376	9.5545
Real	F5R	0.200	0.3	34.8737	30.3235
Real/linear	F5R	0.250	0.3	24.3253	14.8582
Binary	F6	0.050		0.1087	0.0625
Real	F6	0.100	0.3	0.2231	0.1197
Real/linear	F6	0.050	0.1	0.0757	0.0354
Binary	F7	0.050		1.7961	0.6009
Real	F7	0.250	0.3	3.2571	1.2362
Real/linear	F7	0.050	0.2	1.0746	0.2234
Binary	F8	0.040		3.4175	1.8448
Real	F8	0.150	0.3	3.0374	3.5018
Real/linear	F8	0.050	0.1	1.9604	1.6075

9.1 Summary of Experimental Results

The real-coded algorithm with 50% real crossover and 50% linear crossover was better than the real-coded algorithm with 100% real crossover on all problems. (However, the differences for both best and offline performance for F2, and for best performance for F5R are of questionable significance.) The real-coded algorithm with both types of crossover was better than the binary coded algorithm on 7 of the 9 problems, and the real-code algorithm with real crossover was better than the binary-coded algorithm on 5 of the 9 problems.

On problems F4 and F7 the mixed crossover real-coded algorthm did much better than the other two algorithms.

On Problem F5, the Shekel's Foxholes problem, the binary coded algorithm did much better than the real-coded algorithms This problem has periodicities that make it well-suited to a binary-coded algorithm. When the problem was rotated by 30 degrees to get rid of some of these periodicities as function F5R, the real-coded algorithms did better than before, but still not quite as well as the binary coded algorithm.

10 Conclusions

Crossover and mutation in binary-coded genetic algorithms were analyzed as they apply to real-parameter optimization problems. This analysis suggested that binary crossover can be viewed as real crossover plus a perturbation.

Real-coded genetic algorithms with two types of crossover were presented. The above analysis implies that mutation rates for real-coded algorithm should be considerably higher than for binary-code algorithms. For one of these types of crossover (real crossover), Holland's Schema Theorem applies easily.

Experimental results showed that the real-coded genetic algorithm based on a mixture of real and linear crossover gave superior results to binary-coded genetic algorithm on most of the test problems. The real-coded algorithm that used linear crossover in addition to real crossover outperformed the real-coded algorithm that used all real crossover on all test problems.

The strengths of real-coded genetic algorithms include:

(1) Increased efficiency: bit strings do not need to be converted to real numbers for every function evaluation.

(2) Increased precision: since a real-number representation is used, there is no loss of precision due to the binary representation. (See also [Schraudolph, 1991] in these proceedings.)

(3) Greater freedom to use different mutation and crossover techniques based on the real representation. The work of Schwefel, Rechenberg, and other proponents of *Evolutionsstrategie* is relevant here.

However, the greater freedom mentioned in (3) can give practitioner more decisions to make and more parameters to set.

Acknowledgements

The author would like to thank his student Kevin Lohn for conversations, and Digital Equipment Corporation for the loan of a DECStation 3100 on which some computer runs were done.

References

Antonisse, Jim, (1989) A new interpretation of schema notation that overturns the binary encoding constraint. In J. David Schaffer, (ed.), *Proceedings of the Third International Conference on Genetic Algorithms,* San Mateo, CA: Morgan Kaufman Publishers, Inc., 86-91.

Davis, Lawrence, (1989) Adapting operator probabilities in genetic algorithms. In J. David Schaffer, (ed.), *Proceedings of the Third International Conference on Genetic Algorithms,* San Mateo, CA: Morgan Kaufman Publishers, Inc., 61-69.

De Jong, K. A., (1975) Analysis of the behavior of a class of genetic adaptive systems. Ph.D. Dissertation, Department of Computer and Communications Sciences, University of Michigan, Ann Arbor, MI.

Goldberg, David E. (1989). *Genetic Algorithms in Search, Optimization, and Machine Learning.* Reading, Massachusetts: Addison-Wesley.

Holland, J. H. (1975). *Adaptation in natural and artificial systems.* Ann Arbor: The University of Michigan press.

Lucasius, C. B., and G. Kateman, (1989) Applications of genetic algorithms in chemometrics. In J. David Schaffer, (ed.), *Proceedings of the Third International Conference on Genetic Algorithms,* San Mateo, CA: Morgan Kaufman Publishers, Inc., 170-176.

Matyas, J, (1965) Random optimization. *Automation and Remote Control,* **26**, 244-251.

Price, W. L., (1978) A controlled random search procedure for global optimization. *Towards Global Optimization,* edited by L. C. M. Dixon and G. P. Szego, North Holland.

Schaffer, J. David, Richard A. Caruana, Larry J. Eshelman, and Rajarshi Das, (1989) A study of control parameters affecting online performance of genetic algorithms for function optimization. In J. David Schaffer, (ed.), *Proceedings of the Third International Conference on Genetic Algorithms,* San Mateo, CA: Morgan Kaufman Publishers, Inc., 51-60.

Schraudolph, Nicol N., and Richard K. Belew, (1991) Dynamic Parameter Encoding for Genetic Algorithms, These proceedings.

Schwefel, H., (1981) *Numerical Optimization of Computer Models,* (M. Finnis, Trans.), Chichester: John Wiley. (Original work published 1977).

PART 5

FRAMEWORK ISSUES

Fundamental Principles of Deception in Genetic Search

L. Darrell Whitley
Department of Computer Science
Colorado State University
Fort Collins, Colorado 80523
whitley@cs.colostate.edu

Abstract

This paper presents several theorems concerning the nature of deception and the central role that deception plays in function optimization using genetic algorithms. A simple proof is offered which shows that the only problems which pose challenging optimization tasks are problems that involve some degree of deception and which result in conflicting k-arm bandit competitions between hyperplanes. The concept of a *deceptive attractor* is introduced and shown to be more general than the deceptive optimum found in the deceptive functions that have been constructed to date. Also introduced are the concepts of *fully deceptive problems* as well as less strict *consistently deceptive problems.* A proof is given showing that deceptive attractors must have a complementary bit pattern to that found in the binary representation of the global optimum if a function is to be either fully deceptive or consistently deceptive. Some empirical results are presented which demonstrate different methods of dealing with deception and poor linkage during genetic search.

Keywords: deception, tagged bits, GENITOR, uniform crossover.

1 Background

Based on earlier work by Bethke (1980), Goldberg (1987) introduced the concept of *deception* in order to better understand what kinds of situations are likely to create difficulty for a genetic algorithm when performing a function optimization task.

The fundamental theorem of genetic algorithms, the "schema" theory, attempts to explain the ability of genetic algorithms to search complex problem spaces effectively by showing that genetic algorithms change the sampling rates of hyperplanes in an n-dimensional hypercube corresponding to a binary encoding of the solution space (Holland 1975; Goldberg 1989a). Low level building blocks corresponding to large general hyperplanes in the space are recombined to drive the search toward more specific regions in hyperspace that display above average fitness. A problem is deceptive if certain hyperplanes guide the search toward some solution or genetic building block that is not globally competitive (Goldberg 1989a). Most earlier work in this area has concentrated on the construction of difficult or deceptive problems (Goldberg 1987, Goldberg, Korb and Deb 1989, Tanese 1989). More recently this work has expanded to to consider the relationship between deception, hyperplane variance, hyperplane dominance and genetic algorithms as dynamical systems (Grefenstette and Baker 1989, Liepins and Vose 1991, Whitley 1991, Goldberg and Rudnick 1991).

Hyperplanes are represented by bit combinations known as schemata. Consider an order-3 problem involving a search space encoded with 3 bits. This simple space can be represented by a cube, labeled so that adjacent corners in the cube differ by only a single bit. Using * as a don't care symbol, let schema 0** refer to a face in this cube. 0** is a schema of order-1, since it defines a partition of the search space occupied by all strings with a single bit (0) in the position indicated. The value of the hyperplane (or in this case, the simple plane) represented by 0** is denoted f(0**) and is calculated by averaging the fitness value of all the strings that match the template 0** (i.e., the average value of the 4 strings that form the corresponding face of the cube). The "order" of a hyperplane refers to the number of 0 or 1 bits specified in the schema representing that hyperplane. "Order" also relates information about the number of strings contained within the hyperplane represented: a hyperplane of order-1 contains 50% of the strings in the search space. Thus, the schema ***1****0*...* is of order-2 and the corresponding hyperplane contains (i.e., its schema matches) 25% of all the strings in the search space.

In the case of a fully deceptive order-3 function the hyperplane information represented by order-1 and order-2 schemata in the search space leads the search away from the global optimum, and instead directs the search toward what we will refer to as a *deceptive attractor.* Assuming the bits 111 represent the global optimum and the deceptive attractor is 000, full order-3 deception implies that the following relationships hold for the lower-order schemata:

```
f(0**) > f(1**)          f(00*) > f(11*), f(01*), f(10*)
f(*0*) > f(*1*)          f(0*0) > f(1*1), f(0*1), f(1*0)
f(**0) > f(**1)          f(*00) > f(*11), f(*01), f(*10)
```

where f(S) gives the fitness of either a string or the fitness of a schema, S, and is calculated by averaging all of the strings contained in the hyperplane represented by that schema. In other words, all lower-order hyperplane information associated with schemata whose presentation is made up of some subset of these 3 bits leads the search away from the global optimum of 111. A fully deceptive order-3 problem which satisfies the above inequalities has been defined by Goldberg, Korb and Deb (1989) where the bit strings have the following values.

```
Deceptive Function 1    f(000) = 28     f(001) = 26
                        f(010) = 22     f(100) = 14
                        f(110) = 0      f(011) = 0
                        f(101) = 0      f(111) = 30
```

Part of the motivation for building deceptive problems is to better understand what kinds of conditions can inhibit genetic search from finding a globally optimal or near optimal solution. Problems that have the ability to seriously mislead a genetic search have been labeled *GA-hard* problems. As shown later in this paper, fully deceptive subproblems coupled with poor linkage can result in GA-hard problems.

2 Definitions

In attempting to reason about the nature of deception, it is necessary to distinguish different kinds of deceptive situations and to use terms with some precision. I have attempted to keep the definitions offered in this paper as consistent as possible with the existing literature.

First we need to refer to some concepts relating to genetic algorithms in general. In particular we wish to discuss hyperplane competitions, but certain competitions are particularly important. A ***primary hyperplane competition*** of order N involves the complete set of primary competitors, where the ***primary competitors*** in a hyperplane competition of order N are the set of 2^N hyperplanes having schemata with N bit values in the same locations. For example, *0**0, *0**1, *1**0 and *1**1 are competitors in a primary hyperplane competition of order 2.

The "global winner" of a primary hyperplane competition is that hyperplane which has the highest fitness value among a set of competitors, where the fitness value of a hyperplane is the average fitness of all strings that are contained in that hyperplane. This does not necessarily imply that this hyperplane correctly leads to or has bits consistent with the globally optimal solution.

Two more basic concepts are useful for discussing deception. Hyperplane X "contains" hyperplane Y when X is a lower order hyperplane and Y is a higher order hyperplane such that Y has exactly the same bit values in exactly the same locations as X, but also has additional bit values in positions occupied by don't care symbols (*) in X. For example, 0** contains 01*; Note that 0** also contains the same strings as 01* (i.e., 010 and 011) but contains other strings as well (i.e., 001 and 000). This leads to a more specific relationship between hyperplane competitions that is important to deception. Two primary hyperplane competitions of order N and order K (where $K < N$) are "relevant" to one another when collectively the hyperplane competitors of order K "contain" the hyperplane competitors of order N. Thus, the primary hyperplane competition between 11***, 10***, 01*** and 00*** involves a set of hyperplanes that collectively contains the primary hyperplane competition between 111**, 110**, 101**, 100**, 011**, 010**, 001** and 000**. Given some hyperplane competition of order N, "relevant" hyperplanes of order K (where $K < N$) are represented by schemata which have bit values (either 1 or 0) in some proper subset of the locations in which bits occur in the order N hyperplane competition.

Deception implies that the global winner of some hyperplane competition of order N has a bit pattern that is different from the bit pattern of the global winner for some "relevant" lower order hyperplane competition. In other words, deception implies that these lower order competitions have their own "global winners" which do not have the same bit values as the global winner of the relevant hyperplane competition at order N.

A *deceptive problem* is any problem of a specific order N that involves deception in one or more relevant lower-order hyperplane competitions. Deception implies that there exists one or more relevant lower order hyperplane competitions that can potentially guide a genetic search away from the global winner of the hyperplane competition at order N. It does not imply the problem is fully deceptive, or that there is sufficient deception to misguide a genetic algorithm. This usage appears to be consistent with Goldberg's (1987) discussion of the minimal deceptive problem, which involves deception, but is not fully deceptive.

This usage is fairly broad; most problems may involve some degree of deception, since in general we would not expect that *all* lower level hyperplane competitions be consistent with relevant higher order hyperplane competitions. However, as will be shown in the next section, this particular definition separates optimization tasks into two extremely interesting classes: those that are not deceptive (and thus theoretically easy) and those which are deceptive.

A *fully deceptive problem* (or subproblem) of order-N is deceptive when all relevant lower-order hyperplanes lead toward a deceptive attractor. A *deceptive attractor* is a hyperplane of order N other than the true global winner which is supported by relevant lower level hyperplane competitions. For a fully deceptive problem, it can be shown that a deceptive attractor can only be that hyperplane represented by the schema which has a bit pattern that is the complement of the "global winner" of the hyperplane competition at order N (i.e., if the "global winner" of the hyperplane competition of order 3 is **1**1*1**, then the complement and potential deceptive attractor is **0**0*0**). A proof appears later in this paper. This usage of the term "fully deceptive" appears consistent with that found in Goldberg, Korb and Deb (1989).

It turns out that it is useful to distinguish different degrees of deception according to criteria other than order when describing and defining the deceptive attractor.

A *consistently deceptive problem* (or subproblem) of order-N is one in which none of the relevant lower-order hyperplanes lead toward the "global winner" of the primary hyperplane competition at order N we are interested in, but all of the relevant lower-order hyperplanes do not necessarily lead toward the deceptive attractor–except for the order-1 hyperplanes which (as will be shown) can only guide the search toward the "global winner" or the deceptive attractor of the hyperplane competition at order-N. It can be shown that choosing the deceptive attractor to be the complement of the global winner of the hyperplane competition at order N is also a *necessary* condition for the existence of a consistently deceptive problem because of the order-1 hyperplanes involved. Note that a fully deceptive problem is always a consistently deceptive problem, but a consistently deceptive problem is not necessarily a fully deceptive problem.

The term *deceptive function* will refer to a consistently deceptive problem where the number of bits used to encode the solution space corresponds to the order of the deception. A deceptive function is always consistently deceptive and may be fully deceptive. Thus, a consistently deceptive order-3 problem with a 3 bit encoding constitutes a deceptive function.

A *deceptive building block* of order N refers to a situation where some hyperplane H has a higher fitness value than its primary competitors, but all of the relevant lower order hyperplane competitions between schemata composed of subsets of bits in the same locations are misleading. A deceptive building block is similar to the concept of deceptive function, except the deceptive building block may correspond to a particular hyperplane competition that is part of a larger function optimization problem; thus the deceptive building block is in some sense a "subproblem" of some larger function. The concept of a "building block" in genetic algorithms is used to refer to schemata of low order (and often characterized by short defining length) and above average fitness. Building blocks are thought to be important to the process of genetic search. By concentrating on the hyperplane competitions involving building blocks (which should be well represented in sample populations and should be relatively stable under recombination) the genetic algorithm is able to generate good partial solutions and, through recombination, construct higher level schemata and complete strings. Empirical results (such as those presented later in this paper) show that if no building blocks are available, the ability of the genetic algorithm to produce global solutions is impaired. The deceptive building block is in some sense an "anti-building-block" since it represents a hyperplane competition that can mislead the genetic algorithm. A "deceptive building block" does not necessarily have a short defining length; in fact a deceptive building block with a long defining length is one kind of situation that makes the deception more problematic. However, the deceptive building block does involve competitions between hyperplanes with above average fitness.

A *deceptive hyperplane* is a hyperplane that incorrectly leads the search away from some more specific higher order hyperplane that in fact is superior to its competitors. It follows that the lower order hyperplanes which create a deceptive building block are "deceptive hyperplanes" *with respect to that particular building block.*

The notion of a deceptive hyperplane is defined with respect to a particular deceptive building block because the bits that make up the global winner of a hyperplane competition in a deceptive building block may not be the same bits that appear in another hyperplane competition or in the string that corresponds to the global solution of the problem. For example, in an order-4 fully deceptive function, the order-1 hyperplanes are deceptive hyperplanes with respect to the order-4 building block that constitutes the global solution, but the order-1 hyperplanes are not deceptive with respect to any of the order-2 or order-3 competitions, since these all consistently lead to the deceptive attractor. To define "deceptive hyperplanes" with respect to the global solution alone would be too restrictive since this would ignore the forces that drive genetic search. However the definition used here does not prevent us from referring to a hyperplane as being deceptive with respect to the global solution.

3 A Note On Convention

Throughout this paper the following convention will be used. *Let the bit "1" in any binary mapping be interpreted as indicating agreement with the bits in the optimal solution or global winner of a hyperplane competition rather than just a single binary string which maps to a particular value.* This interpretation generalizes the representation so that arguments concerning any specific set of strings hold for other arbitrary strings. This convention can also be operationalized. Liepins and Vose (1990) show that a simple translation function exists which will remap the entire binary space, thus making it possible to reassign the global optimum to an arbitrary bit string. Goldberg (1990) also notes that the function $(x \oplus c')$ translates the optimum to the "all-1s" position and translates all other strings to the correct corresponding positions in Hamming space, where x is each string in the binary encoding and c' is the complement of c, the location of the current optimum. The function $\oplus$ is a bitwise exclusive-or (bitwise addition modulo 2).

4 The Only Challenging Problems are Deceptive

As stated earlier, most problems may involve some deception, since in general we would not expect that *all* lower level hyperplane competitions will be consistent with relevant higher order hyperplane competitions. The following theorem supports the claim that the only challenging optimization tasks are problems involving some degree of deception.

THEOREM 1: *Given a fitness function for a problem representing some optimization task with a binary encoding of length L,* **if** *1) no deception occurs in any of the hyperplanes associated with that particular binary encoding and 2) the winners of the L order-1 hyperplanes can be correctly determined,* **then** *the global optimum of the function is determined by the one string contained in the intersection of the L order-1 hyperplane competition winners.*

PROOF BY CONTRADICTION:

Assume that deception does not occur in any hyperplane of the fitness function and we determine the correct winners of each of the order-1 hyperplanes. The intersection of the order-1 schemata will identify exactly one string as the candidate for the global optimum. If the candidate is not the global optimum, then at least one of the order-1 hyperplanes is deceptive, thus generating a contradiction.

QED

While this theorem makes some idealistic assumptions, it can be operationalized. If no deception occurs, then we can solve a problem in the following fashion. For each bit location we can directly solve a 2-arm bandit problem involving the corresponding two schemata *...*0*...* and *...*1*...*. Alternatively, the competition can be solved using statistical methods. The use of statistical methods potentially provides a means for dealing with the issue of noisy sampling–and thus the problem of determining the appropriate hyperplane competition winners. If no deception is present, such methods will determine (to some desired level of reliability) the probable value of the bits that belong in the corresponding position in the binary string representing the global solution to the problem. Certain difficulties can of course be encountered: what if two competing order-1 hyperplanes have equal or near equal value? Note that if there is no deception, all subpartitions of the order-1 hyperplanes must lead toward the global solution (to do otherwise would imply deception). Thus, if we can solve *any* of the order-1 hyperplanes, we can use the solution to narrow the search space. If there is no deception, it does not matter in what sequence the order-1 hyperplanes are solved. This means that we can break ties and resolve ambiguous competitions if the winners of competing order-1 hyperplanes can be chosen in any partition of the reduced search space. While it may not be possible to always operationalize this idea, it is often possible: we have already found the globally optimal solution to several test problems that have appeared in the genetic algorithm literature by solving the order-1 hyperplanes using statistical methods (Das and Whitley 1991).

This fundamental theory of deception seems almost trivial. Researchers who work with genetic algorithms have long known that hyperplanes at numerous levels play an important role in genetic search. However, this knowledge has not been directly related to deception and to what that implies about Holland's (1975) analogy to the 2-armed or k-armed (multi-armed) bandit problem.

The 2-armed bandit analogy for the order-1 hyperplanes goes as follows. Assume we have a slot machine with 2 arms. Each arm has a different payoff with an associated mean and variance. The problem is to determine which arm has the higher expected payoff. We would like to sample the two arms, while at the same time minimizing our expected loss. To do this, we should allocate exponentially more trials to the observed best. In actual practice, this may involve some initial sampling to determine the observed best. If we are dealing with some order-1 hyperplane competition, then *1*** and *0***, for example, would represent the two arms. If we are dealing with hyperplanes greater than order-1, then this becomes a "k-arm" competition. What Holland was also able to show is that the genetic algorithm can potentially allocate trials to hyperplanes in a near optimal fashion, allocating exponentially more trials to the observed best in numerous hyperplane competitions. The fact that this happens simultaneously for many different hyperplane competitions of different orders is referred to as ***implicit parallelism.***

Deception and the 2-armed bandit analogy directly relate to issues raised by Grefenstette and Baker (1989) concerning the role of implicit parallelism in genetic search. They point out that a genetic algorithm does not solve 2-armed (or k-armed) bandit competitions between hyperplanes because "the genetic algorithm does not perform uniform sampling from the hyperplanes in the population, at least not after the initial generation" (Grefenstette and Baker 1989: 24). As an example, they suggest the following assignment of fitness values for some optimization function:

$$f(x) = \begin{cases} 2 & \text{if } x \in 111*...* \\ 1 & \text{if } x \in 0***...* \\ 0 & \text{otherwise} \end{cases}$$

A uniform sample of strings will indicate $f(0*...*) > f(1*...*)$. However, as genetic search proceeds, the schema 111*...* dominates over its hyperplane competitors and thus, 1*...* will come to be represented in a greater proportion of the strings in the population than 0*...* because the population is no longer uniformly sampling the universe of all strings. Grefenstette and Baker suggest that this represents a flaw in the 2-armed bandit analogy because the competition between 0*...* and 1*...* is not solved correctly. They also point out that the genetic algorithm is not implicitly solving all possible hyperplane competitions correctly, which does in fact show that *implicit parallelism* does not act in a ubiquitous fashion. However, John Holland (personal communication) never intended that implicit parallelism should imply that *all* hyperplane competitions are simultaneously solved through genetic search, only that many competitions are occurring simultaneously.

Having established that the only challenging problems are deceptive problems, there is additional cause for being cautious when characterizing the nature of implicit parallelism. Grefenstette and Baker's point about the limitations of implicit parallelism has serious implications. To see why this is true, first note that Grefenstette and Baker's problem is a deceptive problem. The hyperplane 0*...* is deceptive with respect to the hyperplane competition involving 111*...*. It is not a fully deceptive problem (or subproblem): $f(*1*...*) > f(*0*...*)$ and $f(**1*...*) > f(**0*...*)$. Furthermore, the problem is not GA-hard since, as Grefenstette and Baker correctly point out, a genetic algorithm will quickly converge to strings that reside in the hyperplane represented by 111*...* (Das and Whitley 1991).

The occurrence of a deceptive problem will always produce at least two primary hyperplane competitions (which are analogous to two different k-arm bandit problems), which have solutions that involve different bit patterns. If a genetic algorithm effectively and consistently allocates exponentially more reproductive trials to the true global winner in one of these two hyperplane competitions, such that the schema receiving exponentially more trials becomes ubiquitous in the population *then the genetic algorithm will fail to correctly solve the alternate hyperplane competition.* The genetic algorithm cannot continue to allocate exponentially more trials to the observed best (assuming the observed best is in fact the global winner) in both hyperplane competitions, since the two global winners are incompatible and only one can become ubiquitous in the population.

CONJECTURE 1: *Assuming that sampling noise does not corrupt hyperplane evaluations, the 2-armed (or k-armed) bandit analogy fails to apply to a particular hyperplane competition if and only if deception is involved: a genetic algorithm can only correctly solve one of two competitions that involve a set of hyperplane competitors of order K and some relevant hyperplane competition of order N, where $K < N$ and the global winner of the competition at order K is deceptive with respect to the global winner of the order N competition.*

This conjecture is motivated by the following observations. If no deception occurs in a function optimization task, then implicit parallelism has an opportunity to solve all of the hyperplane competitions correctly and the 2-arm bandit analogy potentially holds. On the other hand, deception will alway create conflicting hyperplane competitions whose global winners have conflicting bit patterns. It may be possible that conflicting winners can be represented in the population in such a way that conflicting patterns can coexist. Such an arrangement, however, would appear to be inconsistent with a critical part of the 2-armed bandit analogy which indicates that the observed best should be allocated an exponential number of reproductive trials. It seems likely that solving all hyperplane competitions correctly would only be reasonable if no deception exists, and as shown, such problems could theoretically be solved without a genetic algorithm. It is disturbing that the same deception that makes global optimization non-trivial also works to undermine implicit parallelism in exactly those hyperplane competitions that would appear to be most critical. The impact of deception on critical hyperplane competitions strikes at the very heart of the theoretical foundations of genetic algorithm.

The empirical evidence suggests that the results of conflicting competitions can go either way. If the higher order schema representing the global winner in a fully deceptive building block has a long defining length, the lower order hyperplanes will typically win their k-armed bandit competitions and the higher order hyperplane will not. On the other hand, if the higher order schema representing the global winner of a fully deceptive building block has a tight linkage and is well represented in the chosen population size, the global winner of the higher order hyperplane competition typically wins its hyperplane competition and the deceptive lower order hyperplanes will not.

These observations lead to a more limited view of implicit parallelism. Many hyperplane competitions are being solved simultaneously so that the search is directed toward the global solution, but this need not imply that all hyperplane competitions are correctly processed by the genetic algorithm. Thus, we might expect that some deception need not be sufficient to mislead a genetic algorithm if the majority of the hyperplane competitions are resolved so as to lead toward global solutions. However, a much better understanding of implicit parallelism is needed to theoretically anchor genetic algorithms.

5 Constructing a Fully Deceptive Function

We now turn our attention to fully deceptive and consistently deceptive functions. Fully deceptive functions can be constructed using Walsh transforms (Goldberg 1989b, 1989c; c.f., Bethke 1980). However, we have found that arbitrary fully deceptive functions of any order greater than 2 can be constructed in the following fashion. (Liepins and Vose (1990) have independently developed a similar algorithm). Sort the binary strings in terms of their relative distance in Hamming space. When the strings are sorted in this fashion (with respect to their distance from the optimal solution or its complement) the distribution of strings into subgroups containing specific numbers of 1s and 0s is binomially distributed. For example, in an order-3 problem such a sorted sequence would be 111 followed by 011, 101 and 110, followed by 001, 010 and 100 and finally ending with 000. Note that there exists more than one such sequence: subgroupings having the same number of "1" bits such as 001, 010 and 100 can be arranged in any sequence. Once an ordering is chosen, number the strings 1 through N. String 1 will be the global optimum; string N will be the deceptive attractor, the complement of string 1. Assign string 2 some positive base value B (in the simplest case this can be zero). For any string X, where $X > 2$, the deceptive function F_d assigns values as follows: $F_d(X) = F_d(X-1)+C$, where C is any positive constant. Finally, the optimum is assigned a value: $F_d(1) = F_d(N)+C$.

The following fully deceptive order-4 function was constructed using the algorithm just outlined. In this case, the base value B is 0 and the constant C is 2. The problem is such that all lower order hyperplane information leads the search away from 1111 and towards 0000.

```
Deceptive Function 2

f(1111) = 30   f(0100) = 22   f(0110) = 14   f(1110) = 6
f(0000) = 28   f(1000) = 20   f(1001) = 12   f(1101) = 4
f(0001) = 26   f(0011) = 18   f(1010) = 10   f(1011) = 2
f(0010) = 24   f(0101) = 16   f(1100) = 8    f(0111) = 0
```

Note that every relevant schema of order 3 or less which is composed exclusively of 0s and *s is superior to all primary hyperplane competitors.

Deceptive Function 2, like all the artificial deceptive functions that have been constructed to date, has a deceptive attractor that is a local optimum in Hamming space since it is superior to any point adjacent to it in Hamming space. The reason that problems such as Deceptive Function 1 and Deceptive Function 2 are fully deceptive is clear: in Hamming space the deceptive optima have an "attracting basin" that covers the entire space except for the single spike that represents the true optimum. However, as will be shown, the deceptive attractors of fully deceptive functions need not always be local optima in Hamming space.

6 The Deceptive Attractor Theorem

The following theorem indicates that for any function mapping a set of strings to a set of values, the resulting assignment will not be fully deceptive at the particular order indicated unless there is a deceptive attractor which is the complement of the global optimum. However, as will be shown, this does not imply that the complement must have a high fitness value that is competitive with the global optimum. This theorem pertains not only to deceptive functions, but also "deceptive building blocks."

The term "attractor" suggests a dynamical system characterized by a stable attractor. The term "deceptive attractor" is used loosely here, but the evidence suggests that 1) if a problem is fully deceptive, 2) the deceptive attractor is the complement of the global optimum, and 3) recombination always disrupts the order-N information which identifies the global optimum, then the deceptive attractor will be a stable point for many populations that sufficiently sample the relevant deceptive hyperplanes at orders less than N. More specifically, if we define some idealized genetic algorithm where the search is dominated by the deceptive hyperplanes at order N-1 or less when search a fully deceptive problem of order N, then the deceptive attractor must be a stable point for any fully deceptive function: all of the order N-1 hyperplane competitions lead toward it (for a specific example, see Whitley 1991).

Nevertheless, the theorems that follows in the reminder of this paper do not rely on notions from dynamics systems, but rather are based on observations about the relationship of binary strings and hyperplanes in an N-dimensional hypercube. These definitions are therefore based on static relationships. If the genetic algorithm is modeled as a dynamical systems, these theorems may no longer hold, but they should at least offer a point of departure.

THEOREM 2: *In order for a function or building block of order N to be consistently deceptive in all relevant lower-order hyperplanes, the deceptive attractor must be the complement of the string which represents the global optimum in the deceptive function, or in the case of a deceptive building block, the deceptive attractor must be the complement of the schema representing the "global winner" of the relevant primary hyperplane competition at order N that is superior to all of its competitors.*

PROOF BY CONTRADICTION:

Assume a deceptive attractor of order N exists such that it shares at least one bit value in common with the "global winner" of the relevant hyperplane competition at order N. Choose one bit which has a value shared by the global winner and the deceptive attractor. Let G represent the bit that agrees with the global winner and let D represent the bit that agrees with the deceptive attractor. These are chosen so that these bits have the same value (0 or 1). The problem is not consistently deceptive at the specific order indicated unless $f(*...*G*...*) < f(*...*D*...*)$, but this is impossible, since the hyperplanes are represented by schemata that have the same bit and thus the hyperplanes have the same fitness.

Let C represent the complement of bits G and D (which are in fact the same bit). To be deceptive, it is necessary that $f(*...*G*...*) < f(*...*C*...*)$, but this also implies that $f(*...*D*...*) < f(*...*C*...*)$. Therefore, if the problem is deceptive in all hyperplanes, this hyperplane competition must be won by an order 1 schema containing the bit C and not D. Therefore the proposed deceptive problem with a deceptive attractor containing the bit D is not consistently deceptive in all hyperplanes at the particular order indicated.

QED

Expressed simply, the relevant order 1 hyperplane competitions can only agree with the global winner at order N or its complement. Therefore, to be fully or consistently deceptive, the only hyperplane that can function as a deceptive attractor is the complement of the global winner of the relevant hyperplane competition at order N. Again, this assumes a static view of the hyperplane relations. When a dynamical point of view is taken the situation becomes much more complex, but certain observation can be made. First, the theorem has stronger implications for fully deceptive functions, since all lower order hyperplanes lead to the same point; this should make it possible to make stronger claims about the status of the deceptive attractor as a true attractor in a dynamical system. Also, note that there will exist other attractors if the problem is not fully or consistently deceptive. There will also be other attractors if the lower order hyperplanes do not dominate the search (due to sampling bias or preservation of higher order schemata); in fact the global solution is one such competing attractor. Nevertheless, in a fully or consistently deceptive problem the order-1 hyperplanes must direct the search toward the deceptive attractor and this will have an impact on the other relevant hyperplane competitions. This leads to the following corollary, again defined from a static perspective.

COROLLARY 2.1: *If the deceptive attractor at order N is not the complement of the global winner of the relevant hyperplane competition at order N, then the problem is neither fully nor consistently deceptive.*

This follows from the proof of the deceptive attractor theorem since that proof considers hyperplanes *...*G*...*, *...*D*...* and *...*C*...*. Since these represent some set of hyperplanes with a single bit specified, if *...*D*...* is in fact not a deceptive hyperplane, then the problem is neither fully deceptive nor consistently deceptive.

7 The Deceptive Attractor and Local Optima

The above argument shows that the deceptive attractor must be the complement of the "global winner" of the relevant hyperplane competition at order N. However, limiting our consideration to deceptive functions for a moment, *it does not follow that the complement of the global winner of the hyperplane competition at order N will necessarily have a high fitness value.* The hyperplanes in a fully deceptive function can lead toward a "deceptive attractor" that is not a local optimum in Hamming space. The assertion also applies to deceptive building blocks, since we can construct such a problem by concatenating several deceptive functions where the deceptive attractor is not the local optimum. Consider the following example:

```
Deceptive Function 3

f(1111) = 30    f(0100) = 27    f(0110) = 5    f(1110) = 0
f(0000) = 10    f(1000) = 28    f(1001) = 5    f(1101) = 0
f(0001) = 25    f(0011) = 5     f(1010) = 5    f(1011) = 0
f(0010) = 26    f(0101) = 5     f(1100) = 5    f(0111) = 0
```

This function was constructed so as to be easy to understand. It is a fully deceptive function but the deceptive attractor has a fitness value that is only 1/3 that of the global optimum. The deceptive attractor is clearly not a local optimum in Hamming space because it is weaker than any of the strings adjacent to it in Hamming space. The deceptive attractor at order N must be hidden by strong strings located adjacent to it in Hamming space if the problem is to be fully deceptive. It turns out this factor is not as critical to consistently deceptive problems. In any case, the example given above shows that the deceptive attractor itself need not be a strong competitive string itself, but it must be cloaked by strong strings adjacent to in Hamming space to hide its weakness. But we can also make some observations about the fitness of the deceptive attractor relative to the strings (or relevant schemata) near it in Hamming space.

THEOREM 3: *A deceptive attractor of order-N for a binary encoded problem cannot maintain full deception at order-N if it is weaker than any string or schema which differs from the deceptive attractor by exactly two bits.*

PROOF:

Consider a fully deceptive problem of order N. Let Z be a binary string (or schema) with bit values of 0 representing the deceptive attractor and $f(Z)$ the value associated with the deceptive attractor. Let X be the string (or schema) that differs from Z by two bits (i.e., it has two 1 bits) and which also has a higher value than any other string that differs from the deceptive attractor by two bits. If it is shown that $f(Z)$

must be greater that *f(X)*, then *f(Z)* must be greater than the value of any string that differs by exactly 2 bits from the deceptive attractor.

Two other strings (or schemata) are relevant. *S1* and *S2* are two unique strings (or schemata) that differ by one bit from both the deceptive attractor and the string (schema) *X*. In other words, they both have a single 1 bit. Assign the names to these strings so that *S1* represents the string (or schema) that has a fitness value greater or equal to *S2*. This also implies a hyperplane competition between these four strings (or schemata) involving schemata of order N-1. Assume *S1* and *X* share a bit we call B1; Also, *S2* and *X* share a different bit in common we will call B2. String *S1* and *X* compete against *S2* and *Z* when we consider schemata of order N-1 that replaces the bit B2 by a don't care operator, *. The remainder of the proof involves only simple algebra. The deceptive attractor theorem and the definition of full deception implies:

$$f(Z) + f(S2) > f(S1) + f(X)$$

which implies

$$f(Z) > f(S1) - f(S2) + f(X).$$

Since $f(S1) \geq f(S2)$, let $f(S1) - f(S2) = K$, where K is a non-negative integer. This implies:

$$f(Z) > f(X) + K, \quad \textbf{therefore } f(Z) > f(X).$$

QED

The obvious difficulty with maintaining full deception at order N occurs in the order N-1 hyperplanes involved. To make the proof more concrete, consider some order-4 deceptive function. Let 0000 be *Z*, let 1001 be *X*, and let 1000 and 0001 be the two uniquely defined strings that differ by one bit from both the deceptive attractor and the string 1001. Pick either string as having the higher fitness value; for illustration purposes assume $f(1000) > f(0001)$. This also implies a relevant hyperplane competition between schemata of order N-1. Full deception implies $f(000*) > f(100*)$, which implies

$$f(0000) + f(0001) > f(1000) + f(1001).$$

The reader can do the remaining math.

COROLLARY 3.1: *In a fully deceptive function, if the deceptive attractor does not reside at a local optimum in Hamming space, a local optimum must exist at some string that is adjacent to the deceptive attractor in Hamming space. In other words, either the deceptive attractor or some string that differs from the deceptive attractor by 1 bit must be a local optimum in Hamming space.*

This follows from Theorem 2. If the deceptive attractor must have a higher fitness value than any string that differs from the deceptive attractor by exactly 2 bits, then the deceptive attractor (and the strings that differ from the deceptive attractor by a single bit) are surrounded by points in Hamming space that are known to be weaker than the deceptive attractor. This implies that either the deceptive attractor is a local optimum, or one of the strings that differs from the deceptive attractor by a single bit is a local optimum in Hamming space.

We now look at a *consistently deceptive problem* that is not fully deceptive. While the requirements for full deception make it necessary that the deceptive attractor not have too low a fitness value, the order N-1 hyperplanes (e.g., f(000*) versus f(100*) in the above example) where deception breaks down first are not necessarily those that we would expect to be most important to genetic search. If we relax the requirements for deception at some order N, we can build problems which are consistently deceptive but not fully deceptive; of particular interest are problems that are consistently deceptive where many, but not all of the lower-order hyperplanes lead toward the deceptive attractor, but where none lead toward the global winner of the hyperplane competition at order N. Using the set of values given in Deceptive Function 3, if the value of the string "0000" is changed from 10 to 7, the problem is no longer fully deceptive because $f(000*) < f(100*)$, $f(000*) = f(010*)$ and finally $f(00 * 0) = f(10 * 0)$, but all other hyperplane information still leads toward 0000, the deceptive attractor. However in such cases it still makes sense to talk about order-4 deception that is less than fully deceptive, because the order-4 deceptive problem does not decomposed into an order-3 deceptive problem. In fact, several order-3 hyperplanes continue to be deceptive with respect to the correct order-4 building block. The hyperplanes *000 or 0*00 continue to have a higher fitness value than all of their competitors and therefore continue to support deception at the order-4 level. Even though the function is not fully deceptive since all of the lower order hyperplanes do not lead toward the deceptive attractor, the problem is nevertheless misleading in all hyperplanes. In none of the hyperplane competitions does a hyperplane win that directs the search toward the correct building block. We have reduced the degree of deception, but there are shades of gray between fully deceptive order-3 problems and fully deceptive order-4 problems. It is this kind of situation that motivated the distinction between fully deceptive problems and consistently deceptive problems. At this point there is no empirical evidence regarding the difficulty of consistently deceptive problems.

8 The Deceptive Attractor Remapping Strategies

What practical significance does all this have? Knowing that the deceptive attractor must be the complement of the global winner is useful if deception is occurring in a known or predictable location (ie: within a parameter setting); problem recodings that significantly change the Hamming distance between the global winner and the deceptive attractor of a hyperplane competition will reduce the degree of deception that currently exists. Small changes may not reduce the level of deception; for example, in Deceptive Function 3 reversing the mapping between 1000 and 0000 does not reduce the level of deception–in fact it strengthens the deception. But remapping many of the strings so that the "global winner" is moved closer in Hamming space to the deceptive attractor *and those strings that help to maintain the deception* will in many cases reduce the level of deception.

It would also seem possible that a remapping could create deception in other parts of the search space, although the possibility of this happening does not imply anything concerning the likelihood of such an event. The problem with recoding or remapping strategies is that knowing the exact location of the deception is not in general practical, since it involves finding a binary mask indicating the locations where a particular case of deception is occurring; *finding this binary mask involves a search space as large as the function optimization space.* It is also unclear if a single mask for deception would suffice: is it necessary to know how many distinct cases of deception exist and where they are located? The difficulty of locatings the bits that compose a deceptive building block problem also precludes such simple fixes for deception as inverting the final solution or inverting strings during search and evaluating their complements. This would appear to be bad news for simple remapping strategies, unless we can simply remap arbitrary portions of the encodings and hope that we somehow manage to remap part of the deception.

9 Deception and Linkage Problems

Deception is only part of the problem. The other factor that works to make a problem GA-hard involves the "linkage" between bits in a deceptive building block. Deception is a more serious problem if it occurs across some random (and therefore, unknown) combination of bits in the encoding. Having a poor linkage implies that the deceptive bits are highly separated on the encoding. Hyperplanes represented by bits that are distributed over a long encoding are inadequately sampled by the genetic algorithm because of a higher rate of disruption during recombination. Put more simply, if the deceptive bits are close together, then strings that sample some favorable combination of bits are more likely to pass on appropriate schemata to their offspring intact (because crossover disrupts critical schemata less often) and the genetic algorithm will not be misguided. Consider when deception occurs in a building block of 3 bits. We would expect the correct combination (111) to occur 12.5% of the time in the initial random population; if the critical bits tend to be passed on as a single block, then they will spread in the population and the deception that occurs in the lower order hyperplanes will not have a significant impact on search. However, if the same 3 bits are highly separated on the encoding, they are less likely to be inherited intact by offspring (due to the disruptive effects of crossover) and thus, the lower level hyperplane information becomes critical. When this happens, the genetic algorithm is likely to converge toward an incorrect solution.

Consider Deceptive Function 1. Also, assume the 3 bits that make up Deceptive Function 1 are highly separated in a binary encoding made up of several deceptive functions. As the competition between 000 (with an evaluation of 28) and 111 (with and evaluation of 30) heats up, there is not enough selective pressure to overcome the disruptive effects of recombination. Thus, the competition falls back to the level of the relevant lower order hyperplanes, and here 000 wins because of the deception built into the problem.

To build a suitable test function that is likely to be difficult for a genetic algorithm to solve, Goldberg, Korb and Deb (1989) defined a 30 bit function made up of 10 copies of a fully deceptive 3-bit function. The problem is difficult (or "ugly") when the bits are arranged so that each of the 3 bits of the subfunctions are uniformly and maximally distributed across the encoding. Thus each 3 bit subfunction, i, has bits located at positions i, i+10 and i+20. Goldberg et al. show that a standard genetic algorithm consistently converges to an incorrect solution on this problem, with each subfunction converging to 000 instead of 111. Again, this is consistent with the notion of hyperplane sampling, and biases known to exist in genetic algorithms against schemata with long defining lengths.

10 A Description of the Experiments

All of the following experiments used population sizes of 200 or 2000 as indicated. The experiments with population sizes of 200 used a total of 10,000 evaluations while the experiments with population sizes of 2000 used a total of 50,000 evaluations. These results do not necessarily represent our "best efforts" at solving these problems; rather, they are intended as comparative results. A rank based selection scheme was used with a linear selective bias of 1.5. Test problems include "ugly" versions of fully deceptive order-3 problems made up of Deceptive Function 1. A "random" version of the problem has also been defined by Goldberg, Korb and Deb (1989) so that the bits of the 10 subproblems are distributed randomly, such that some of the subproblems are highly distributed across the string of 30 bits, while other subproblems have bits that are closer together in the encoding. Since each test problem is actually composed of 10 fully deceptive subproblems, results are reported in terms of the number of subproblems correctly solved.

Additional tests were carried out comparing uniform crossover and 1-point crossover on an "ugly" fully deceptive order-4 problem. We constructed a 40 bit function composed of 10 4-bit fully deceptive subfunctions (specifically, Deceptive Function 2), with the 10 subfunctions distributed at positions i, i+10, i+20, i+30, for i = 1 to 10. No tests were done using a "random" arrangement of bits on this problem.

All of the reported results are based on 30 independent runs of the genetic algorithm. The genetic algorithm used was GENITOR which keeps copies of the best strings found so far. Because of this, GENITOR produces more selective pressure than would be the case with a standard genetic algorithm using a linear selective bias (Goldberg and Deb 1991). The additional selective pressure and conservation of previous "best" strings allows it to solve some of the deceptive subproblems.

10.1 GENITOR With 1-point Crossover

The 1-point crossover operator was a reduced surrogate operator. No mutation was used in these experiments. These experiments provide a baseline for comparative purposes, since they represent the operation of the algorithm in its normal mode of operation. Results for this and other operators are given in Table 1.

10.2 GENITOR With Uniform Crossover

Uniform crossover involves choosing each bit independently from the two parents, producing a random assortment of bits from the two parents. The potential advantage is that now the location of the bits on the encoding is irrelevant; the "ugly" version of the problems is exactly the same as the "random" version of the problem. Thus, there is no bias against schemata with long defining lengths. On the other hand, the advantages of tight linkage (i.e., functionally related bits that are close in the encoding have a tight linkage) are lost, since they no longer can be inherited as a block. One might expect uniform crossover to be competitive on problems such as the "ugly" deceptive problem, since there is not a bias against schemata with long defining length. But when this is considered more carefully, it seems clear that uniform crossover breaks apart higher level schemata during recombination and thereby shifts the burden of search to the order-1 and order-2 schemata. If the problem is deceptive, this action will increase the probability that the deceptive hyperplanes will be able to mislead the genetic algorithm. This suggests that Syswerda's tests (1989) using uniform crossover were not sufficiently deceptive to make them challenging tests problems (Das and Whitley 1991).

10.3 Tagged Bits and Random Orderings

The difference between the results on the "ugly" and "random" versions of the fully deceptive order-3 problem clearly illustrates that the location of the bits composing the subproblems in the overall problem encoding is important. This implies that using "tags" to randomize the location of the specific bits in the overall encodings might be effective in improving the performance of the genetic algorithm. In the "random" version of the test function, there is one fixed random ordering that is used by all strings. When tags are used each string has its own individual ordering. Normally, the interpretation of a binary string is positionally dependent. If we tag the bits, for example, so that the eighth bit in some binary strings is denoted (8 0), then we know the eighth bit is a "0" regardless of its location. The binary 11011001 for example, could be represented using the following LISP-like notation

((1 1) (2 1) (3 0) (4 1) (5 1) (6 0) (7 0) (8 1))

This kind of scheme can of course be implemented any number of ways; the point is that the bits can now be allowed to move and different "linkages" can developed in different strings.

The key issues in the implementation used here is that in the initial population both the bit strings and the sequence of bit tags were randomly generated. This means that each string started with its own randomly generated set of tags. In order for recombination to occur, two parent strings must align so that they have the same sequence of bits. To accomplish this one parent is randomly chosen to be the "template" and the other parent is rearranged so that its sequence of bits match this template. Recombination is then applied using 1-point crossover. Also, note that the offspring inherits the same sequence of bit tags as the parent providing the template. Thus, as search proceeds, the distribution of bit tag sequences will not remain random.

Goldberg and Bridges (1990) have also considered the use of tagged bits; the main difference between the work presented here and previous work is that the genetic search presented here began with a random distribution of different bit tag sequences, whereas in Goldberg's studies (which were in part analytical), it was assumed that the population started with all strings having the same sequence and inversion was applied to strings in an effort to try to improve the linkage.

As is the case with uniform crossover, using a tagged bit representation means that there is no difference between the "ugly" and "random" versions of this problem. The tagged bit scheme consistently produced the best results.

10.4 A Distributed Genetic Algorithm

A distributed genetic algorithm was used that employed 10 subpopulations. Each subpopulation contained 200 strings and was allowed 5,000 evaluations; these restrictions made the total amount of work comparable to the runs using a single population of 2000 and a total of 50,000 evaluations. On the "ugly" fully deceptive order-3 problem it solved 55% of the subproblems correctly. In other tests we had allowed each subpopulation of 200 strings a total of 50,000 recombinations; given the additional recombinations it solved 100% of the subproblems. The success of the distributed algorithm lies in the fact that each subpopulation solves a different subset of the deceptive subproblems; migration of strings between subpopulations allows these different solutions to be recombined, thereby solving all of the subproblems. We should also note however, that the distributed algorithm used a mutation operator and the serial algorithm did not; this could also account for part of the results.

In related experiments we used 1, 2, 5, 10, 50 and 100 subpopulations with and without mutation. The total population size and number of recombinations was held constant in the various experiments; the single population was composed of 5000 strings and allowed 500,000 recombinations while the run with 100 subpopulations, where each subpopulation had 50 members (100 * 50 = 5000) and was allocated 5000 recombinations (100 * 5000). Both the runs with and without mutation showed the same trend: the more distributed the population, the better the results on ugly deceptive problems (Starkweather, Whitley and Mathias 1991).

11 Summary of Experimental Results

The following table summarizes the experimental results. The test problems are the ugly order 3 fully deceptive problem (Ugly,O-3), the random version of the same problem (Random,O-3), of the ugly order 4 fully deceptive problem (Ugly,O-4). The "1-point" versions are direct application of GENITOR, while the "Tagged" and "Uniform," versions should be self explanatory. The "Parallel" version used 10 subpopulations of 200 (*total population, 2000) and 5000 evaluations in subpopulation (*total evaluations, 50,000).

Problem	Approach	Pop Size	Solved	Evals
Ugly,O-3	1-point	200	27%	10,000
Ugly,O-3	Uniform	200	27%	10,000
Random,O-3	1-point	200	46%	10,000
Ugly,O-3	Tagged	200	53%	10,000
Ugly,O-3	Uniform	2000	35%	50,000
Ugly,O-3	1-point	2000	38%	50,000
Random,O-3	1-point	2000	52%	50,000
Ugly,O-3	Parallel	2000*	55%	50,000*
Ugly,O-3	Tagged	2000	64%	50,000
Ugly,O-4	Uniform	200	3%	10,000
Ugly,O-4	1-point	200	7%	10,000
Ugly,O-4	Tagged	200	16%	10,000

TABLE 1: Summary of Results

While we have not done any statistical tests on this data, three hypotheses are suggested by these results. First, the use of "tagged bits" appears to help when the initial linkage represents a worse case situation. Second, uniform crossover does not outperform 1-point crossover in this domain and may be inferior. Third, a distributed algorithm produces considerable improvement in performance. However, one should be careful not to generalize from these results; these are highly artificial problems and may not be typical of other optimization problems.

12 Conclusions and Future Directions

The theoretical results and empirical tests offered in this paper are meant to 1) stimulate further debate on the issue of deception and its relationship to hyperplane sampling and implicit parallelism and 2) to establish a foundation for further research. The paper has probably raised more questions than have been answered.

There has been a great deal of debate recently in the genetic algorithm community about the problems we use as test functions. What does it really mean when one variant of genetic algorithm outperforms another on some test suite? The theoretical results presented in this paper concerning deception and the 2-armed bandit analogy strongly support a position which has been advocated by David Goldberg: test problems must be used that involve some degree of deception. This also places increased importance on work relating Walsh transforms to genetic algorithm and deception (Goldberg 1989b; 1989c). This does not imply, however, that the only reasonable test problems are "ugly" fully deceptive problems. These problems are highly artificial in two ways: 1) they decompose into subproblems that can be solved independently and 2) deception occurs in every position in every subproblem so that the complement of the "deceptive solution" is always the global solution. I would very strongly argue that methods which utilize information about *where* the deception exists will only solve these very artificial problems–however, in general this information is not available.

We have already obtained empirical evidence that simple problems which are not deceptive in the order-1 hyperplanes can be solved by solving the order-1 hyperplanes (Das and Whitley 1991). The methods appear to be faster than genetic search (since much less information is being processed) and robust even when given noisy hyperplane sampling. Theorem 1 on solving nondeceptive problem by solving the order-1 hyperplane competitions is somewhat idealistic (since it assumes no sampling noise and accurate determination of order-1 winners) but nevertheless points toward a workable method for deciding if problems are significantly deceptive. This potentially represents an important contribution to the field of genetic algorithms.

This paper has established a description of various kinds of deception. The concepts of "fully deceptive" and "consistently deceptive" also suggest a new need for empirical tests to evaluate the difficultly of problems displaying different degrees of deception. Are problems that are consistently deceptive as hard or nearly as hard as fully deceptive problems? Are fully deceptive problems as hard to solve if the deceptive attractor is not a local optimum in Hamming space and does not have a fitness value competitive with the correct building block? We are already working on the answers to these questions.

In general, the experiments described in this paper are only a first step toward arriving at a better understanding of deceptive problems. Our results suggest that the use of "tagged bits" might be helpful in solving the linkage problems. Other have argued against the use of "tagged bits" because finding the optimal sequence of bits while also finding the optimal bit values greatly increases the size of the search space (Liepins and Vose 1991; Goldberg and Bridges 1990). However, it may be that we do not need to find optimal sequences, but rather just avoid particularly "bad" or "ugly" sequences. On the other hand, these test problem started with the worst bit arrangement we could devise; on real problems where parameters are coded as a contiguous block of bits the use of tagged bits may actually make the initial arrangement worse, not better. Thus, I would argue that this issue is still open and no firm conclusions should be drawn at this time.

Approaches that have *not* been examined include Goldberg, Korb and Deb's (1989) "Messy GAs" which were designed in part to handle deceptive problems; also not examined are crowding techniques. Crowding techniques would seem like a natural solution to deception, since crowding forces the development of "niches" based on the similarity of the binary strings competing for space in a population. Given the results presented in this paper concerning the relationship of a deceptive attractor to the relevant global winner of a hyperplane competition, we might expect the global winner to occupy a different niche than that occupied by the deceptive attractor and the strings or schemata close to the deceptive attractor in Hamming space. In some sense the results using the distributed genetic algorithm also is produced by a kind of "niching," but one based implicitly on genetic drift and sampling errors rather than explicit forced niching.

Acknowledgements

My thanks to David Goldberg for insightful comments on an earlier draft; the remaining flaws are, of course, my own. The research was supported in part by a grant from the National Science Foundation, grant number IRI-9010546, and by a grant from the Colorado Institute of Artificial Intelligence (CIAI). CIAI is sponsored in part by the Colorado Advanced Technology Institute (CATI), an agency of the State of Colorado. CATI promotes advanced technology education and research at universities in Colorado for the purpose of economic development.

References

Bethke, A. (1980) Genetic Algorithms as Function Optimizers. Ph.D. Dissertation, Computer and Communication Sciences, University of Michigan, Ann Arbor.

Das, R. and Whitley, D. (1991) The Only Challenging Problems are Deceptive: Global Optimization by Solving Order-1 Hyperplanes. To appear: *Fourth International Conf. on Genetic Algorithms.*

Goldberg, D. (1987) Simple Genetic Algorithms and the Minimal, Deceptive Problem. In, *Genetic Algorithms and Simulated Annealing,* L. Davis, ed., pp: 74-88, Morgan Kaufmann, Pubs.

Goldberg, D. (1989a) *Genetic Algorithms in Search, Optimization and Machine Learning.* Reading, MA: Addison-Wesley.

Goldberg, D. (1989b) Genetic Algorithms and Walsh Functions: Part I, A Gentle Introduction. *Complex Systems* 3:129-152.

Goldberg, D. (1989c) Genetic Algorithms and Walsh Functions: Part II, Deception and its Analysis. *Complex Systems* 3:153-171.

Goldberg, D. (1990) Construction of High-order Deceptive Functions Using Low-order Walsh Coefficients. IlliGAL Report No. 90002. Department of General Eng. Univ. of Illinois at Urbana-Champaign.

Goldberg, D. and Bridges, C. (1990) An Analysis of a Reordering Operator on a GA-Hard Problem. *Biological Cybernetics* 62: 397-405.

Goldberg, D. and Deb, K. (1991) A Comparative Analysis of Selection Schemes. *Foundation of Genetic Algorithms,* G. Rawlins, ed. Morgan Kaufmann.

Goldberg, D., Korb, B., and Deb, K. (1989) Messy Genetic Algorithms: Motivation, Analysis, and First Results. *Complex Systems* 4:415-444.

Goldberg, D. and Rudnick, M. (1991) Genetic Algorithms and the Variance of Fitness. IlliGAL Report No. 91001. Department of General Eng. Univ. of Illinois at Urbana-Champaign.

Grefenstette, J. and Baker, J. (1989) How genetic algorithms work: a critical look at implicit parallelism. *Proceeding of the Third International Conference on Genetic Algorithms, 1989.* Washington, D.C., Morgan Kaufmann, Publishers.

Holland, J. H. (1975) *Adaptation in Natural and Artificial Systems.* Ann Arbor, MI: University of Michigan Press.

Liepins, G. and Vose, M. (1990) Representation Issues in Genetic Optimization. *J. Experimental and Theoretical Art. Intell.* 2(1990)101-115.

Liepins, G. and Vose, M. (1991) Deceptiveness and Genetic Algorithm Dynamics. *Foundation of Genetic Algorithms,* G. Rawlins, ed. Morgan Kaufmann.

Starkweather, T. Whitley, D., and Mathias, K. (1991) Optimization Using Distributed Genetic Algorithms *Parallel Problem Solving from Nature,* Springer-Verlag, Publishers.

Syswerda, G. (1989) Uniform Crossover in Genetic Algorithms. *Proc. Third International Conf. on Genetic Algorithms,* Morgan Kaufmann, Publishers.

Tanese, R. (1989) Distributed Genetic Algorithms. *Proc. Third International Conf. on Genetic Algorithms,* Morgan Kaufmann, Publishers.

Whitley, D. (1991) Deception, Dominance and Implicit Parallelism in Genetic Search. Technical Report. Department of Computer Science, Colorado State University.

Isomorphisms Of Genetic Algorithms

David L. Battle & Michael D. Vose
Computer Science Dept., 107 Ayres Hall
The University of Tennessee
Knoxville, TN 37996-1301
(615) 974-5067 @cs.utk.edu

Abstract

We begin with the premise that the role of Holland schemata in directing genetic search should be granted both as a matter of empirical fact and as a natural consequence of the Schema Theorem. From this we conclude that schemata more general than Holland's can also be made to direct genetic search, and that a duality exists between problem representations and which schemata are relevant for their optimization. This duality provides a theoretical framework in which to interpret problem representations.

Keywords: schema, representations.

1 Introduction

It is a common practice to choose among alternate representations. For example, consider the objective function

$$f(x) = 4 + \frac{11}{6}x - 4x^2 + \frac{7}{6}x^3$$

It may readily be checked that f is a type – II minimal deceptive problem (Goldberg 1989) over the integer interval $[0, 3]$ when integers are given their natural binary encoding:

string	fitness
00	4
01	3
10	1
11	5

However, if a Gray code is used then the problem gets easier; it becomes type – I:

value	string	fitness
0	00	4
1	01	3
2	11	1
3	10	5

The point is that values in the domain of f may be represented by binary strings in several ways, and some are better than others (we do not suggest that Gray coding is on average better than natural coding; it is an instructive exercise to prove that all *fixed* coding schemes do equally well when *all* functions are averaged over).

It was noticed by Liepins and Vose (1990) that an entire class of deceptive problems would be totally easy for genetic optimization if an appropriate representation were used. Of course, this by itself is trivial; the nontrivial result of Liepins and Vose is that a suitable representation is induced by a matrix.

Matrices over the finite field of integers mod 2 are quite powerful as a means of altering representation. Since permutation matrices have binary entries, it follows that the inversions studied by several researchers[1] are implementable by matrices. In fact, Gray coding is also induced by a suitable matrix (see appendix).

From one perspective, a change in representation produces an alternate function to optimize. However, this paper develops an alternate interpretation, which we refer to as *the duality principle*, which explains representational transformations and their effects in terms of schemata.

2 Schemata

J.H. Holland (1975) has defined a schema to be a subset of strings with similarities at certain string positions[2] and has used them to analyze the behavior of Genetic Algorithms (GAs). For example, consider the two strings

011

111

which are identical in the second and third positions. Regarding $*$ as a symbol which may be interpreted as either 0 or 1, these two strings are represented by

$*11$

In this way, strings over the alphabet $\{*, 0, 1\}$ correspond to schemata of Holland.

The schema $*11$ can be thought of as a function which maps binary strings of length three into the set {true, false} according to:

$$*11\,(x) = \text{true} \quad \text{iff } x \text{ matches } *11 \text{ at every position not containing } *$$

[1] Bagley (1967), Cavicchio (1970), Frantz (1972), Goldberg & Bridges (1988), etc.

[2] In this paper we are only concerned with fixed length binary strings.

The most general definition of a schema might therefore be an arbitrary predicate. This abstract approach has been developed by Vose (1991) who has shown that the fundamental theorem of Genetic Algorithms (the Schemata Theorem) generalizes to predicates. Our paper also generalizes Holland schemata but in a way which preserves their algebraic structure. The schemata we introduce may be thought of as a concrete example (or special case) of those investigated by Vose.

Let Ω be the set of all length ℓ binary strings, and let $N = 2^\ell$. We regard Ω as the vector space

$$\mathcal{Z}_2 \times \ldots \times \mathcal{Z}_2$$

where $\mathcal{Z}_2$ denotes the finite field of integers modulo 2. The standard basis vectors are $\{u_1, \ldots, u_N\}$ where u_j differs from the zero vector in that its j th component is 1.

Note that, for $\ell = 3$, the subspace generated by u_1 is the set of vectors $\{000, 100\}$. Translating this subspace by the vector 011 produces a coset which is identical to the set of vectors represented by the Holland schema $*11$. In general, Holland schemata correspond to cosets of those subspaces generated by subsets of the basis vectors $\{u_1, \ldots, u_N\}$.

We use this observation to represent Holland schemata with a matrix. Let M be an invertible ℓ by ℓ matrix[3] and let $\mathcal{S}_M$ denote the collection of cosets of those subspaces generated by subsets of the column vectors of M. In this notation, Holland schemata correspond to $\mathcal{S}_M$ when M is the identity matrix I.

3 Operators And Corresponding Schemata

Let $\mathcal{R}$ denote an operator defined over Ω such as fitness, crossover, or mutation, and let $\mathcal{R}_M$ be essentially the composition $\mathcal{R}_M = M \circ \mathcal{R} \circ M^{-1}$. More precisely, $\mathcal{R}_M$ represents the operator obtained by:

1. Mapping the arguments of $\mathcal{R}$ by M^{-1}
2. Applying $\mathcal{R}$
3. When the results are vectors, mapping them by M

Consider a GA with t th generation P_t, fitness function f, crossover operator χ, and mutation operator μ. The growth or decline in the population of a particular schema from $\mathcal{S}_I$ may be viewed as a consequence of an operator $\mathcal{G}$ (encompassing f, χ, and μ) which moves the GA from one generation to the next. The corresponding operator for the GA with fitness function f_M, crossover operator χ_M and mutation operator μ_M is $\mathcal{G}_M$.

Theorem 1 The influence of $\mathcal{G}$ on $\mathcal{S}_I$ is identical to the influence of $\mathcal{G}_M$ on $M\,\mathcal{S}_I$.

Proof: From the commutativity of the diagram

[3] All of our matrices have entries in $\mathcal{Z}_2$.

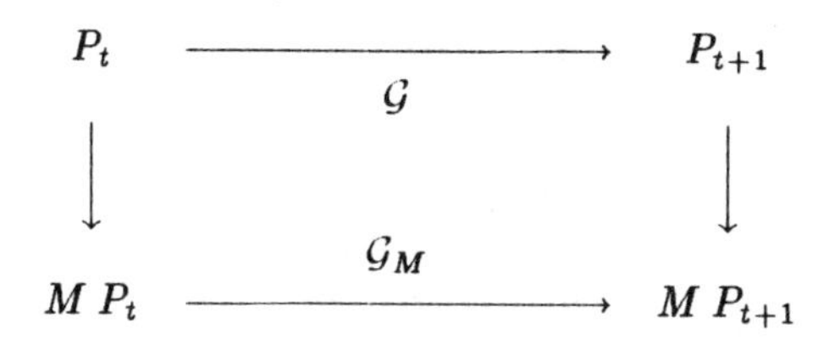

it follows that the action of $\mathcal{G}$ on P_t is isomorphic via M to the action of $\mathcal{G}_M$ on $M\,P_t$. □

Theorem 2 If L is invertible, then $L\,\mathcal{S}_M = \mathcal{S}_{LM}$

Proof: Let $\{m_{i_1}, \ldots m_{i_k}\}$ be a subset of the columns of M, let the subspace they generate be denoted by $< m_{i_1}, \ldots m_{i_k} >$, and let $s \in \Omega$. A typical element of $L\,\mathcal{S}_M$ is

$$\begin{aligned} L\,\{<m_{i_1}, \ldots m_{i_k}> + s\} &= <Lm_{i_1}, \ldots Lm_{i_k}> + L\,s \\ &= <c_{i_1}, \ldots c_{i_k}> + L\,s \end{aligned}$$

where $\{c_{i_1}, \ldots c_{i_k}\}$ are columns of LM. Hence $L\,\mathcal{S}_M \subseteq \mathcal{S}_{LM}$. It follows that

$$L\,\mathcal{S}_M \subseteq \mathcal{S}_{LM} = LL^{-1}\mathcal{S}_{LM} \subseteq L\,\mathcal{S}_M$$ □

Theorem 3 The schemata $\mathcal{S}_M$ satisfy the Schemata Theorem when the genetic operators χ_M and μ_M are used.

Proof: Let $s_I \in \mathcal{S}_I$ be a Holland schema. By Theorem 2, there ia s schema $s_M \in \mathcal{S}_M$ corresponding to it,

$$s_M = M\,s_I$$

Note that

$$\sum_{x \in s_I \cap P_t} f(x) = \sum_{x \in M(s_I \cap P_t)} f(M^{-1}x) = \sum_{x \in s_M \cap M P_t} f_M(x)$$

Hence the utility of the Holland schema s_I with respect to the GA $\mathcal{G}$ is the same as the utility of the schema s_M with respect to the GA $\mathcal{G}_M$. Moreover, Theorem 1 implies that $|\, s_I \cap P_{t+1} \,| = |\, s_M \cap M P_{t+1} \,|$. Since the Schemata Theorem relates the utility of a schema s_I to its representation in the next generation, the diagram

$$\sum_{x \in s_I \cap P_t} f(x) \qquad\qquad |\, s_I \cap P_{t+1} \,|$$

$$\updownarrow \qquad \longleftarrow \text{Schemata Theorem} \longrightarrow \qquad \updownarrow$$

$$\sum_{x \in s_M \cap M P_t} f_M(x) \qquad\qquad |\, s_M \cap M P_{t+1} \,|$$

shows the identical relationship holds with respect to the schema s_M and the GA $\mathcal{G}_M$. □

Theorems 1 2 and 3 show that whatever relation Holland schemata ($\mathcal{S}_I$) have to the direction of genetic search corresponding to $\mathcal{G}$ (i.e., $\mathcal{G}_I$), the same relation holds between the schemata $\mathcal{S}_M$ and the GA corresponding to $\mathcal{G}_M$. Abusing our notation slightly, we say *the isomorphism M transfers the suitability of the Holland schemata $\mathcal{S}_I$ for the GA $\mathcal{G}_I$, to the suitability of $\mathcal{S}_M$ for the GA $\mathcal{G}_M$.*

One of the open questions in genetic algorithm research has been whether there are subsets other than Holland schemata that, under appropriate genetic operators, behave according to the Schemata Theorem. Vose (1991) had answered this question abstractly by generalizing the schemata theorem to arbitrary predicates. We have answered this question concretely by using an isomorphism (M) to exhibit explicit schemata ($\mathcal{S}_M$) and their corresponding operators ($\mathcal{G}_M$).

4 Duality

Continuing our abuse of notation, let $\mathcal{G}$ be a GA with random initial population P and fitness function f. Consider the influence of changing the objective function to $h = f \circ M$.

From one point of view, this change simply produces $\mathcal{G}'$, a GA which optimizes a different function. However, the isomorphism between $\mathcal{G}'$ and $\mathcal{G}'_M$ provides an alternate interpretation. The fitness function of $\mathcal{G}'_M$ is

$$h_M = h \circ M^{-1} = f \circ M \circ M^{-1} = f$$

Moreover, since the initial population P of $\mathcal{G}$ was random, so also is the initial population MP of $\mathcal{G}'_M$. The only difference between $\mathcal{G}$ and $\mathcal{G}'_M$ is in their crossover and mutation operators. In other words, the difference is in their schemata; $\mathcal{G}$ explores f with respect to the schemata $\mathcal{S}_I$, while $\mathcal{G}'_M$ explores f with respect to the schemata $\mathcal{S}_M$.

If we regard isomorphic GAs as being the same, then the isomorphism M between $\mathcal{G}'$ and $\mathcal{G}'_M$ proves the following:

Theorem 4 Genetic optimization of $f \circ M$ when guided by $\mathcal{S}_I$ (Holland schemata) is equivalent to the genetic optimization of f if directed by $\mathcal{S}_M$ (the images under M of Holland schemata).

This duality between changing objective functions and changing schemata sheds light on the representational transformations proposed by Liepins and Vose (1990). They regard f and h as the same function but with different binary encodings, and view the matrix M as altering its representation. Our duality principle shows their choice of representation matrix M is equivalent to a choice of schemata $\mathcal{S}_M$.

For example, the function

String	Fitness
000	29
001	26
010	22
011	0
100	14
101	1
110	2
111	30

is completely deceptive with respect to $\mathcal{S}_I$, but computing the utilities of $\mathcal{S}_M$ for the matrix

$$M = \begin{pmatrix} 1 & 1 & 1 \\ 0 & 1 & 1 \\ 1 & 0 & 1 \end{pmatrix}$$

gives

Schema	Utility	Schema	Utility
{111, 001, 010, 100}	92	{011, 100}	14
{000, 110, 101, 011}	32	{101, 010}	23
		{110, 001}	28
		{000, 111}	59
{110, 001, 011, 100}	42	{010, 100}	36
{000, 111, 101, 010}	82	{101, 011}	1
		{111, 001}	56
		{000, 110}	31
{101, 010, 011, 100}	37	{001, 100}	40
{000, 111, 110, 001}	87	{110, 011}	2
		{111, 010}	52
		{000, 101}	30

Since each schema containing 111 has greatest utility, using the schemata $\mathcal{S}_M$ to direct genetic search optimizes f easily. By duality, arranging for $\mathcal{S}_M$ to direct genetic search is equivalent to optimizing $f \circ M$.

5 Conclusion

We have generalized Holland schemata in a manner which preserves algebraic structure. The resulting schemata, $\mathcal{S}_M$, correspond to cosets of subspaces generated by the columns of an invertible matrix M.

We have demonstrated that the schemata $\mathcal{S}_M$ behave according to the Schemata Theorem under appropriate genetic operators by using an isomorphism M to show how their properties may be inferred from those of the Holland schemata.

The correspondence we develop between schemata $\mathcal{S}_M$ and genetic operators $\mathcal{G}_M$ leads to an equivalence between problem representations and which schemata are relevant for their optimization. We use the resulting duality between objective functions and schemata to interpret transformations of problem representations.

We have considered isomorphisms (and their corresponding schemata) induced by matrices so as to directly interpret the representational transformations proposed by Liepins and Vose (1990). The results of this paper are easily generalized from linear to arbitrary invertible maps (i.e., from vector space isomorphisms to set isomorphisms) where schemata generalize to the arbitrary predicates of Vose (1991).

While not the focus of this paper, we have also noted that inversions (permutations of string positions) and Grey coding are both special cases of representational transformations induced by matrices (see appendix).

6 Appendix

Here we establish that Gray coding is a special case of the transformations of problem representations induced my matrices.

Let binary strings be represented as column vectors. We prove that the matrix M with 1's down the main and subdiagonal implements Gray coding, and that the lower triangular matrix N consisting of 1's implements the decoding (recall that all arithmetic is done in the finite field of integers modulo 2).

Definition 1 Let M_{k+1} be the matrix defined by

$$M_{k+1} = \left[\begin{array}{c|ccc} 1 & 0 & \cdots & 0 \\ \hline 1 & & & \\ 0 & & M_k & \\ \vdots & & & \\ 0 & & & \end{array}\right] \quad ; \quad M_1 = [\,1\,]$$

Definition 2 Let C_{k+1} be the matrix defined by

$$C_{k+1} = \left[\begin{array}{ccc|ccc} 0 & \cdots & 0 & 1 & \cdots & 1 \\ \hline & C_k & & & C_k & \end{array}\right] \quad ; \quad C_1 = [\,0 \quad 1\,]$$

Note that C_{k+1} has the binary integers 0 through $2^{k+1} - 1$ as columns. This is clear by induction: the left half of C_{k+1} agrees with C_k, and so contains the integers from 0 to $2^k - 1$; the right half adds 2^k to each column of C_k.

Definition 3 Let A_{k+1} be the matrix defined by

$$A_{k+1} = \left[\begin{array}{ccc|ccc} 0 & \cdots & 0 & 1 & \cdots & 1 \\ \hline & A_k & & & A_k^* & \end{array}\right] \quad ; \quad A_1 = [0 \quad 1]$$

where a matrix superscripted with "*" has its first row (binary) complemented.

Theorem 5 $A_k = M_k C_k$.

Proof: Base $k = 1$: $[1][0 \quad 1] = [0 \quad 1]$.
Induction:

$$M_{k+1}C_{k+1} = \left[\begin{array}{c|ccc} 1 & 0 & \cdots & 0 \\ \hline 1 & & & \\ 0 & & M_k & \\ \vdots & & & \\ 0 & & & \end{array}\right] \left[\begin{array}{ccc|ccc} 0 & \cdots & 0 & 1 & \cdots & 1 \\ \hline & C_k & & & C_k & \end{array}\right]$$

$$= \left[\begin{array}{ccc|ccc} 0 & \cdots & 0 & 1 & \cdots & 1 \\ \hline & M_k C_k & & & (M_k C_k)^* & \end{array}\right]$$

$$= \left[\begin{array}{ccc|ccc} 0 & \cdots & 0 & 1 & \cdots & 1 \\ \hline & A_k & & & A_k^* & \end{array}\right]$$

$$= A_{k+1}$$

□

Theorem 5 says that if consecutive integers modulo 2^k are transformed by the matrix M_k, then their images form the columns of A_k (column indices also mod 2^k).

Theorem 6 Consecutive columns of A_k differ in exactly one bit (column indices modulo 2^k).

Proof: Base $k = 1$: $A_k = [0 \quad 1]$.
Induction:

$$A_{k+1} = \left[\begin{array}{ccc|ccc} 0 & \cdots & 0 & 1 & \cdots & 1 \\ \hline & A_k & & & A_k^* & \end{array}\right]$$

The induction hypothesis covers all adjacent columns except the two middle and two outer columns. Since in either pair the columns differ in the first bit, we must show them identical elsewhere. This is equivalent to showing the last column of A_k is identical to the first column of A_k^*. By the inductive hypothesis, the first and last columns of A_k differ in exactly one bit. By construction, this bit is in the first row. □

Theorem 6 says that the columns of A_k are Gray coded (column indices mod 2^k). Hence Theorems 5 and 6 together prove that the transformation

$$v \longrightarrow Mv$$

maps an integer v (represented as a column vector) to its Gray code. We conclude by identifying the inverse transformation N.

Definition 4 Let N_{k+1} be the matrix determined by

$$N_{k+1} = \left[\begin{array}{c|ccc} 1 & 0 & \cdots & 0 \\ \hline 1 & & & \\ \vdots & & N_k & \\ 1 & & & \end{array}\right] \quad ; \quad N_1 = [1]$$

Theorem 7 $M_k^{-1} = N_k$.

Proof: Base $k = 1$: $[1][1] = [1]$.
Induction:

$$\begin{aligned} M_{k+1}N_{k+1} &= \left[\begin{array}{c|ccc} 1 & 0 & \cdots & 0 \\ \hline 1 & & & \\ 0 & & M_k & \\ \vdots & & & \\ 0 & & & \end{array}\right] \left[\begin{array}{c|ccc} 1 & 0 & \cdots & 0 \\ \hline 1 & & & \\ \vdots & & N_k & \\ 1 & & & \end{array}\right] \\ &= \left[\begin{array}{c|ccc} 1 & 0 & \cdots & 0 \\ \hline 0 & & & \\ \vdots & & M_k N_k & \\ 0 & & & \end{array}\right] \\ &= \left[\begin{array}{c|ccc} 1 & 0 & \cdots & 0 \\ \hline 0 & & & \\ \vdots & & I_k & \\ 0 & & & \end{array}\right] \\ &= I_{k+1} \end{aligned}$$

□

Acknowledgements

This research was supported by the National Science Foundation (IRI-8917545).

References

Bagley, J. D. (1967). The Behavior of Adaptive Systems which Employ Genetic and Correlational Algorithms. (Doctoral dissertation, University of Michigan). *Dissertation Abstracts International* 28(12), 5106B. (University Microfilms No. 68-7556)

Cavicchio, D. J. (1970). *Adaptive Search Using Simulated Evolution.* Unpublished doctorial dissertation, University of Michigan, Ann Arbor.

Frantz, D. R. (1972). Non-linearities in genetic adaptive search. (Doctoral dissertation, University of Michigan). *Dissertation Abstracts International,* 33(11), 5240B–5241B. (University Microfilms No. 73-11 116).

Goldberg, D. E., & C. L. Bridges (1988) *An analysis of a reordering operator on a GA-hard problem* (TCGA Report No. 88005). Tuscaloosa: University of Alabama, The Clearinghouse for Genetic Algorithms.

Goldberg, D. E. (1989) *Genetic Algorithms in Search, Optimization, and Machine Learning.* Addison-Wesley.

Holland, J. H. (1975). *Adaptation in natural and artificial systems.* Ann Arbor: The University of Michigan Press.

Liepins & Vose (1990) *Representational Issues in Genetic Optimization.* Journal of Experimental and Theoretical Artificial Intelligence, 2 (1990), 101-115.

Vose, M. D. (1991) *Generalizing The Notion Of Schema In Genetic Algorithms.* Artificial Intelligence, In press.

Conditions for Implicit Parallelism

John J. Grefenstette
Navy Center for Applied Research in Artificial Intelligence
Code 5514
Naval Research Laboratory
Washington, DC 20375-5000
E-mail: GREF@AIC.NRL.NAVY.MIL

Abstract

Many interesting varieties of genetic algorithms have been designed and implemented in the last fifteen years. One way to improve our understanding of genetic algorithms is to identify properties that are invariant across these seemingly different versions. This paper focuses on invariants among genetic algorithms that differ along two dimensions: (1) the way user-defined objective function is mapped to a fitness measure, and (2) the way the fitness measure is used to assign offspring to parents. A genetic algorithm is called *admissible* if it meets what seem to be the weakest reasonable requirements along these dimensions. It is shown that any admissible genetic algorithm exhibits a form of implicit parallelism.

Keywords: Implicit parallelism, invariants, k-armed bandits

1 Introduction

Whenever a new variation of genetic algorithm is proposed, it is reasonable to expect that the designer will provide an analysis of how the new algorithm compares with previous algorithms. Often this takes the form of empirical studies, but such studies are usually difficult to perform, and the generality of the results is often difficult to assess. Theoretical comparisons offer more robust insights, but our existing theory is sparse. The seminal result used in the analysis of genetic algorithms is of course the Schema Theorem (Holland, 1975), which has also provided the foundation for a number of attempts to characterize problems thought to be especially difficult for genetic algorithms (Bethke, 1981; Goldberg, 1987). We take a slightly different approach to understanding genetic algorithms. This paper offers an extension to the analysis begun in James Baker's dissertation (Baker, 1989) and carried forward in (Grefenstette and

Baker, 1989). This approach aims to identify properties that are invariant across different versions of the genetic algorithm. Such results provide a sense of coherence to the field, in that commonalities are exposed among superficially different versions of the genetic algorithm. These results can also serve to spotlight the features which distinguish broad classes of genetic algorithms from one another.

As in (Grefenstette and Baker, 1989), this paper focuses on invariants among genetic algorithms that differ along two dimensions: (1) the way user-defined objective function is mapped to a fitness measure, and (2) the way the fitness measure is used to assign offspring to parents. A genetic algorithm is called *admissible* if it meets what seem to be the weakest reasonable requirements along these dimensions. It is shown that any admissible genetic algorithm exhibits a form of *implicit parallelism*, in the sense that trials are allocated in an exponentially differentiated way to a large number of subsets based on implicit competitions.

The remainder of the paper is organized as follows: Section 2 summarizes previous developments in this approach, and discusses two important genetic algorithm design parameters: the fitness function and the selection algorithm. Section 3 offers a new definition of admissible genetic algorithms and describes a form of implicit parallelism that is invariant for this class. Section 4 presents some observations on the prevailing views of implicit parallelism, based on the k-armed bandit analogy. Section 5 offers some ideas for extensions to this approach, and the final section summarizes the paper.

2 Background

Throughout this paper, we will adopt the generational model of genetic algorithms, shown in Figure 1.

procedure *GA*
begin
  t = 0;
  initialize $P(t)$;
  evaluate **structures in** $P(t)$;
  while termination condition not satisfied do
  begin
    t = t + 1;
    select $P(t)$ **from** $P(t-1)$;
    apply *genetic operators* **to structures in** $P(t)$;
    evaluate **structures in** $P(t)$;
  end
end.

Figure 1: A Genetic Algorithm Model

Although the model considered here is a full scale genetic algorithm, with all the usual genetic operators including crossover and mutation, we focus the discussion on the effects of the selection algorithm and defer consideration of the effects of operators such as crossover and mutation. There are two reasons for this approach. First, the effects of the genetic operators can be viewed as orthogonal to the effects of selection. That is, the choice of recombination and mutation operators represents another dimension along which genetic algorithms differ. The analysis of genetic operators is usually stated in terms of the disruption to the effects of selection, so it seems natural to characterize the effects of selection first, and then to factor in how crossover and mutation impact those effects. Second, by restricting the analysis to selection at this stage, we can temporarily ignore representation issues, since a structure's internal representation has no direct effect on how many times that structure is selected for reproduction. Consequently, our results can be stated more generally, in terms of arbitrary subsets of the search space, rather than being restricted to hyperplanes induced by a particular binary representation.

The most useful results will be those that relate the behavior of the genetic algorithm, in terms of how much effort is allocated to various regions of the search space, to the measure of interest to the end user, that is, the objective function. A few definitions will help clarify the discussion. We define the *growth rate (gr)* (expected number of offspring) for an individual x as a composition of three functions:

$$gr(x) \equiv select(u(f(x)))$$

The *objective function* f is normally determined by the intended application of the algorithm. We assume that the objective function captures some figure of merit, such as cost or payoff, that a user is interested in optimizing.[1] In general, we assume that the definition of the objective function is not under the control of the GA programmer. In contrast, the next two functions in the composition leading to $gr(x)$ are design parameters of the genetic algorithm. The *fitness function* u is defined by the GA programmer to map the range of the objective function into a non-negative interval, say [0,1]. The fitness function might be a simple linear transformation, or may be more exotic (logarithmic, exponential, polynomial) and might also be time-varying in order to scale the problem (Grefenstette, 1986; Goldberg, 1989). The *selection algorithm* is also chosen by the GA programmer to map fitness values into (an expected) number of offspring. A wide variety of selection algorithms have been studied, including proportional methods, ranking methods, and threshold methods, among others (see (Baker, 1989) for an exhaustive survey).[2]

The aim of this work is to identify aspects of the dynamic behavior of genetic algorithms that are invariant across classes of fitness functions and selection algorithms. Let A be an arbitrary subset of the search space. We focus on the growth rate of subsets, where the growth rate of a subset A at time t, $gr(A,t)$, is defined as:

$$gr(A,t) \equiv \frac{1}{n} \sum_{i=1}^{n} gr(x_i)$$

where $\{ x_1, x_2, \cdots, x_n \}$ are the representatives of A in the population at time t.[3]

As an example of the kind of results of interest to us, we first consider a common class of genetic algorithms. A *dynamic linear fitness function* has the form $u(x) = c(t) f(x) + d(t)$, for some (perhaps, time-varying) coefficients c and d, where $c(t) > 0$. A *proportional selection algorithm* computes the function $gr(x) = u(x) / \overline{u}(t)$, where $\overline{u}(t)$ is the population mean fitness at time t. The following characterization of an invariant in the growth rate appeared in (Grefenstette and Baker, 1989).

---

[1] Without loss of generality, we assume throughout this paper that the task is to maximize the objective function.

[2] The final step in the *select* phase of Figure 1 is to apply a sampling algorithm to assign an integer number of offspring to each parent (Baker, 1987). This step will not concern us here.

[3] The reader should feel free to interpret the results in terms of hyperplanes, but should bear in mind that the results apply to arbitrary subsets of the search space.

Theorem 1. In any genetic algorithm using a dynamic linear fitness function and proportional selection algorithm, for any pair of subsets $< A, B >$ represented in $P(t)$,

$$\text{if } f(A,t) \leq f(B,t) \text{ then } gr(A,t) \leq gr(B,t).$$

The theorem has the following interpretation. Suppose one is given any arbitrary, fixed population of structures from the search space. Imagine enumerating all subsets of the search space that are represented in the population, and then obtaining a list **L** by sorting these subsets by decreasing observed mean value under the objective function. Theorem 1 says that, for any genetic algorithm with a dynamic linear fitness function and proportional selection algorithm, any subset in the list **L** grows at least as fast (during this generation) as any subset occurring later in the list. If we change the genetic algorithm by choosing a different dynamic linear fitness function (that is, by choosing different transformation coefficients $c(t)$ and $d(t)$), we will change the relative *magnitudes* of the of the growth rate of the subsets in **L**, but not the relative *order* within **L**. The order within **L** is invariant for all genetic algorithms in this class.

As shown in (Grefenstette and Baker, 1989), there are many classes of genetic algorithms of practical interest, such as genetic algorithms that use a rank-based selection, to which Theorem 1 does not apply. It would be interesting to partition the space of genetic algorithms into interesting classes and find invariants for theses classes like the one in Theorem 1. The remainder of this paper develops this idea further, starting with the class of all ''reasonable'' genetic algorithms.

3 Implicit Parallelism in Admissible Genetic Algorithms

How can we approach the task of characterizing common features of all genetic algorithms? We begin by trying to identify the weakest acceptable conditions for genetic algorithms. Certainly, the notion of survival of the fittest is central to genetic algorithms. That is, a genetic algorithm should assign a fitness value to individuals that is consistent with the objective function, but consistency can be defined in a variety of ways. The following definitions try to capture the weakest reasonable form of consistency. We say a fitness function is *monotonic* iff

$$u(x_i) \leq u(x_j) \textit{ iff } f(x_i) \leq f(x_j)$$

That is, a monotonic fitness function does not reverse the sense of any pairwise ranking provide by the objective function. A slightly stronger form of consistency with the objective function is reflected in the following definition: A fitness function is *strictly monotonic* iff it is monotonic and

$$\textit{if } f(x_i) < f(x_j) \textit{ then } u(x_i) < u(x_j)$$

A strictly monotonic fitness function preserves the relative ranking of any two points in the search space with distinct objective function values.

Exactly analogous definitions can be applied to selection algorithms. A selection algorithm is *monotonic* iff

$$gr(x_i) \leq gr(x_j) \textit{ iff } u(x_i) \leq u(x_j)$$

A selection algorithm is *strictly monotonic* iff it is monotonic and

$$\textit{if } u(x_i) < u(x_j) \textit{ then } gr(x_i) < gr(x_j)$$

We say that a genetic algorithm is *admissible* if its fitness function and selection algorithm are both monotonic. A genetic algorithm is *strict* if its fitness function and selection algorithm are both strictly monotonic.[4]

These definitions give us some new tools for classifying genetic algorithms. For example, a genetic algorithm that uses a dynamic linear fitness function and a proportional selection algorithm is strictly monotonic, since a better individual always has a higher growth rate than a worse individual. Similarly, a genetic algorithm that uses a ranking selection scheme in which the growth rate varies linearly from, say, 2 for the best individual to 0 for the worst, would also be strictly monotonic. On the other hand, consider a genetic algorithm with a *threshold* selection algorithm, such as

> *All individuals with less than average fitness are eliminated, and all remaining individuals have equal growth rate.*

This selection algorithm is monotonic (since a better individual is never eliminated in favor of a worse one), but not strictly monotonic (since better individual may have the same growth rate as a worse one, as long as they both survive the cut). The interesting question is whether these classes of genetic algorithms have any common behaviors. We begin be considering the most general class, the admissible genetic algorithms.

In order to characterize the implicit parallelism of the admissible genetic algorithms, we need to borrow some terminology from the field of multi-objective decision making (Schaffer, 1984), specifically, the notion of one set partially dominating another. Consider two arbitrary subsets of the solution space, A and B. Let the representatives of subset A at time t be

$$A(t) = < a_1 , a_2 , \cdots , a_n >,$$

sorted such that $f(a_i) \geq f(a_{i+1})$ for $1 \leq i < n$. Let the representatives of subset B at time t be

$$B(t) = < b_1 , b_2 , \cdots , b_m >,$$

also sorted in order of decreasing f. Since $A(t)$ and $B(t)$ may have different cardinalities, we define two auxiliary lists, $\hat{A}$ and $\hat{B}$, of size $k = \max(n, m)$, as follows:

Case 1: $n \leq m$. Let $\hat{B} = B(t)$ and let

$$\hat{A} = < \hat{a}_1 , \hat{a}_2 , \cdots , \hat{a}_k >,$$

where $\hat{a}_i = a_1$ for $1 \leq i < m-n$, and $\hat{a}_i = a_{m-n+i}$ for $m-n+1 \leq i \leq k$.

Case 2: $m < n$. Let $\hat{A} = A(t)$ and let

$$\hat{B} = < \hat{b}_1 , \hat{b}_2 , \cdots , \hat{b}_k >,$$

where $\hat{b}_i = b_i$ for $1 \leq i \leq m$, and $\hat{b}_i = b_m$ for $m+1 \leq i \leq k$.

That is, if $A(t)$ is smaller than $B(t)$, we augment $\hat{A}$ by adding copies of the best representative of A. If $B(t)$ is smaller than $A(t)$, we augment $\hat{B}$ by adding copies of the worst representative of B. Finally, we say *B partially dominates A* ($A <_p B$) at time t iff

$$f(\hat{a}_i) \leq f(\hat{b}_i) \quad \text{for } 1 \leq i \leq k$$

and at least one inequality is strict. Intuitively, if $A <_p B$, then B is "better" than A in

[4] Baker (1989) called admissible genetic algorithms *reliably consistent.*

the sense that each representative of B is at least as good as the corresponding representative of A. This gives a preference criterion that all admissible genetic algorithms can agree on, as shown by the following:

Theorem 2. In any admissible genetic algorithm, for any pair of subsets $< A, B >$ represented in $P(t)$,

$$\text{if } A <_p B \text{ then } gr(A,t) \leq gr(B,t).$$

Proof. Consider an admissible genetic algorithm G and a pair of subsets $< A, B >$ that each have at least one representative in $P(t)$. Let $\hat{A}$, $\hat{B}$, and k be defined as above. If $A <_p B$ at time t then

$$f(\hat{a}_i) \leq f(\hat{b}_i) \text{ for } 1 \leq i \leq k.$$

Since G is admissible, the fitness function is monotonic, so

$$u(\hat{a}_i) \leq u(\hat{b}_i) \text{ for } 1 \leq i \leq k.$$

Since the selection algorithm is monotonic,

$$gr(\hat{a}_i) \leq gr(\hat{b}_i) \text{ for } 1 \leq i \leq k.$$

So,

$$\frac{1}{k} \sum_{i=1}^{k} gr(\hat{a}_i) \leq \frac{1}{k} \sum_{i=1}^{k} gr(\hat{b}_i)$$

By the construction of $\hat{A}$ and $\hat{B}$,

$$gr(A,t) \leq \frac{1}{k} \sum_{i=1}^{k} gr(\hat{a}_i) \text{ and } \frac{1}{k} \sum_{i=1}^{k} gr(\hat{b}_i) \leq gr(B,t),$$

so the result follows. □

A slightly stronger result can be obtained for strict genetic algorithms:

Theorem 3. In any strict genetic algorithm, for any pair of subsets $< A, B >$ represented in $P(t)$,

$$\text{if } A <_p B \text{ then } gr(A,t) < gr(B,t).$$

Proof. Similar to the previous proof. □

To help illustrate this result, consider again an enumeration **L** of all subsets of a search space represented in a given population. The relation $<_p$ induces a partial order over this list of subsets. Consider any two subsets A and B in **L** such that $A <_p B$. Theorem 3 says that in any strict genetic algorithm the dominating subset B grows at least as fast as the dominated subset A. The effect of changing the selection algorithm from, say, a proportional selection algorithm to a rank-based selection algorithm, is to alter the magnitudes of the growth rates of the two subsets, but the relative order is invariant for all strict genetic algorithms.

The following corollary makes some contact between the previous result and the exponential allocation of trials heuristic advocated in (Holland, 1975).

Corollary 3.1. If B completely dominates A in the strict sense that every representative of B is strictly better than every representative of A, then B grows exponentially faster than A in any strict genetic algorithm (ignoring the effects of recombination and mutation).

Proof. Since B completely dominates A, B has a higher growth rate than A in any population in which each subset is represented. The compounding effect of continually higher growth rate gives B an exponentially larger allocation of trials than A (modulo the effects of crossover and mutation). □

The following figure offers an intuitive illustration of this result. Consider an arbitrary fitness function, as shown in Figure 2. Draw a horizontal line at any height y. The corollary says that, for *any* strict genetic algorithm, the number of representatives in the population from the set $B = \{ x \mid f(x) > y \}$ (the shaded region in Figure 2) grows exponentially faster than the number of representatives from the complement of B.

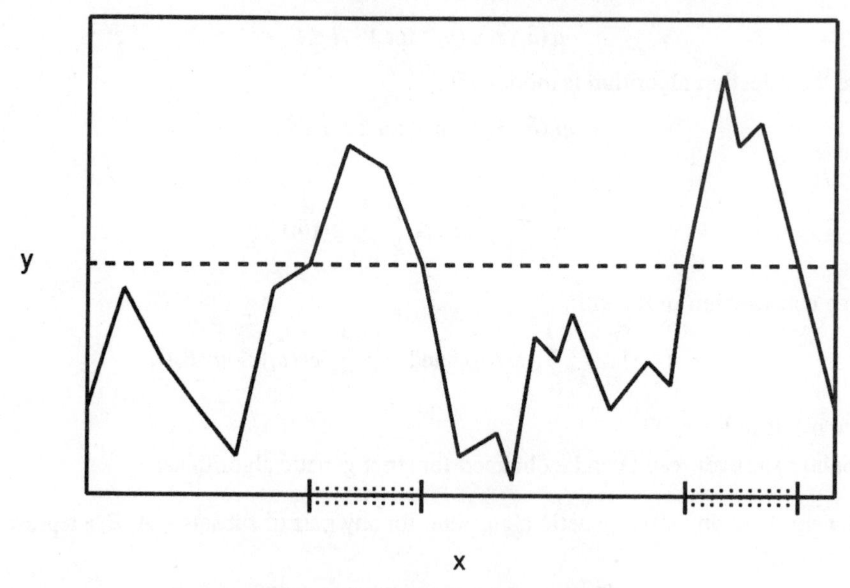

Figure 2: Shaded Region is $B = \{ x \mid f(x) > y \}$

This provides an intuitive explanation of the power of genetic algorithms, and it has the advantage that it is expressed in the terms most meaningful to the end user, that is, in terms of the objective function, rather than in terms that are properly internal to the genetic algorithm, such as fitness.

4 The K-Armed Bandit Paradox

Much of our current understanding of implicit parallelism in genetic algorithms derives from Holland's analysis of the k-armed bandit problem. Holland (1975) shows that, given a k-armed bandit problem with unknown payoff distribution (of a certain class), a nearly optimal adaptive strategy is to allocate trials to the best arm at an exponentially increasing rate. Holland draws an analogy between this strategy and the behavior of genetic algorithms: an n-order hyperplane partition defines a k-armed bandit problem, with $k=2^n$. However, there are some subtleties about the correspondence between the payoff in the k-armed bandit sense and the performance of the corresponding hyperplane.

In order to take the analogy seriously, it must be recognized that the payoff associated with a given "arm" is *not* an estimate of the mean value of the objective function for that hyperplane. In fact, there are simple examples that show that the genetic algorithm does *not* solve the k-armed bandit problem in the sense of choosing the hyperplane with the best mean objective function value within a given hyperplane partition (Grefenstette and Baker, 1989). Rather, the payoff on an "arm" is the mean of the observed fitness, which is of course a highly biased estimate of the mean fitness after the first generation. A possible source of confusion is that, unlike the player of a k-armed bandit, additional trials by the genetic algorithm (after the very first generation) are not unbiased samples from the hyperplanes within a given hyperplane partition. Instead, the samples are highly biased toward the most promising regions within each hyperplane.

According to Holland [private communication], the proper interpretation of the k-armed bandit analysis is to think of k as specifying the size of the finest grain hyperplane partition for which the current distribution of payoffs in the competing hyperplanes is reasonably uniform. Additional samples, taken uniformly from a given hyperplane, will then reduce the variance in performance associated with that hyperplane, and the variance of coarse (low-order) hyperplanes will be reduced at rates that are exponentially higher the than higher-order hyperplanes. By this argument, low-order hyperplanes are typically confirmed early on, and the solution proceeds toward increasing k as more samples are accumulated.

But this involves a paradox: The exponential allocation strategy is most useful when competing arms have relatively high variance; if the variance is low, a few samples from each arm gives a reliable indication of the best arm. However, in the case of competing hyperplanes having relatively high variance, the genetic algorithm rapidly shifts its focus of attention to more specific hyperplane competitions involving hyperplanes with lower performance variance. That is, the genetic algorithm abandons the near-optimal allocation of trials just when it seems to be most appropriate. This strategy does indeed reduce the variance in the observed payoff of the hyperplane, but the observed payoff may be arbitrarily far from the true average payoff. This is where the analogy with the k-armed bandit problem breaks down. Because of biased sampling within a hyperplane, the genetic algorithm's view of a given hyperplane competition may bear little relation to the mean objective function value of the corresponding hyperplanes. This rather subtle point is often obscured in discussions of the "building block" hypothesis that imply genetic algorithms proceed by solving 2-armed bandit problems, followed by $k = 4$, $k = 8$, and so on (Goldberg, 1989). This statement of the building-block hypothesis, which underlies much of the work on "deceptive problems" (Bethke, 1981; Goldberg, 1987), needs to be re-examined in light of the discussion above.

It also follows from these considerations that the notion that implicit parallelism can be quantified by counting the number of hyperplane partitions in a random population (Goldberg, 1985; Fitzpatrick and Grefenstette, 1988) is probably misguided. It is always necessary to consider the dynamic distribution of samples within the population in order to accurately assess how many hyperplanes are being processed by the genetic algorithm. It is not clear how this can be done without explicitly considering the objective function, the fitness function, and the selection algorithm as part of the analysis.

5 Directions for Further Work

There are many promising directions for continuing this line of work. One direction is to try to characterize the implicit parallelism in genetic algorithms with various degrees of sensitivity to the selection algorithm. Intuitively, genetic algorithms with linear fitness functions and proportional selection are highly sensitive to the objective function. That is, large differences in objective function values are reflected as large differences in growth rates. The use of dynamic scaling fitness functions reducing sensitivity. Rank-based selection schemes reduce sensitivity to the objective function even more. Figure 3 sketches out the genetic algorithm landscape with respect to sensitivity. The vertical axis represents a subjective set of expectations about the speed of convergence of various types of genetic algorithms. The horizontal axis represents a rough measure of how sensitive the algorithms are to the objective function. The point is that there seems to be some correlation between convergence and sensitivity that should be explored in more formal detail.

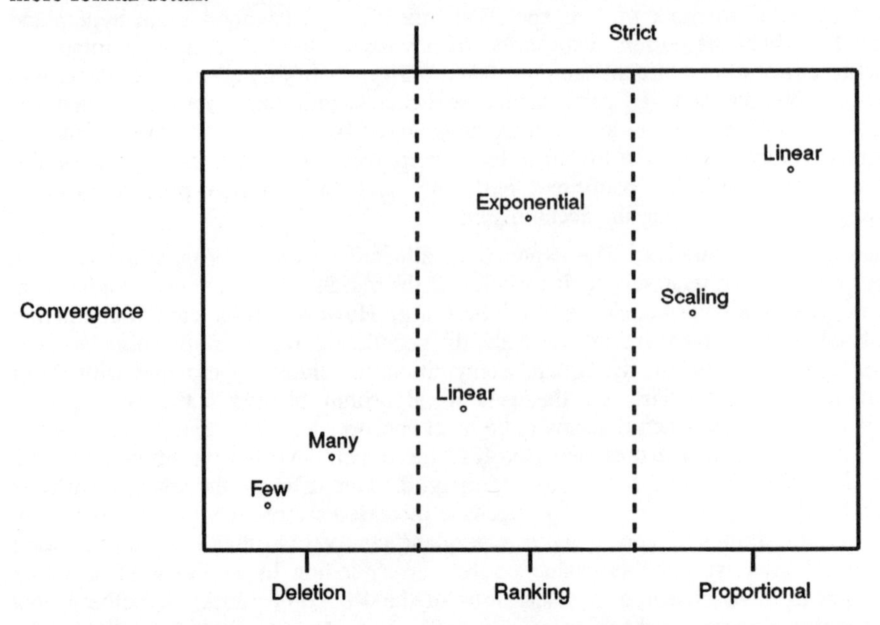

Figure 3: Convergence Rate as a Function of Sensitivity to Objective Function

It is not clear that more sensitivity is necessarily better. For example, consider the effects of the user's decision to change the objective function from $f(x) = x^2$ to $f(x) = x^{10}$. With a proportional selection algorithm, a genetic algorithm is likely to suffer premature convergence with the new objective function. With a ranking selection algorithm, the genetic algorithm explores the identical sequence of points in either case. Given that genetic algorithms will often be used to optimize a surface with unknown properties, the genetic algorithm designer should be prepared to use algorithms with the appropriate sensitivity for the application at hand. We conjecture that, given appropriate formal definitions of sensitivity, theorems similar to Theorem 2 and 3 could be developed for various sub-classes of admissible genetic algorithms distinguished by the sensitivity of the selection algorithm.

6 Summary

It is unlikely that a single version of the genetic algorithm will be equally useful for all applications, so it is important to try to understand the similarities and differences among various classes of genetic algorithms. This paper aims to improve our understanding of genetic algorithms by identifying properties that are invariant across different fitness functions and selection algorithms. A genetic algorithm is called admissible if it meets what seem to be the weakest reasonable requirements along these dimensions. It is shown that any admissible genetic algorithm exhibits a form of implicit parallelism.

The results presented here illustrate some invariants based on one way to partition the class of genetic algorithms. Other partitions could be explored using similar methods. For example, classifying selection algorithms by sensitivity to the objective function seems to be a promising direction for further research along these lines.

We hope this work is seen as making at least two very positive statement about genetic algorithms: that the desirable results we often obtain in the simulations are less dependent than we thought on particular selection or recombination operators, and that there are many unexplored directions for extending the theory of genetic algorithms.

References

Baker, J. E. (1987). Reducing bias and inefficiency in the selection algorithm. *Proceedings of the Second International Conference Genetic Algorithms and Their Applications* (pp. 14-21). Cambridge, MA: Erlbaum.

Baker, J. E. (1989). *Analysis of the effects of selection in genetic algorithms,* Doctoral dissertation, Department of Computer Science, Vanderbilt University, Nashville.

Bethke, A. D. (1981). *Genetic algorithms as function optimizers,* Doctoral dissertation, Department Computer and Communication Sciences, University of Michigan, Ann Arbor.

Fitzpatrick, J. M. and J. J. Grefenstette (1988) Genetic algorithms in noisy environments. *Machine Learning, 3(2/3),* (pp. 101-120).

Goldberg, D. E. (1985). *Optimal initial population size for binary-encorded genetic algorithms.* TCGA Report No. 85001, Tuscaloosa: University of Alabama.

Goldberg, D. E. (1987). Simple genetic algorithms and the minimal, deceptive problem. In *Genetic algorithms and simulated annealing.* D. Davis (ed.), London: Pitman Press.

Goldberg, D. E. (1989). *Genetic algorithms in search, optimization, and machine learning.* Reading: Addison-Wesley.

Grefenstette, J. J. (1986). Optimization of control parameters for genetic algorithms. *IEEE Transactions on Systems, Man, and Cybernetics, SMC-16(1),* 122-128.

Grefenstette, J. J. and J. E. Baker (1989). How genetic algorithms work: A critical look at implicit parallelism. *Proceedings of the Third International Conference Genetic Algorithms* (pp. 20-27). Fairfax, VA: Morgan Kaufmann.

Holland, J. H. (1975). *Adaptation in natural and artificial systems.* Ann Arbor: University Michigan Press.

Schaffer, J. D. (1984). *Some experiments in machine learning using vector evaluated genetic algorithms* Doctoral dissertation, Department of Electrical Engineering Vanderbilt University, Nashville.

PART 6

VARIATION AND RECOMBINATION

The CHC Adaptive Search Algorithm: How to Have Safe Search When Engaging in Nontraditional Genetic Recombination

Larry J. Eshelman

Philips Laboratories,
North American Philips Corporation,
345 Scarborough Road,
Briarcliff Manor, New York 10510

Abstract

This paper describes and analyzes CHC, a nontraditional genetic algorithm which combines a conservative selection strategy that always preserves the best individuals found so far with a radical (highly disruptive) recombination operator that produces offspring that are maximally different from both parents. The traditional reasons for preferring a recombination operator with a low probability of disrupting schemata may not hold when such a conservative selection strategy is used. On the contrary, certain highly disruptive crossover operators provide more effective search. Empirical evidence is provided to support these claims.

Keywords: cross-generational competition, elitist selection, implicit parallelism, uniform crossover, incest prevention, restarts.

1 Introduction

Any search algorithm that operates via a reproduction-recombination cycle on a population of structures can be called a genetic algorithm (GA) in the broad sense of the term. Since Holland's seminal work (1975), however, in order for any such algorithm to qualify as a genetic algorithm in a more restricted sense, it must be shown that its search behavior displays what Holland calls *implicit parallelism*. Thus, it is incumbent upon anyone who claims that some new population-based search algorithm is a genetic algorithm in the more restricted sense to make a case for its implicit parallelism.

In this paper I argue that CHC, a nontraditional genetic algorithm, does indeed display implicit parallelism. Furthermore, I argue that some of the features that on the surface seem to disqualify it as a true GA not only do not so disqualify it, but in fact make it more powerful than the traditional GA.

I shall use the following outline of a genetic algorithm as a point of reference to describe how CHC differs from a traditional GA (based on Grefenstette and Baker, 1989, but generalized):

```
procedure GA
begin
   t = 0;
   initialize P(t);
   evaluate structures in P(t);
   while termination condition not satisfied do
   begin
      t = t + 1;
      select_r C(t) from P(t-1);
      recombine structures in C(t) forming C'(t);
      evaluate structures in C'(t);
      select_s P(t) from C'(t) and P(t-1);
   end
end.
```

By a *traditional GA* I mean a GA for which the following is assumed: (1) The initialization of the population P(0) (of fixed size M) is random. (2) The selection for reproduction (*select*$_r$) is biased toward selecting the better performing structures. (3) The selection for survival (*select*$_s$) is unbiased, typically replacing the entire parent population P(t-1) with the child population C′(t) generated from P(t-1). (4) The recombination operator is either one- or two-point crossover. (5) A low rate of mutation is used in the recombination stage to maintain population diversity.

CHC differs from a traditional GA on all but the first of these points. First, it is driven by survival-selection (*elitist selection*) rather than reproduction-selection. In other words, the bias in favor of the better performing structures occurs in survival-selection rather than reproduction-selection. Second, a new bias is introduced against mating individuals who are similar (*incest prevention*). Third, the recombination operator used by CHC is a variant of uniform crossover (*HUX*), a highly disruptive form of crossover. Finally, mutation is not performed at the recombination stage. Rather, diversity is maintained (or more precisely, reintroduced) by partial population randomization whenever convergence is detected (*restarts*).

In the remainder of this paper I describe CHC in detail, contrast it with the traditional GA, give a theoretical justification for the differences, and provide some empirical comparisons. Sections 2 and 3 describe and analyze the essential features of CHC — a conservative selection algorithm working in conjunction with a radical (i.e., highly disruptive) recombination operator. Sections 4 and 5 describe mechanisms for avoiding premature convergence and maintaining diversity. Sections 6 and 7 present strong empirical evidence that CHC outperforms the traditional GA. Section 8 shows how the philosophy of CHC can be extended to permutation problems. Section 9 offers some concluding remarks. An appendix at the end of this paper provides pseudocode for CHC.

2 Elitist Selection

CHC replaces "reproduction with emphasis" with "survival of the fittest". More precisely, during selection for reproduction, instead of biasing the selection of candidates C(t) for reproduction in favor of the better performing members of the parent population P(t-1), each member of P(t-1) is copied to C(t), and randomly paired for reproduction. (In other words, C(t) is identical to P(t-1) except that the order of the structures has been shuffled.) During survival-selection, on the other hand, instead of replacing the old parent population P(t-1) with the child population C′(t) to form P(t), the newly created children must compete with the members of the the parent population P(t-1) for survival — i.e., competition is *cross-generational.* More specifically, the members of P(t-1) and C′(t) are merged and ranked according to fitness, and P(t) is created by selecting the best M (where M is the population size) members of the merged population. (In cases where a member of P(t-1) and a member of C′(t) have the same fitness, the member of P(t-1) is ranked higher.) I shall call this procedure of retaining the best ranked members of the merged parent and child populations *population-elitist selection* since it guarantees that the best M individuals seen so far shall always survive.

Several other GA's use fitness-biased survival-selection — Whitley's GENITOR (1989), Syswerda's Steady State GA (SSSGA) (1989), and Ackley's Iterated Genetic Search (IGS) (1987). CHC differs from all three of these algorithms in that the competition for survival is cross-generational — a child only replaces a member of the parent population if it is better. Furthermore, unlike these three algorithms, CHC operates in generational cycles with many matings rather than one mating per cycle. CHC's sole reliance upon survival-selection for its bias in favor of the better performing individuals also distinguishes it from GENITOR and SSSGA, but not IGS. Finally, CHC's deterministic, rank-based method of doing selection distinguishes it from SSSGA and IGS, but not GENITOR.

In the remainder of this section I shall address the question whether a GA using only population-elitist selection (i.e., relying solely upon cross-generational, deterministic survival-selection for its bias in favor of the better performing individuals) will produce exponential sampling of the better performing schemata (building blocks defined by combinations of bit-values), and thus exhibit implicit parallelism. The answer to this question depends upon what is meant by implicit parallelism. It will be argued that although CHC does not exhibit implicit parallelism as it was originally defined by Holland, it does exhibit a weaker version of implicit parallelism.

First, it is easy to show that elitist selection does not exhibit strong implicit parallelism, where implicit parallelism is understood to mean: *If the observed performance of schema H_j is consistently higher than the observed performance of H_i, the market share of H_j (fraction represented in the population) grows at an exponentially greater rate than the market share of H_i* (Grefenstette and Baker, 1989). The reason that elitist selection fails to demonstrate strong implicit parallelism is that at most one additional copy of an individual is made in each generation. This means that for any schema H_j that has a high average observed performance but also a high performance variance, CHC cannot compensate in one generation for the performance of the poor individuals by making extra copies of the good individuals. If the poor performing individuals containing H_j are replaced, but only one additional copy is made of the high performing individuals, the net result may be fewer instances of schema H_j even though its average performance was relatively high compared to the population average.

Elitist selection, however, does permit CHC to exhibit Grefenstette's and Baker's (1989) weak version of implicit parallelism: *If schema H_j completely dominates schema H_i, then the market share of H_j grows at least as fast as that of H_i in any generation.* (Schema H_j is said to *completely dominate* schema H_i if every possible representative of H_j is at least as good as every possible representative of H_i.)

In order to see how elitist selection satisfies the conditions for weak implicit parallelism, consider what happens to a single individual. Ignoring for the moment the disruptive effects of recombination, at most two copies of any member of the parent population can survive into the next generation. Furthermore, an individual can never have more copies survive than another individual if it is worse than that other individual. (The converse does not follow: if it is better, it does not necessarily produce more copies.) Thus, the market share of those schemata that completely dominate other schemata will grow at least as fast as the market share of the latter (again ignoring the effects of disruption due to recombination).

To illustrate how this can lead to exponential growth of the number of instances of the better schemata let us first consider a degenerate case in which the recombination operator is the identity operator and simply returns the two parents. In this case C′(t) will not be changed by recombination and will be the same as P(t-1). After survival-selection, however, the best M/2 individuals in P(t-1) will have duplicate copies in P(t), replacing the worst M/2 individuals in P(t-1). Thus, in $\log_2 M$ generations the best individual will have taken over the population. Even though no individual can produce more than a single offspring per generation, good individuals, since they are ranked higher, survive at least as long as, and usually longer than, lower ranked individuals, and, therefore, tend to produce more offspring. Since these offspring, in turn, produce surviving offspring, exponential growth will result. Those individuals that are in the top half of the population will be growing at the same rate (doubling in number each generation). In the long run, however, some of these individuals will be squeezed out of the top half of the population, and the better individuals will increase their market share at the expense of the lower ranked individuals.

Now let us reintroduce recombination into the analysis. (I shall assume that the recombination operator produces two new strings by recombining material from two parents.) Suppose schema H_j completely dominates schema H_i, and instances of both H_j and H_i appear in the parent population. Remember that this means that each instance of H_j performs at least as well as any instance of H_i. This is enough to guarantee that under elitist selection the individuals containing H_j are safe from being replaced by those offspring containing H_i. Further, if the recombination operator sometimes preserves instances of H_j, they will eventually increase their market share at the expense of instances of H_i (unless every sample of H_j has the same performance value as every sample of H_i). In other words, in so far as individuals containing H_j tend to be better than individuals containing H_i, the market share of H_j will grow exponentially for a longer period of time.

Note that in the above account nothing was assumed about using a recombination operator with a low probability of disrupting schemata. The interesting thing about elitist selection is that it is not necessary to make such an assumption in order to insure exponential growth of the market share of the better schemata. In traditional selection (proportional or linear rank selection) good performers get more copies than bad performers. Elitist selection achieves the same thing by preserving good performers for more generations and thus over time generating more copies. The effect of a disruptive crossover operator, however, is quite different for elitist selection than for traditional selection. Consider, for example, the extreme case where the crossover operator is so disruptive that often all the children are worse than the worst member of the parent population. This would be disastrous for a traditional GA. On the other hand, if elitist selection is used, the parent population will survive intact. Elitist selection is much more robust than traditional selection because it compensates for those occasions in which it generates many inferior children by replacing fewer of the parents.

The essential requirement for any recombination operator used in conjunction with elitist selection is that it not be strongly biased against the better schemata. It does not matter if the operator is highly disruptive provided the offspring of the relatively poor individuals do not have a better chance of surviving a mating than the offspring of the better individuals. Given this proviso it follows that if schema H_j completely dominates schema H_i, then the market share of H_j will be expected to grow at least as fast as H_i in any generation. Furthermore, to the extent that there are instances of H_j that are strictly better than instances of H_i the exponential growth of the market share of H_j can be expected to eventually overtake that of H_i.

Since elitist selection's claim to even weak implicit parallelism depends upon the proviso that the recombination operator is not biased against the better schemata, we must turn our attention to whether there are any highly disruptive recombination operators for which this proviso holds. But before taking up this question, we need to address an even more fundamental question: Why should anyone want to use a highly disruptive recombination operator? I shall address both questions in the next section.

3 Uniform Crossover

It is important to keep in mind that what we are looking for in a GA is an operator that provides productive recombinations and not simply preservation of schemata. There is a tradeoff between effective recombination and preservation. A recombination operator that always crossed over a single differing bit, for example, would create new individuals while being minimally disruptive, but it would not be a very useful operator. More generally, this tradeoff can be seen by examining the formula for the minimum number of schemata preserved via crossover (i.e., when the parents are complementary on all loci):

$$2^x + 2^{(L-x)}$$

where L is the string length and x is the number of bits crossed over. Note that the number of schemata guaranteed to be preserved is greatest when the operator does no recombining, i.e., when $x = 0$ or L. On the other hand, the number of schemata guaranteed to be preserved is the lowest when $x = L/2$. Thus, if we are mainly interested in preservation, we would prefer that x be low. But what if our concern is with recombination?

The intuitive idea behind recombination is that by combining features from two good parents we may produce even better children. What we want to do is copy high valued schemata from both parents, simultaneously instantiating them in the same child. Clearly the more bits that we copy from the first parent the more schemata we copy and thus the more likely we are to copy, without disruption, high valued schemata. On the other hand, the more bits we copy over from the first parent the fewer we can copy from the second parent, thus increasing the likelihood of disrupting high valued schemata from the second parent. Consequently, a crossover operator that crosses over half the bits (or even better, half the differing bits) will be most likely to combine valuable schemata simply for the reason that the maximum number of schemata is combined from each parent.

Combining many schemata from both parents, however, is just half the story. Every low order schema is defined at a proper subset of the defining positions of at least one higher order schema.[1] Although this is a truism, it has important consequences for search. If the best performing individual in the population is not the global optimum, then certain instances of lower order schemata that are critical to its good performance are combined with other lower order schemata that are hindering performance. The job of recombination is to test these associations by vigorously trying a variety of recombinations. As an illustration of a crossover operator that fails in this regard, consider the one-point crossover operator, H1X, which always cuts the strings at the mid-point into two equal segments. One would not expect this to be a very effective crossover operator for a search algorithm. There is one set of schemata that are always disrupted and another set that are always preserved. Although in any crossover operation H1X will recombine a maximum number of schemata, on subsequent crossovers it will always operate on the same schemata. The case is similar, but less extreme, with a two-point crossover operator, H2X, that crosses over half the material — i.e., one point is chosen at random anywhere on the string, and the second point is chosen so as to be $L/2$ loci from the first point. H2X has the same low probability ($2/L$) of disrupting two adjacent bits as traditional two-point crossover (2X). (It is assumed in the following discussion that the parents differ on the defining loci.) As the defining length is increased, however, the probability of disrupting order-2 schemata increases more rapidly for H2X than for 2X. If the defining length is $L/2$, the probability of disruption for 2X is 0.5 whereas for H2X it is 1.0 (assuming an even number of bits in the string). In fact, any schema (a) whose defining length is greater than or equal to $L/2$ and (b) which has no "gap" of adjacent undefined loci greater than or equal to $L/2$ will always be disrupted by H2X, whereas there will always be some probability that 2X will not disrupt such a schema, provided there are some undefined loci. (It should be kept in mind, however, that the only reason 2X has such a low probability of disruption is that it often does not do much recombination. For example, the probability that it will exchange exactly one bit is $2/L$.) More generally, certain schemata tend to get passed through repeated applications of H2X

[1] The order of a schema is the number of defining loci. Its defining length is the difference between the first and last defined loci.

without being disturbed whereas other schemata rarely or never survive intact. This is the result of the strong positional bias of H2X that privileges certain schemata, those that have a short defining length, at the expense of other schemata. In order for a valuable schema to be recombined with schemata from another individual, the valuable schema has to be preserved. Likewise, in order for the performance of a valuable schema to be distinguished, it must be pried apart from other schemata. Privileging some schemata at the expense of others hinders this process. (See Schaffer, Eshelman and Offutt (1991) for a more detailed discussion of this point.)

In summary, neither preserving schemata by not doing much recombination (e.g., 2X) nor doing a lot of recombination but always of the same schemata (e.g., H2X) is a very satisfactory alternative. In the light of these considerations, let us now examine the actual recombination operator used by CHC — a variant of uniform crossover. Uniform crossover (UX) exchanges bits rather than segments. For each position in the string the bits from the two parents are exchanged with fixed probability p (typically 0.5). CHC uses a modified version of UX, HUX, that crosses over exactly half the non-matching alleles (what Booker (1987) calls the reduced surrogate), where the bits to be exchanged are chosen at random without replacement. HUX guarantees that the children are always the maximum Hamming distance from their two parents. The flip side of HUX's disruptiveness is that it maximizes the chance of two good schemata, one from each parent, being combined in a child since half the material from each parent is chosen. Furthermore, all schemata of the same order have an equal chance of being disrupted or preserved.

We are now ready to address the question raised at the end of the previous section. Granted that elitist selection will allocate trials exponentially to the better schemata, provided the disruptiveness is not strongly biased against the better schemata, what reason is there to believe that HUX will not be strongly biased against the better schemata? In so far as better individuals are better because the performance reflects the discovery of some good high order schema, it might seem that it would be much harder to propagate this high order schema than other lower order schemata that are not so good. Furthermore, if this critical schema does not completely dominate its competitors, there is the possibility, in spite of the preservative powers of elitist selection, of losing this critical schema from the parent population — e.g., if other parents can introduce better individuals in the parent population without this schema, even though they might be better yet if they contained this critical schema.

Admittedly, under some circumstances HUX will be biased against the better schemata. It should be kept in mind, however, that the same qualification applies to the traditional GA. In fact, any recombination operator will be biased in favor of certain types of building blocks and against others. Just as the traditional GA works with building blocks with short defining lengths, CHC works with low order building blocks. HUX is especially good at prying apart large (high order) building blocks into small (low order) building blocks and recombining them. Under some circumstances this feature may be a weakness, just as two-point crossover's bias in favor of short defining length building blocks will, under certain circumstances, be a weakness.

Granted that no recombination operator can under all circumstances satisfy the proviso that it not be biased against the better schemata, it might still be objected that the circumstances under which HUX and, more generally, uniform crossover fail are much more serious. After all, the probability of preserving a building block whose defining length is 10 using two-point crossover in a string of 100 bits is of a different order of magnitude than preserving a building block whose order is 10 using uniform crossover. It might seem that uniform crossover cannot be much more successful than mutation on problems that require the manipulation of schemata of high order. The probability of preserving any schema using uniform crossover (assuming that two children are produced and that the mate contains the compliment of the specified schema) is only twice as great as randomly generating this schema.

It should be kept in mind, however, that although HUX is highly disruptive, it is, unlike mutation, a true recombination operator. It cannot introduce new alleles, but simply recombines schemata contained in the parents. Consequently, loci that are converged cannot be disturbed by uniform crossover. In other words, the more converged the two parents are, the less disruptive uniform crossover is. Furthermore, convergence is not some *deus ex machina* brought in at the last minute to rescue uniform crossover. There is no way a GA can allocate more trials to the schemata of better performing individuals without converging on the loci defining those schemata. Convergence can be a sign of progress as well as a sign of stagnation.

HUX and, more generally, uniform crossover can only be made to look as bad as random search by assuming a worst-case analysis. It is true that if there is one instance of an order-10 schema in the population, and all other individuals in the population are converged on its complimentary bits, then the probability of preserving it is $1/2^9$ (if two children are produced). Why, however, assume this worst case? If, for example, there is a only one instance of a schema H_k of order k, in a population of size M, then the probability that a child will be created by uniform crossover that preserves this schema is:

$$\sum_{i=0}^{i=k-1} \frac{Instances(H_k, i)}{(M-1)\; 2^{(k-i)-1}}$$

where $Instances(H_k, i)$ is the number of individuals in the population that share the same value for exactly i bits of the k bits defining schema H_k. $Instances(H_k, i)/(M-1)$, then, is the probability of choosing a mate that matches on i of the k defining bits of H_k. The worst case occurs whenever for *all* $i > 0$, $Instances(H_k, i) = 0$. On the other hand, if $Instances(H_k, i)$ is relatively large for i near k, then crossover is likely to preserve schema H_k.

In order to better understand how likely CHC is to propagate any high order schema H_k we need to consider three cases. In the first case, there are lower order building blocks that contribute to the performance of the individuals independently of finding the high order building block. In this case, there is likely to be some convergence in the population making it easier to find and propagate the critical high order schema H_k. In the second case, all the low order schemata that might serve as building blocks for H_k have similar values to their competitors, and so are no help in finding H_k. On the other hand, there is no pressure for the bits in the loci defining H_k to converge on any particular set of values. Values for these bits will tend to wander randomly. This means that the probability of propagating H_k is not the worst case of $1/2^{(k-1)}$ but closer to $1/2^{(k/2-1)}$. The reason is that given a randomly initialized population and no pressure for convergence, half the bits on average will match those in H_k. Finally, the worst case occurs when the problem is a deceptive problem (Goldberg, 1987), and there is pressure for the lower order building blocks to converge on values that are compliments of the bit-values defining H_k. In this case, if the order of H_k is high, the schema will be unlikely to propagate $(1/2^{(k-1)})$. Furthermore, if there are several such high order deceptive schemata instantiated in different individuals, then it will be highly unlikely that they will ever be successfully combined intact in a single offspring.

Even in this latter case the traditional GA does not necessarily have an advantage over CHC. The traditional GA can do relatively well on such deceptive problems only if two additional conditions hold. First, the GA must use a population size that is large enough to make it likely that these high order deceptive schemata will occur in the initial, randomly generated population. Secondly, the high order deceptive schemata should have short defining lengths, and, more importantly, their defining loci should not be interspersed. The first condition can easily be met provided one is willing to pay the cost — large populations require more evaluations to converge to a solution. The trouble is that typically one does not know whether or not the problem is a high order deceptive problem. If it turns out not to be such a problem, then a much smaller population is likely to be more efficient for search. The second condition, however, is the critical one. What reason is there for assuming that the problem will be so well behaved that the deceptive schemata will tend not to overlap? It is easy to contrive problems where this assumption is met, but there is no reason to believe that real problems are going to be so benign. In fact, GAs are usually advertised as being most useful for large, complex, poorly understood problems, where there is little or no a priori knowledge about how the bits interact. Under such conditions the traditional GA will have no advantage over CHC with regard to deceptive schemata.

4 Avoiding Incest

The exponential growth of instances of good schemata is of little value if it leads to premature convergence. One of the effects of crossing over half the differing bits between the parents is that the danger of premature convergence is lessened. Even if at each generation the most recent descendant mates with one of its original ancestors (the same one each time), it will take $\log_2 h$ generations to converge (within one bit) to the original ancestor where h is the Hamming distance between the original parents. In the case of two-point crossover, on the other hand, each of the two children will differ from its nearest (measured by Hamming distance) parent by an amount ranging from one bit to no more than half the length of the string L. Thus, the longest time it can take to converge within one bit of its ancestor is $\log_2 h$ generations and the shortest is one generation.[2] Of course, a child is unlikely to be repeatedly mated with one of its remote ancestors, but since better individuals have more descendents, it is fairly likely that a individual will be mated with one of its near relatives. In so far as this leads to crossing over of individuals that share a lot of alleles, exploration via recombination quickly degenerates. Although always crossing over half the differences (using HUX) slows this process, sometimes individuals are paired that have few differences. If one or both children survive this mating, it will be even more likely that such an event will occur the next generation.

[2] If, instead of always choosing the child that is nearest (in terms of Hamming distance) to the parent that is its original ancestor, a child is chosen at random, then convergence may take longer than $\log_2 h$ generations, but the expected value will still be less than $\log_2 h$.

CHC has an additional mechanism to slow the pace of convergence — a mechanism for helping avoid *incest*. During the reproduction step, each member of the parent population is randomly chosen without replacement and paired for mating. Before mating, however, the Hamming distance between potential parents is calculated, and if half that distance (the Hamming distance of the expected children from their parents) does not exceed a difference threshold, they are not mated and are deleted from the child population. (The difference threshold is set at the beginning of the run to L/4 — half the expected Hamming distance between two randomly generated strings.) Thus, typically only a fraction of the population is mated to produce new offspring in any generation. Whenever no children are accepted into the parent population (either because no potential mates were mated or because none of the children were better than the worst member of the parent population), the difference threshold is decremented. The effect of this mechanism is that only the more diverse potential parents are mated, but the diversity required by the difference threshold automatically decreases as the population naturally converges. The number of survivors for each generation stays remarkably constant throughout the search because when CHC is having difficulty making progress, the difference threshold drops faster than the average Hamming distance, so that more individuals are evaluated. Conversely, when CHC is finding it easy to generate children that survive, the difference threshold drops at a slower rate, and the number of matings falls. (See Eshelman and Schaffer (1991) for a more detailed discussion of the benefits of incest prevention.)

5 Restarts

Nothing has been said so far concerning mutation. This is because in the reproduction-recombination cycle CHC does not use any mutation. The use of HUX and incest prevention in conjunction with a population size large enough to preserve a number of diverse structures (e.g., 50) enables CHC to delay premature convergence, and thus do quite well without any mutation. But these various mechanisms cannot guarantee that no allele will prematurely converge. Some sort of mutation is needed.

Mutation, however, is less effective in CHC than in the traditional GA. Since CHC is already very good at maintaining diversity, mutation contributes very little early on in the search. On the other hand, late in the search, when the population is nearly converged, mutation, combined with elitist selection, is not very effective in reintroducing diversity. At this stage of the search mutation will rarely produce an individual who is better than the worse individual in the population, and consequently, very few new individuals will be accepted into the population. In contrast to CHC, a traditional GA, by replacing the parent population each generation, insures that new variations will constantly be introduced.[3]

CHC's way out of this impasse is to introduce mutation only when the population has converged or search has stagnated (i.e., the difference threshold has dropped to zero and there have been several generations without any new offspring accepted into the parent population). More specifically, whenever the reproduction-recombination cycle

[3] This might explain why the Evolution Strategy work in Germany (Hoffmeister and Bäck, 1990), which initially used a form of cross-generational competition, later abandoned it, opting for a selection strategy more closely resembling that used by the traditional GA.

achieves its termination condition, the population is reinitialized (diverged) and the cycle is repeated. The reinitialization, however, is only partial. The population is reinitialized by using the best individual found so far as a template for creating a new population. Each new individual is created by flipping a fixed proportion (e.g., 35%) of the template's bits chosen at random without replacement. One instance of the best individual is added unchanged to the new population. This insures that the next search cannot converge to a worse solution than the previous search. This outer loop, consisting of reinitialization followed by genetic search, is iterated until its termination condition is met (either reaching a fixed number of reinitializations, or repeated failure to find any better structures than the retained structure).[4]

It should be pointed out that for CHC there is no danger that the best individual found so far, even though included in the reinitialized population, will rapidly take over the population. CHC's incest preventing mechanism (the dropping difference threshold), in combination with elitist selection and disruptive recombination prevents this. This does not mean, however, that saving the best individual has no effect on search. Crossover will still recombine the schemata represented in this surviving best individual with those of the other individuals in the population, but mating is prevented whenever two individuals are paired that are relatively more similar than the population as a whole as reflected by the difference threshold.

An advantage of partial reinitializations over chronic mutation is that CHC can do quite well on a large range of problems using the same parameter settings. Restarts provide many of the benefits of a large population without the cost of slower search. On easy problems the optimal solution is found in the first initialization cycle, whereas on hard problems the optimal solution is found only after repeated restarts.

6 Empirical Results

CHC's performance has been compared with that of a traditional GA on a number of functions. The most extensive comparisons are based on a test suite of 10 functions, F1-F10, previously studied by Schaffer, et al. (1989). They tested a traditional GA[5] on functions F1-F10 with 840 different parameter settings including 6 population sizes, 10 crossover rates, 7 mutation rates, and two crossover operators (one- and two-point). Each search was run for 10000 evaluations and repeated 10 times. The test suite consisted of the five De Jong functions (1975) as well as five other multi-modal functions, including two sine-wave-based functions, a FIR digital filter optimization task, a 30-city traveling salesman problem (with a sort order representation), and a 64-node graph partition task.

In order to identify how well a GA can perform on each of these 10 functions, I chose for each function the 5 parameter sets that worked best (i.e., found the best solutions

[4] Ackley's (1987) IGS algorithm also includes an outer, reinitialization loop, but IGS still retains mutation in the inner loop, and fully randomizes the population when it is reinitialized.

[5] The GA in Schaffer's study used Baker's (1987) stochastic universal sampling, dynamic linear fitness scaling (scaled to the worst individual) (Grefenstette and Baker, 1989), the individual elitist strategy (Grefenstette, 1986) (not to be confused with the population-elitist selection discussed in this paper), a generation gap of 1.0 (i.e., the entire parent population is replaced by the children), and Gray coding for the numeric parameters (Caruana and Schaffer, 1988).

in the least number of evaluations) from among the 840 parameter sets tested and conducted runoffs. In this second series of experiments, the search was continued until the global optimum was found or until a maximum of 50000 evaluations was reached. (Because F4 is a noisy function, it was treated somewhat differently. The search was halted whenever the performance value before adding noise was less than two standard deviations of the noise function from the optimum (i.e., F4(x) < 2).) Each search was repeated 50 times. CHC was also tested on F1-F10 under the same conditions except that a single set of parameters was used for all the functions (a population size of 50 and a divergence rate of 35%). Note that since a single parameter set was used for CHC whereas the parameter settings for the GA were selected independently for each function, there is a methodological bias in favor of the traditional GA.

Table 1: Mean Number of Evaluations to Find the Global Optimum

Func	Traditional GA			CHC		
	#opt[a]	mean[b]	sem[c]	#opt	mean	sem
F1	50	805	48	50	1089	25
F2	50	9201	703	50	9065	591
F3	50	1270	100	50	1169	27
F4	50	2228	135	50	1948	97
F5	50	1719	96	50	1396	38
F6	37	9272	1291	50	6496	725
F7	50	8688	738	50	3634	291
F8	11	30986	3712	50	7279	725
F9	1	24402	NA	29	24866	2404
F10	2	5520	1087	33	10217	1107

[a] The number of searches out of 50 that the algorithm succeeded in finding the optimum value.

[b] Mean number of evaluations to find the optimum in those searches where it did find the optimum.

[c] Standard error of the mean.

Table 1 compares the performance of CHC with the best GA for each of the functions. In spite of the methodological bias in favor of the GA, CHC does better on nine of the ten functions. (An algorithm performs better on a function if it finds the global optimum more often, or it finds the global optimum the same number of times as its competitor but in fewer evaluations.) For five of the six functions in which both algorithms found the optimum in all 50 searches, CHC, on average, found the optimum in fewer evaluations, and for the remaining four functions, CHC found the optimum more often than the GA. Furthermore, the only function on which the traditional GA does significantly better than CHC is F1, a smooth, unimodal function that is the easiest for most traditional optimization techniques. On the other hand, CHC does significantly better on all the multi-modal functions (F5-F10).

Since the traditional GA was rarely able to find the optimum on the last three functions, perhaps it is only fair to consider how well CHC does when the number of evaluations is limited to 10000 as was the case in the study done by Schaffer, et al.

The GA was able to find the optimum once for F8 and twice for F10 within 10000 evaluations whereas CHC was able to find the optimum 42 times for F8 and 20 times for F10. In the case of F9, a 30-city traveling salesman problem, neither algorithm was very successful at finding the optimum (i.e., best known tour of 421) within 10000 evaluations — CHC found it three times and the GA not at all. If we look, however, at the mean of the shortest tours found over 50 runs, it is obvious that CHC is doing significantly better. The mean performance for CHC was 461 after 10000 evaluations and 429 after 50000 evaluations, whereas for the GA it was 538 after 10000 evaluations and 482 after 50000 evaluations. (The standard error for all four means was less than 5.)

Finally, something should be said about the role played by restarts. The divergence rate used in these experiments (35%) is a rate that works well over a wide range of functions; however, it is not necessarily the best rate for functions F1-F10. In fact, performance can be significantly improved on F10 (so that it will find the optimum in all 50 runs) without significantly affecting performance on the other functions, by using a divergence rate of 35%, but completely randomizing the population whenever there is a total of three restarts that do not result in improvements. In should also be pointed out that CHC rarely had to resort to restarts when searching the easier functions, F1, F3-F5. The functions that required the most restarts were the most difficult functions, F9 and F10 — an average of 6.2 restarts per run for F9 and 9.6 per run for F10. This is not to imply that it is simply the use of restarts that accounts for CHC's better performance on these more difficult functions. For both F9 and F10, CHC's performance was significantly better at the point of the first restart (7798 and 4302 mean evaluations, respectively) than the GA had achieved after 50000 evaluations.

7 Deceptive Problems

Although the results of the previous section indicate that CHC is a robust search algorithm, it should be noted that there are two identifiable classes of functions that are consistently more difficult for CHC to optimize than the traditional GA: (1) functions that are easy for a robust hillclimber (e.g., a stochastic hillclimber) and (2) tightly ordered (benign) deceptive functions. These are functions for which CHC's usual assets are liabilities. CHC's mechanisms for slowing convergence, especially incest prevention, enable it to perform a coarse-to-fine search of the space. The traditional GA's tendency to quickly converge on a promising area of the search space can be to its advantage, however, if no additional insights are to be gained from knowledge of the problem's macrostructure.

In the case of tightly ordered deceptive functions — functions in which the bits of the deceptive subproblems are adjacent — the traditional GA's positional bias — favoring schemata of short defining length — gives it an advantage over CHC, provided the GA uses a very low mutation rate (e.g., zero) and a large population size (e.g., 500). As was argued in section 3, however, there is no reason to believe that real problems are going to be so benign. After all, GAs are supposed to be most useful for poorly understood search spaces where there is little or no a priori knowledge about which bits interact, and so little likelihood that the problem will be represented so that bits constituting deceptive subproblems will be tightly ordered.

In order to get a better handle on CHC's ability to deal with deceptiveness, CHC was run on Goldberg's order-3 deceptive problem (Goldberg, Korb and Deb, 1989) and Liepins' order-n deceptive problems for n of 3, 4 and 5 (Liepins and Vose, 1991). All four problems consisted of 10 deceptive subproblems. Liepins' fully deceptive function, f, for a string of length n is defined as follows:

$$f(x) = \begin{cases} 1 - \dfrac{1}{2n}, & \text{if } o(x) = 0; \\ 1, & \text{if } o(x) = n; \\ 1 - \dfrac{1+o(x)}{n}, & \textit{otherwise}, \end{cases}$$

where $o(x)$ is the number of 1's in the string. Goldberg's fully deceptive order-3 function is defined in Table 2.

Table 2: Goldberg's order-3 deceptive problem

f(000) = 28		
f(001) = 26	f(010) = 22	f(100) = 14
f(011) = 0	f(101) = 0	f(110) = 0
f(111) = 30		

Table 3 shows CHC's performance for the four functions based on 50 replications of each search. The first line shows the results using the "standard" parameter settings — population size of 50 and divergence rate of 0.35. Results for two other divergence rates (but the same population size) are also shown.

Table 3: CHC's Performance on Deceptive Functions:
Percent of Subproblems Solved in 50000 Evaluations

div rate	Goldberg's order-3	Liepins' order-3	Liepins' order-4	Liepins' order-5
0.35	85.8	100	79.6	19.2
0.25	99.2	100	92.2	25.6
0.15	100	100	99.6	17.4

Goldberg (1989) reports that for his order-3 problem a traditional GA was able to find the optimum in about 40000 trials if the subproblems are tightly ordered, but converges to the local optimum (all 0's) if the subproblems are loosely ordered (the first 3-bit subproblem is located at positions 1, 11, and 21, the second subproblem is located at positions 2, 12 and 22, etc.). His messy GA was able to find the optimum for the loosely ordered problem in about 40000 evaluations. Since CHC's crossover operator is not position biased, its performance is not affected by the ordering. CHC was able to find the optimum in 20960 function evaluations (mean of 50 runs, with a minimum of 9933, a maximum of 36297 evaluations, and a standard error of 980) when the divergence rate was 0.15. This makes CHC a worthy competitor to Goldberg's messy GA (at least for low-order, loosely ordered deceptive problems). For the Liepins' order-3 problem the mean number of evaluations to find the optimum were 7196, 6207, and 6170, respectively, for the three divergence rates of 0.35, 0.25, and 0.15. It should also be pointed out that if we don't restrict CHC to using HUX but allow CHC to use two-point, reduced surrogate crossover and a divergence rate of 0.5, CHC is able to find the global optimum for Goldberg's order-3 deceptive problem in 3162 evaluations and for Liepins order-3, -4 and -5 deceptive problem in 2143, 4929, 11230 evaluations, respectively (means for 50 runs).

8 Extending CHC to Permutation Problems

This section illustrates how it is possible to extend the philosophy of CHC to a GA that manipulates permutations of integers to solve traveling salesman problems. The basic algorithm is the same as before except that it uses different operators for mutation (when it does a restart) and crossover, and each individual is improved by a hillclimber before it is evaluated, thus limiting the search to the space of local minima. The operators are similar to those used by Muhlenbein's ASPARAGOS/PGA (Gorges-Schleuter, 1989, Muhlenbein, 1990). The crossover operator creates a single child by preserving the edges that the parents have in common and then randomly assigning the remaining edges in order to generate a legal tour. Whenever the population converges, the population is partially randomized for a restart by using the best individual found so far as a template and creating new individuals by repeatedly swapping edges until a specific fraction of the edges (e.g., 30%) differ from those of the template. Whenever an individual is created for the initial population, by mutation for a restart, or by crossover, a hillclimber is used to improve the tour. The hillclimber is a variant of Lin's two-opt (1973). Subtours are systematically inverted by swapping two edges. Whenever an improvement is found, the tour is updated, and the hillclimber continues until it can make no more improvements. If the individual is created by crossover or mutation (if a restart), the hillclimber is constrained to swapping the new edges — i.e., the edges that were created randomly by crossover or by mutation in the case of a restart. There is a tremendous improvement in efficiency by restricting the edges that can be swapped by the hillclimber. This is especially apparent when used in conjunction with crossover — the time devoted to hillclimbing drops exponentially as the population converges and the parents have more edges in common.

Incest prevention works the same as it does for the binary version of CHC, except that instead of counting the number of bits that the two parents have in common, the number of edges that they share is counted. If the number of common edges is not above the incest threshold, the potential parents are not mated. Only one child is created per mating. No changes were made in CHC's selection procedure.

The algorithm was tested on Padberg's 532-city traveling salesman problem (1987). (Most standard problems of less than 100 cities are not much of a challenge.) Table 4 reports the performance for 10 runs. Each run was allowed 4 restarts with a divergence rate of 30%. The population size was 50.

Table 4: 532-City-TSP Tour Lengths

mean	sem	best	worst	#runs <27800	#runs <27750
27747	11.8	27710	27849	9	9

It should be noted that the best value (27710) found by CHC is within 0.1% of the optimum (27686). CHC's performance on this problem is comparable to that reported for ASPARAGOS — a mean of 27770 for ten replications with the best run having a tour of 27715 (Gorges-Schleuter, 1989). More recently, Muhlenbein has reported that his parallel GA has been able to find a tour of 27702 after 3 hours of search using 64 processors (1990). This result outclasses any previous GA on a large TSP problem. Although CHC's performance is not quite as good as that reported by Muhlenbein, it should be pointed out that CHC was able to find a tour of 27710 in under 3 hours (the average run time was 2.5 hours per replication) on a *single* processor — a Sun SPARCstation 1+.

9 Conclusion

CHC outperforms the traditional GA as a function optimizer over a wide range of functions. Given a fixed number of evaluations or the goal of finding the optimum in a minimum number of trials, a GA that can maintain population diversity while using a small population size (e.g., 50) will have an advantage, since large population sizes mean fewer generations and, thus, less search. CHC is able to maintain population diversity without sacrificing implicit parallelism by combining a highly disruptive crossover operator with a conservative selection procedure that preserves any progress made.

The evolution of CHC has been a story in which, once elitist selection was in place, every change that increased the recombination power of crossover, and consequently its disruptiveness, also improved CHC's performance across a wide variety of functions. The switch from traditional crossover to UX, and then to HUX, and finally the addition of the downward adjusting difference threshold to prevent "incest" have all led to dramatic improvements. Moving mutation from inside to outside the reproduction-recombination loop also significantly improved performance, especially on the more difficult problems, and was instrumental in allowing CHC to perform well on a number of functions using the same set of parameter values.

Acknowledgements

Dave Schaffer, Dan Offutt, and Rich Caruana have contributed many useful ideas throughout the evolution of CHC. I would also like to thank Dave Schaffer, Dan Offutt and Benjamin Zhu for their helpful comments on various drafts of this paper.

References

D. H. Ackley. (1987) *A Connectionist Machine for Genetic Hillclimbing.* Boston, MA: Kluwer Academic Publishers.

J. E. Baker. (1987) Reducing Bias and Inefficiency in the Selection Algorithm. *Proceedings of the Second International Conference on Genetic Algorithms and Their Applications, 14-21. Hillsdale, NJ: Lawrence Erlbaum Associates.*

L. Booker. (1987) Improving Search in Genetic Algorithms. In L. Davis (ed.), Genetic Algorithms and Simulated Annealing, 61-73. San Mateo, CA: Morgan Kaufmann.

R. A. Caruana and J. D. Schaffer. (1988) Representation and Hidden Bias: Gray vs. Binary Coding for Genetic Algorithms. *Proceedings of the 5th International Conference on Machine Learning*, 153-161. San Mateo, CA: Morgan Kaufmann.

K. A. De Jong. (1975) Analysis of the Behavior of a Class of Genetic Adaptive Systems. Ph.D. Dissertation, Department of Computer and Communication Sciences, University of Michigan, Ann Arbor, MI.

L. J. Eshelman and J. D. Schaffer. (1991) Preventing Premature Convergence in Genetic Algorithms by Preventing Incest. *Proceedings of the Fourth International Conference on Genetic Algorithms*. San Mateo, CA: Morgan Kaufmann.

D. E. Goldberg. (1987) Simple Genetic Algorithms and the Minimal, Deceptive Problem. In L. Davis (ed.), *Genetic Algorithms and Simulated Annealing*, 74-88. San Mateo, CA: Morgan Kaufmann.

D. E. Goldberg, B. Korb and K. Deb. (1989) Messy Genetic Algorithms: Motivation, Analysis, and First Results. TCGA Report No. 89003. To appear in *Complex Systems*.

M. Gorges-Schleuter. (1989) ASPARAGOS An Asynchronous Parallel Genetic Optimization Strategy. *Proceedings of the Third International Conference on Genetic Algorithms*, 422-427. San Mateo, CA: Morgan Kaufmann.

J. J. Grefenstette. (1986) Optimization of Control Parameters for Genetic Algorithms. *IEEE Transactions on Systems, Man & Cybernetics* **SMC-16** (1): 122-128.

J. J. Grefenstette and J.E. Baker. (1989) How Genetic Algorithms Work: A Critical Look at Implicit Parallelism. *Proceedings of the Third International Conference on Genetic Algorithms*, 20-27. San Mateo, CA: Morgan Kaufmann.

F. Hoffmeister and T. Bäck. (1990) Genetic Algorithms and Evolution Strategies: Similarities and Differences. *Proceedings of the First International Workshop on Parallel Problem Solving from Nature*. Dortmund, Germany: University of Dortmund.

J. H. Holland. (1975) *Adaptation in Natural and Artificial Systems*. Ann Arbor, MI, University of Michigan Press.

G. E. Liepins and M. D. Vose. (1991) Representational Issues in Genetic Optimization. *Journal of Experimental and Theoretical AI* (May).

S. Lin and B. W. Kernighan. (1973) An Effective Heuristic Algorithm for the Traveling Salesman Problem. *Operations Research* 21: 498-516.

H. Muhlenbein. (1990) Parallel Genetic Algorithms and Combinatorial Optimization. Symposium on Parallel Optimization. Madison, WS.

W. Padberg and G. Rinaldi. (1987) Optimization of a 532-City Symmetric TSP. *Operation Research Letters* 6: 1-7.

J. D. Schaffer, R. A. Caruana, L. J. Eshelman and R. Das. (1989) A Study of Control Parameters Affecting Online Performance of Genetic Algorithms for Function Optimization. *Proceedings of the Third International Conference on Genetic Algorithms*, 51-60. San Mateo, CA: Morgan Kaufmann.

J. D. Schaffer, L. J. Eshelman and D. Offutt. (1991) Spurious Correlations and Premature Convergence in Genetic Algorithms. In G. J. E. Rawlins (ed.), *Foundations of Genetic Algorithms and Classifier Systems.* San Mateo, CA: Morgan Kaufmann.

G. Syswerda. (1989) Uniform Crossover in Genetic Algorithms, *Proceedings of the Third International Conference on Genetic Algorithms*, 2-9. San Mateo, CA: Morgan Kaufmann.

D. Whitley. (1989) The GENITOR Algorithm and Selection Pressure: Why Rank-Based Allocation of Reproductive Trials is Best. *Proceedings of the Third International Conference on Genetic Algorithms*, 116-121. San Mateo, CA: Morgan Kaufmann.

Appendix: Pseudocode for CHC

```
procedure CHC
begin
    t = 0;
    d = L/4;
    initialize P(t);
    evaluate structures in P(t);
    while termination condition not satisfied do
    begin
        t = t + 1;
        select_r C(t) from P(t-1);
        recombine structures in C(t) forming C'(t);
        evaluate structures in C'(t);
        select_s P(t) from C'(t) and P(t-1);
        if P(t) equals P(t-1)
            d--;
        if d < 0
        begin
            diverge P(t);
            d = r × (1.0 - r) × L;
        end
    end
end.
```

```
procedure select_r
begin
   copy all members of P(t-1) to C(t) in random order;
end.

procedure select_s
begin
   form P(t) from P(t-1)
      by replacing the worst members of P(t-1)
      with the best members of C'(t)
   until no remaining member of C'(t)
      is any better than any remaining member of P(t-1);
end.

procedure recombine
begin
   for each of the M/2 pairs of structures in C(t)
   begin
      determine the Hamming_distance
      if (Hamming_distance/2) > d
         swap half the differing bits at random;
      else
         delete the pair of structures from C(t);
   end
end.

procedure diverge
begin
   replace P(t) with M copies of the best member of P(t-1);
   for all but one member of P(t)
   begin
      flip r × L bits at random;
      evaluate structure;
   end
end.

variables
   M  population size
   L  string length
   t  generation
   d  difference threshold
   r  divergence rate
```

Genetic Operators for Sequencing Problems

B.R. Fox and M.B. McMahon
Planning and Scheduling Group
McDonnell Douglas Space Systems Co.
16055 Space Center Blvd
Houston, Tx 77062

Abstract

The traditional recombination operators which work well when solutions are coded as bit strings do not work when the solutions are coded as sequences. Little research has been devoted to the understanding of how recombination operators can be applied to sequences. A thorough understanding of recombination operators for sequences will open up the possibility of applying genetic algorithms to a wide variety of problem domains, including scheduling. In this paper, several recombination operators are evaluated and a model for sequences and two new recombination operators are introduced.

Keywords: scheduling, sequences, genetic operators, recombination

1 Introduction

The Planning and Scheduling Group at MDSSC-Houston is responsible for the development of generic scheduling technology that can be applied to a wide variety of scheduling problems within NASA, including Space Shuttle and Space Station Freedom operations. There are two basic goals: to develop basic technology that can be used in the implementation of a domain independent interactive scheduler [Fox 1989, 1990] and to incorporate advanced optimization methods developed at universities and NASA research centers. Candidate optimization algorithms are evaluated on the basis of both their theoretical properties and empirical performance.

Given the intractable nature of scheduling problems, it is natural to consider recent research in Genetic Algorithms which may offer some advantages over traditional search techniques. A preliminary survey of the literature reveals a small number of GA papers devoted to the Traveling Salesman Problem (TSP) [Goldberg 1985, Oliver 1987, Whitley 1989]. Unfortunately, most scheduling problems within NASA are best characterized as Job-shop Scheduling Problems (JSP). Under most circumstances, algorithms that are based upon specific properties of the TSP cannot be applied to the JSP. However, the domain independent nature of GA's suggests that an effective means for generating an optimal sequence of cities in a traveling salesman problem can be used to generate an optimal sequence of scheduling operations in a job-shop scheduling problem.

The application of GA's to the TSP presents specific difficulties because of the constraints imposed upon the symbol string that represents a tour of cities. While GA's traditionally assume that the symbols within a genotype can be independently modified and rearranged, the string that represents a tour of cities must contain exactly one instance of each city label. Any omission or duplication of labels constitutes an error. Clearly, the traditional cross-over and mutation operators have a tremendous capacity for producing non-viable off-spring because of omission and duplication of city labels. However, the motivating principles behind these operators still seem applicable in this domain. The cross-over operator is a mechanism for incorporating the best attributes of two parents into a new individual, and the mutation operator is a mechanism for introducing necessary attributes into an individual when those attributes do not already exist within the current population. Hence, GA research related to the TSP is concerned with creating analogs of the traditional operators that satisfy these general principles and respect the constraints imposed upon the symbol strings that represent city tours.

2 Statement of the Problem

Evaluation of genetic reordering operators proposed in the literature is complicated by several factors. First, there is no common framework for analyzing sequences, and the information they contain, that can be used for comparing the theoretical properties of the operators. Second, the proposed operators are closely coupled with other elements of the traditional GA: fitness functions, mating strategies, selective pressure, etc. Hence, it is difficult to identify the factor or factors most responsible for good performance. But most important, comparisons made in the literature provide only partial information. An objective assessment of proposed operators must consider code size and complexity, actual execution time, number of generations, best solution, initial rate of convergence, long-term rate of convergence, etc.

The problem, addressed in this paper, is to perform an objective assessment of genetic reordering operators and to determine if the underlying principles of genetic algorithms, as applied to bit-string problems, can be used to find solutions for scheduling problems. This involved creating a generic GA testbed (GENOA) so that proposed operators could be compared in a context where the operator was the only variable part and it involved developing a representation for sequences that clearly showed how portions of each parent may be used to construct children. Serendipitously, the representation for sequences used in this investigation enabled the development of two new GA reordering operators, and the generic testbed enabled the investigation of other factors, such as evaluation functions and mating strategies, that have effect on performance.

3 Survey of Related Research

Genetic operators for reordering a sequence of symbols may be divided into two classes: unary operators, which require one parent to produce one or more children, and binary operators, which require two parents.

Unary operators take an initial sequence of symbols and reorder a part of that sequence, keeping part of the original sequence in tact. The most primitive reordering operator, swap, simply exchanges two symbols in a sequence. Other unary operators include splice, which removes the symbols between two cutpoints and splices them onto the end of the sequence, and inversion, which reverses the order of the symbols between two cutpoints. Although splice and inversion can be implemented by repeated application of the swap operator, for sake of comparison, they are considered separately in the empirical results.

Binary operators are more sophisticated in theory and implementation. Since they draw information from two parents, binary operators have a greater opportunity to reproduce common or valuable attributes. Four operators found in literature include order crossover (OX) [Davis 1985, Oliver 1987], cycle crossover (CX) [Oliver 1987], partially mapped crossover (PMX) [Goldberg 1989],and edge recombination (ER) [Whitley 1989].

OX creates children which preserve the order and position of symbols in a subsequence of one parent while preserving the relative order of the remaining symbols from the other parent. It is implemented by selecting two random cut points which define the boundaries for a series of copying operations. First, the symbols between the cutpoints are copied from the first parent into the child, into the same positions as they appear in the parent. Then, starting just after the second cutpoint, the symbols are copied from the second parent into the child, omitting any symbols that were copied from the first parent. When the end of the second parent sequence is reached, this process continues with the first symbol in the second parent until all of the symbols have been copied into the child. A second child can be constructed by switching the roles of the parent sequences. For example:

```
Parent 1:  A B C D |  E F G |  H I  J K
Parent 2:  K A G F |  B D H |  I  J C E

 Child 1:  A B D H |  E F G |  I  J C K
 Child 2:  C E F G |  B D H |  I  J K A
```

CX creates children so that the position of each symbol in a child is determined by one of its parents. It is implemented by dividing the set of symbols into two subsets such that the positions of symbols of the first subset are disjoint from the positions of symbols in the second subset. Given two subsets that satisfy this requirement, a child is constructed by copying symbols from the first subset into the child, into the same positions as they appear in the first parent, and similarly, by copying symbols from the second subset into the child, into the same positions as they appear in the second parent. A second child can be constructed by switching the roles of the parent sequences. A set of symbols can be divided into two subsets that satisfy this requirement by a set closure algorithm. First choose a single symbol to serve as the seed of the subset under construction. Then, repeatedly apply the following rule: if symbol X is in the subset under construction and if X conflicts with symbol Y, then symbol Y should be added to the subset under construction. Two symbols X and Y conflict with each other if X holds the same position in one parent as Y holds in the other parent. An example of the CX operator is:

```
Parent 1:  H K C E F D B L A I G J
Parent 2:  A B C D E F G H I J K L

Subset 1:  H             L A I   J
Subset 2:    K C E F D B       G

 Child 1:  H B C D E F G L A I K J
 Child 2:  A K C E F D B H I J G L
```

Unfortunately, if two sequences represent the same cycle of cities but have a different starting point, this algorithm will include all of the cities in the first subset, leaving none for the second. In such circumstances, the children will be identical to the parents.

PMX creates children which preserve the order and position of symbols in a subsequence of one parent while preserving the order and position of many of the remaining symbols from the other parent. It is implemented by selecting two random cut points which define the boundaries for a series of swapping operations. The first child begins as an exact copy of the first parent. The goal is to modify the child so that the symbols between cut points are an exact copy of the symbols between cutpoints in the second parent. This is accomplished by a series of swap operations where each symbol must appear between the cutpoints is swapped with the symbol that occupies its desired position. (The method is slightly more complicated when one of the subject symbols already occupies a position between the cutpoints.) A second child can be constructed by switching the roles of the parent sequences. For example:

```
Parent 1:  H G A | B C G | I E D F
Parent 2:  I H D | E F G | A C B J

 Child 1:  H J A | E F G | I B D C
 Child 2:  I H D | B C J | A F E G
```

ER creates children which preserves edges, or immediate predecessor/successor relationships found in the parent sequences. It is implemented by constructing an edge map that lists the neighbors of each symbol in the sequence. The first symbol to be placed into the child is chosen at random. The next symbol is chosen from the list of the first symbol's neighbors. The process of choosing symbols from the previous symbol's list of neighbors is continued until all symbols are chosen or until there are no neighbors to choose from but there are symbols left. In this case, any of the remaining symbols is chosen at random. In the following example, A is chosen first, followed by B, after that the choices are arbitrary. The child produced in this example consists entirely of edges from its parents. For example:

Parent 1: A B C D E F
Parent 2: D F B A C E

Edge Map: A: B,C
B: A,C,F
C: A,B,D,E
D: C,E,F
E: C,D,F
F: A,B,D,E

Child : A B F D C E

4 Formal Model of Sequences

An alternate representation of sequences was developed that coherently explains properties preserved by these various operators. The representation chosen for this investigation is easily discovered when some familiar mathematics are juxtaposed with traditional GA's. Ideally, the representation for an individual should be a bit-string where each bit represents an independent attribute of the individual. Typically the representation for a city tour with a TSP is a sequence of city names or labels. Converting, the labels to bit-strings and sequences to the concatenation of these bit-strings would only complicate the representation and any operators that might be applied to it. However, a bit-string representation of a sequence is easily derived by another method. Briefly summarized, a sequence of cities in a TSP is an instance of a totally ordered set, a totally ordered set is a special case of a partially ordered set, and partially ordered sets are frequently represented by boolean (0-1) matrices that represent predecessor and successor relationships. For example, the sequence [B,A,C,E,D] can be represented by the matrix shown in Figure 1. Matrix element [X,Y] contains a one, if and only if symbol X occurs before symbol Y in the given sequence. When completed, the column sums of the matrix reveal the number of predecessors of a symbol, which in turn, determines position within the sequence. Similarly the row sums reveal the number of successors.

On first inspection it might appear that the boolean matrix representation obscures the information contained in the sequence. On closer inspection it might appear that the matrix contains a lot of redundant information. In reality, the matrix accurately represents all of the information contained in the sequence in exactly the form required by traditional GA's. The complication, which remains true regardless of representation, is that a city tour must contain exactly one instance of each city label with no omissions or duplications. In the matrix representation this translates into three requirements. First, the matrix for n cities must contain exactly $n(n-1)/2$ ones. Second, the transitive nature of order must be respected: if matrix element [i,j] is one and matrix element [j,k] is one then matrix element [i,k] must also be one. Third, the matrix must not contain any cycles: matrix element [i,i] must be zero for all i. If a matrix contains fewer than $n(n-1)/2$ ones but respects the other constraints then the cities are partially ordered and there is at least one way the

	A	B	C	D	E	
A	0	0	1	1	1	3
B	1	0	1	1	1	4
C	0	0	0	1	1	2
D	0	0	0	0	0	0
E	0	0	0	1	0	1
	1	0	2	4	3	

Figure 1: Matrix representation for sequence: B A C E D

matrix can be completed to produce a totally ordered sequence of cities.

The boolean matrix representation of a sequence encapsulates all of the information about the sequence, including both the micro-topology of individual city-to-city connections and the macro-topology of predecessors and successors. The boolean matrix representation can be used to understand existing operators and to develop new operators that can be applied to sequences to produce desired effects while preserving the necessary properties of the sequence.

5 The Precedence Matrix and Unary Reordering Operators

In general, any permutation of symbols within a sequence can be achieved by a series of operations which exchange two symbols in the sequence. For example, the inversion operator is a selective sequence of swap operations. The head of a subsequence is exchanged with the tail of the subsequence. This exchange continues, moving the head and tail one element closer to the the center of the subsequence, until the two middle elements are swapped or until the head and tail both point to the same element. All of unary operators described in this paper can be implemented as a series of swap operations.

Using the precedence matrix the micro operations that are performed by the swap operation become apparent. The exchange of symbols X and Y in a sequence corresponds to the operation which exchanges rows X and Y and columns X and Y in the corresponding matrix. Exchanging row X with row Y, enumerates the fact that X is taking Y's successors and Y is taking X's successors. By exchanging the columns, each element is exchanging predecessors. Swapping predecessors and successors is an obvious side effect when swapping elements, but it is worthwhile to look at the swap operation within the context of the precedence matrix because it explicitly shows how each relationship between two elements changes as ones get moved from one column to another or from one row to the other. In addition, understanding the effects of moves on the matrix caused by the swap operator will help in understanding the results of more complicated operators.

The matrix analog of this operator illustrates the desirable properties of a matrix operator. First, it preserves the number of ones in the matrix and the transitive relationships that they imply. Second, it does not introduce cycles. The matrix will have these qualities, regardless of the number of times the swap operation is performed. This verifies that the swap operator and series of swaps will always produce viable off-spring.

6 The Precedence Matrix and Binary Reordering Operators

The matrix representation provides partial explanation for the binary operators that have been proposed. Each operator extracts some information from the parent sequences, recombines that information in a child sequence, and possibly introduces some new information not found in either parent. All of the information that defines parent and child sequences is explicitly represented in the matrix representation. For instance, the edges tabulated by the edge recombination operator, are explicitly represented by ones within the matrix representation. However, the matrix stores not only information about the direct predecessor and successor of a symbol, but also stores information about all predecessors and successors. Order crossover and cycle crossover both divide the set of symbols into two subsets and then recombine these two subsets to form a child sequence. In both cases, the operators specifically preserve the relative order of symbols within each subset. This information is explicitly encoded in the matrix representation and can be extracted by selecting the appropriate rows and columns of the parent matrices and can be recombined by relatively simple boolean operations. The relative ordering of two subsets can be enforced by mapping ones into specific elements of the resulting child matrix. These ideas are generalized by two new operators introduced in the following section.

7 Two New Operators

The challenge of a binary operator is to satisfy the general principle, i.e. to incorporate attributes from both parents, and to satisfy the specific constraints that the result must be a sequence of cities with no omissions or duplications. The matrix representation suggests ways that this might be accomplished. Given the constraints imposed upon the boolean matrices it is unlikely that a random selection of bits from two parents would produce viable off-spring. The bits must be chosen in a way that guarantees consistency when interpreted as a sequence.

Two general principles enable the development of operators that satisfy these constraints. The first principle is that the selection of a subset of bits from a single matrix is guaranteed to be consistent, although incomplete. By natural extension, the selection of the subset of bits that are common to two parents will also be consistent. This forms the basis of an intersection operator. The second principle is that a subset of bits from one matrix can be safely combined with a subset of bits from a second matrix if these two subsets have no intersection. This forms the basis of a union operator.

8 The Intersection Operator

The intersection operator was motivated by the principle that the characteristics that are common to two good solutions should be passed on to the children. On first inspection, it may appear difficult to extract all of the characteristics common to two sequences. That is, given two sequences, [A,B,C,D,E,F,G] and [A,E,F,D,B,C,G], determine all of the predecessor/successor relationships that the two have in common. The precedence matrix makes this task trivial.

This concept of using the precedence matrix to find common predecessor/successor relationships between two sequences is the basis of the intersection operator. A descendant sequence is constructed from two parents in three steps. Given the boolean matrix representation of the parent sequences, first create the matrix which is the logical AND (intersection) of the two parents. The result represents exactly all of the predecessor and successor relationships that are common to the two parents, including micro and macro topological information. For example, the boolean matrix representation of the sequences [A,B,C,D,E,F,G] and [A,E,F,D,B,C,G] and their intersection is shown in Figure 2. The intersection requires that A precede all of the other cities, G follow all of the other cities, B precede C, and E precede F, but B, C, D, E, and F can be interleaved in any fashion that satisfies any of the constraints.

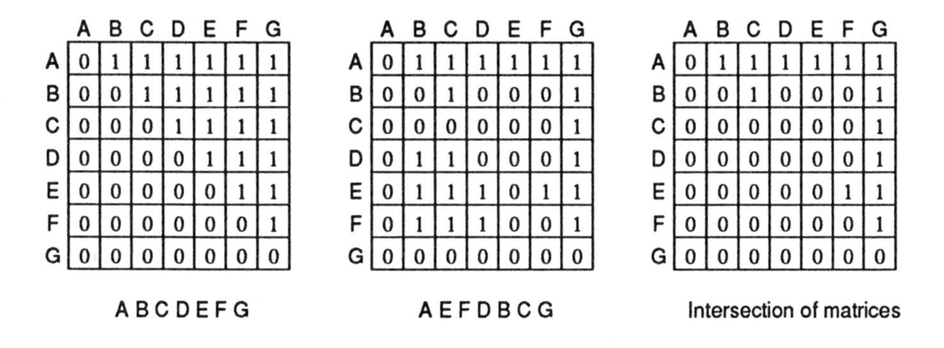

Figure 2: Taking the Intersection of Two Sequences

Second, add into that matrix a subset of the ones that are unique to one of the parents. This results in a matrix that strongly favors both parents with some attributes unique to one parent, but with some attributes still undefined. The intersection operator is reminiscent of Whitley's edge recombination operator because it includes all of the edges actually in common to both sequences along. But, it also includes the macro ordering relationships (such as the relationships of A and G to the others) and it does not include the inverse connections used in Whitley's operator. Finally, convert this underconstrained matrix to a sequence through an analysis of the row/column sums. This is analogous to completing the matrix without the necessity of actually performing the matrix manipulations. For example, the sequence [A,B,E,D,F,C,G] is one possible sequence derived from the intersection. Preliminary results on the performance of this operator compared with other proposed operators is provided in a later section.

9 The Union Operator

The union operator is reminiscent of the traditional crossover operator but is careful to avoid the introduction of transitive links that might result in cycles. A descendant sequence is constructed from two parents in four steps. Given the boolean matrix representation of the parent sequences, first partition the set of symbols (cities) into two disjoint sets. Second, construct the matrix which contains the bits from the first parent that define the relationships within the first subset of cities and construct the matrix which contains the bits from the second parent that define the relationships within the second subset of cities. Third, perform the logical OR (union) of these two matrices resulting in a matrix that contains unique attributes from both parents but with some attributes still undefined. Finally, like the intersection operator, convert this underconstrained matrix to a sequence through an analysis of the row/column sums. For example, the boolean matrix representation of the sequences [A,B,C,D,E,F,G] and [A,E,F,D,B,C,G] and their union when partitioned into subsets A, B, C and D, E, F, G is shown in Figure 3.

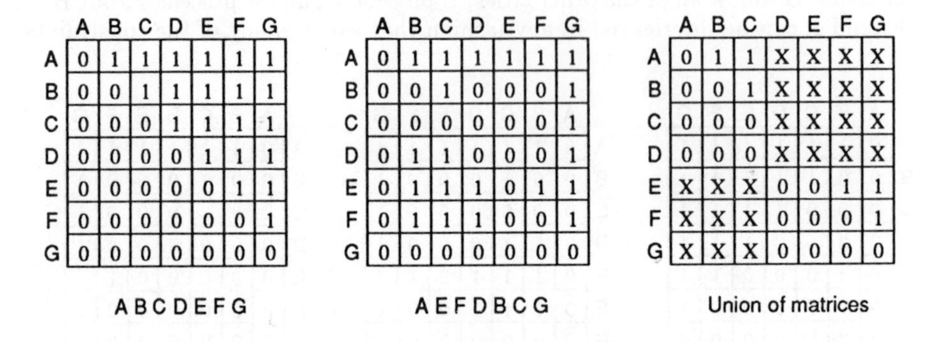

Figure 3: Taking the Union of Two Sequences

(Equivalent operations can be performed even if the partition does not so conveniently divide the matrix into quadrants. This partition was chosen because it clearly illustrates the principle.) Notice that the undefined portions of the matrix are precisely those quadrants that define the relationships between the two subsets. For example, the sequence [E,A,F,B,D,C,G] is one possible sequence derived from the intersection. Notice, the A, B, and C respect the ordering from the first sequence and D, E, F, and G respect the ordering from the second sequence. In practice, it is safe to populate one of the two remaining quadrants with information from one parent or the other.

Experimental results indicate that the method used to partition the set of symbols into two subsets has a significant effect on performance. If symbols are divided into two subsets by purely random means then the probability that these subsets span long subsequences in the parents is very small. Hence, the matrix that results from the union operator will contain a relatively small amount of information about subsequences found in the parents.

In order to compensate for this inadequacy, a special method can be used to divide the set of symbols into two subsets. One parent sequence is chosen as the basis for the subset decomposition. The first symbol of that parent is copied into the first subset. Thereafter, each successive symbol of that parent is copied into either the first or second subset according to the position of a two-way switch. The switch is not set randomly at each step, instead a random process is used to flip the switch from one position to the other. As long as the switch stays in one position, the algorithm will copy complete subsequences into one subset. Subsequences are broken only when the switch is flipped between positions. As a Markov process, the average subsequence length is inversely proportional to the transition probability. Hence, using this method, the set of symbols can be divided into two subsets that contain a controllable amount of subsequence information.

10 Architecture of the GENOA Testbed

The Genetic Operator Analysis (GENOA) Testbed is a testbed which has been developed for the sole purpose of analyzing reordering operators. It was written in Ada in support of NASA Contract NAS9-17885 and is freely available to anyone working on a government project. Ada provides packaging and the separation of specification and implementation. This allows the easy modification and addition of new genetic operators and related techniques.

In analyzing or comparing operators, it is of primary concern to evaluate operators under similar terms. But in the literature, the operator descriptions are tightly coupled with related techniques, like mating pool selection or elitism. To understand the behavior of the operator itself, it is necessary to separate these techniques from the operator. But, since these techniques play an integral part in providing optimal results, they are implemented as options.

An important concern is the applicability of each genetic operator to a variety of sequencing problems. For example, in the Traveling Salesman Problem, the value of a sequence is equivalent to the value of that sequence in reverse order. This trait is not true of all sequencing problems. In scheduling problems, this is a gross error. To test each genetic operator fairly, the testbed is not limited to any particular problem. Instead, a problem description, like a genetic operator, is an interchangeable part. This allows the operators to be tested over a broad spectrum of problems.

11 Empirical Results

Seven reordering operators were tested on the Traveling Salesman Problem using a variety of city topologies. In addition, a "random" operator was implemented which creates a population of random sequences for each new generation. The motive for the random operator was to set a benchmark for the other operators and to provide insight to the possibility of letting the computer blindly hash out a solution given enough time. If the reordering operators do not find a better solution than the best found using the random operator, then there is no reason to pursue them.

The operators were tested on four different city topologies. In addition they were run once preserving the best solutions each generation (elitism) and once not preserving the best solutions. A generational replacement rather than steady state method was used to preserve solutions. Plots were generated which showed the score of the best solution along the Y-axis and the number of generations along the X-axis. Another set of plots were run based on time. The four city topologies tested were:

1) Cities whose distances were randomly generated
2) Clusters of cities (to test micro and macro topologies)
3) Cities arranged to form concentric circles
(optimal solutions are easily visualized)
4) The 30 city problem whose coordinates are given in Oliver's paper [Oliver 1987].

To test the operators fairly, a standard set of parameters was selected by which all tests were run. These parameters settings were:

Sequence Length = 30
Population Size = 900
Selection Method = Tournament Selection (Tournament size = 2)
No mutation.

Figures 4 through 7 show the results of each of the operators using Oliver's 30 City problem. The graphs represent a typical run. The best (shortest path) from each generation is plotted. Figure 4 shows the results of 100 generations without preserving good solutions. Figure 5 shows what happens when elitism is used. Figure 6 shows a timed run of 45 minutes without elitism and Figure 7 shows the timed run using elitism. The purpose of these figures is not to display specific values at given times, but to give the characteristic behavior of each of the operators. For this reason, each graph will not be described in detail, rather the graphs are presented in sets so that the behavior of the operators may be observed.

Some of the characteristics to look for in these graphs include: 1) the rate of progress - evident by looking at the grade of the graph; 2) continual progress - evident by the direction of the graph, especially on longer runs; 3) the amount of noise - evident by the variation between generations; 4) the execution speed - evident by the number of points plotted on the timed runs.

In general, the unary operators were very effective in finding good solutions. All three of the operators (Swap, Splice, and Invert) showed this similar behavior, regardless of the problem topologies. This may be in part due to the fact that minor changes produce little disruption. A large amount of disruption can prevent the GA from stabilizing, while just the right amount can help the GA find the global optimum.

The binary operators did not perform as well as the unary operators. This may be in part due to their sensitivity to the selection of parameters, such as population size and selection methods, or it may be that they naturally create more disruption.

Both binary and unary operators do remarkably better when good solutions are preserved. It is interesting to note that preserving good solutions increases the potential for the GA to prematurely converge. This does not appear to be a problem

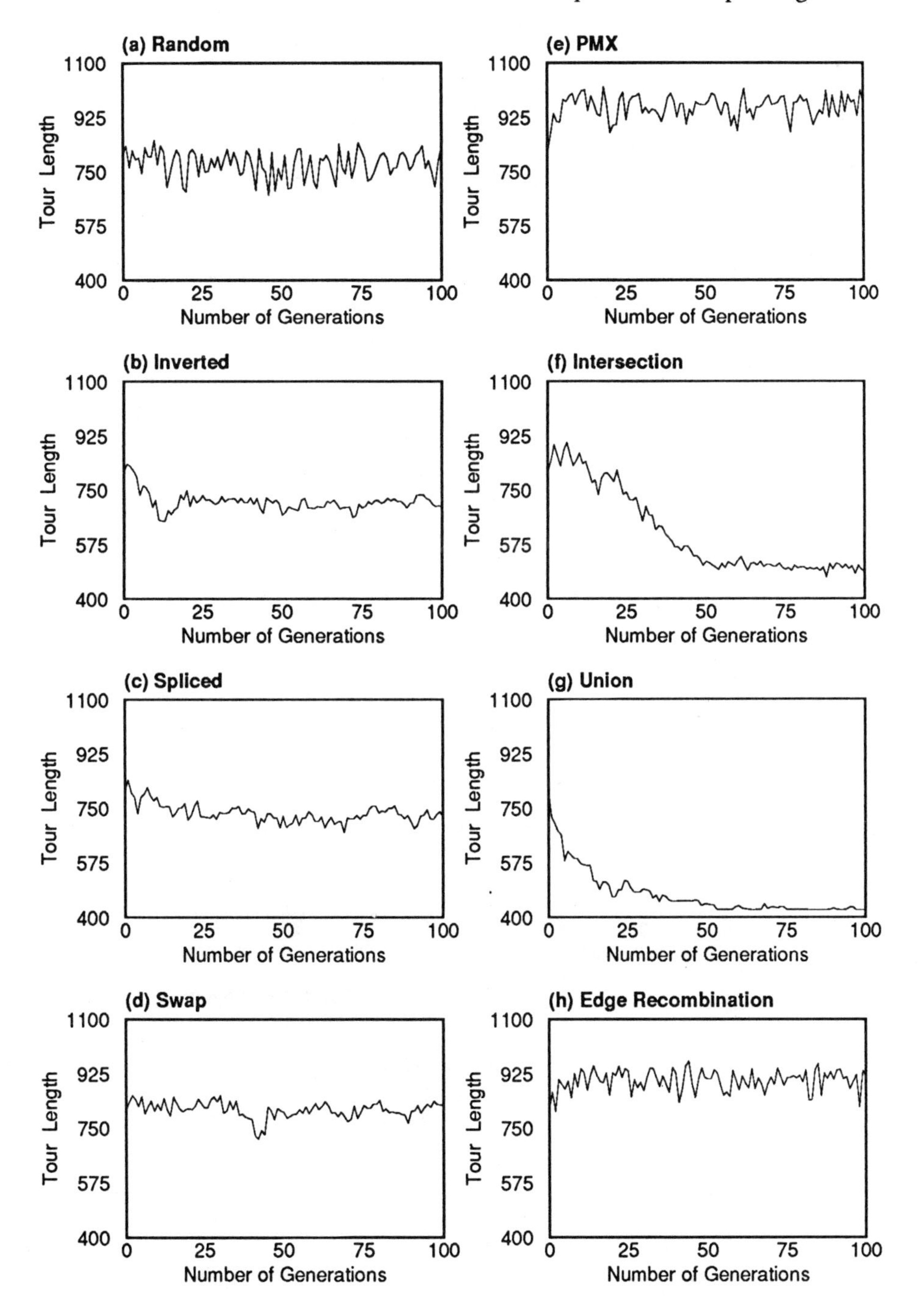

Figure 4: Oliver's 30 City Problem - 100 generations without preserving best solutions of each generation.

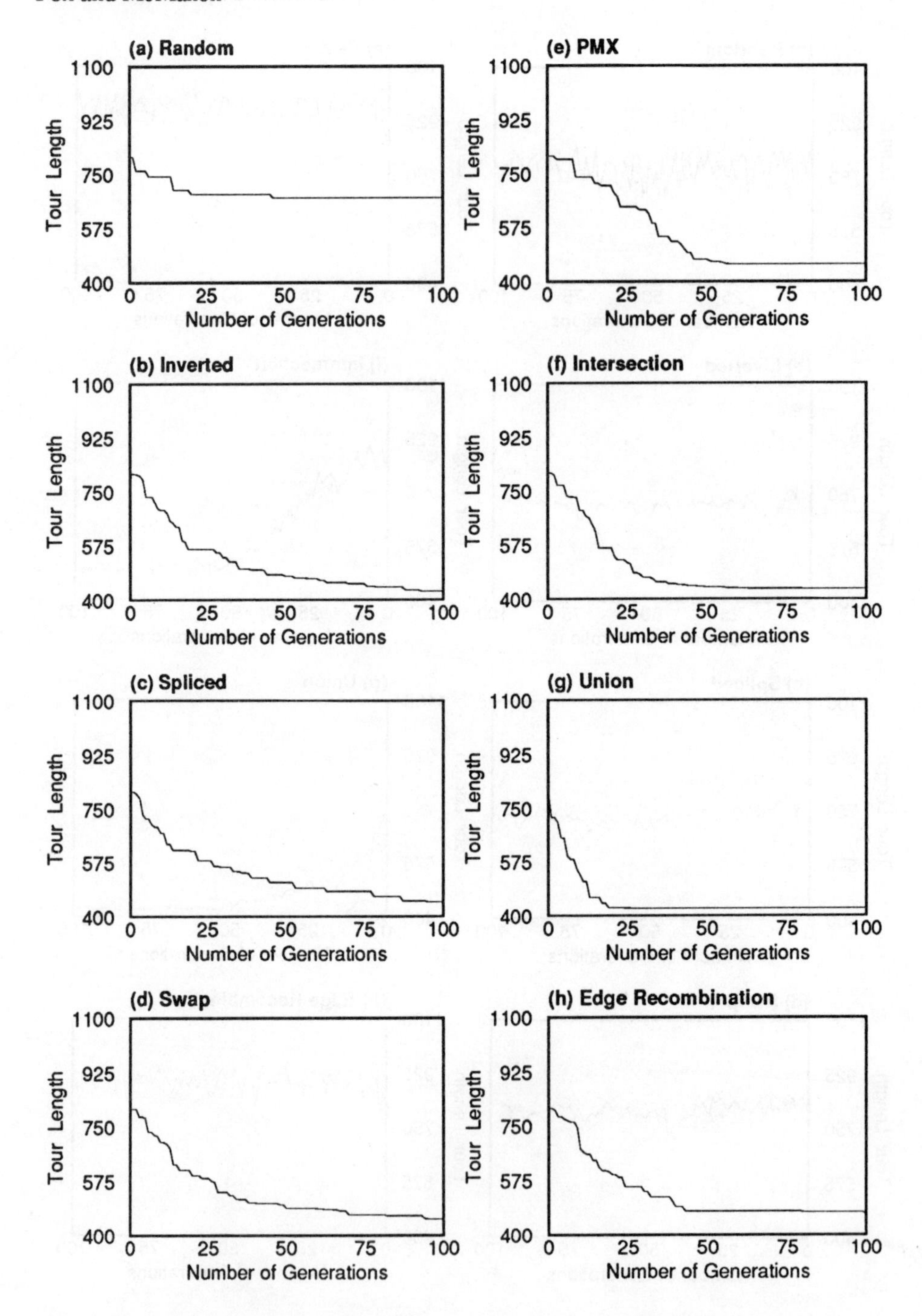

Figure 5: Oliver's 30 City Problem - 100 generations preserving best solutions of each generation.

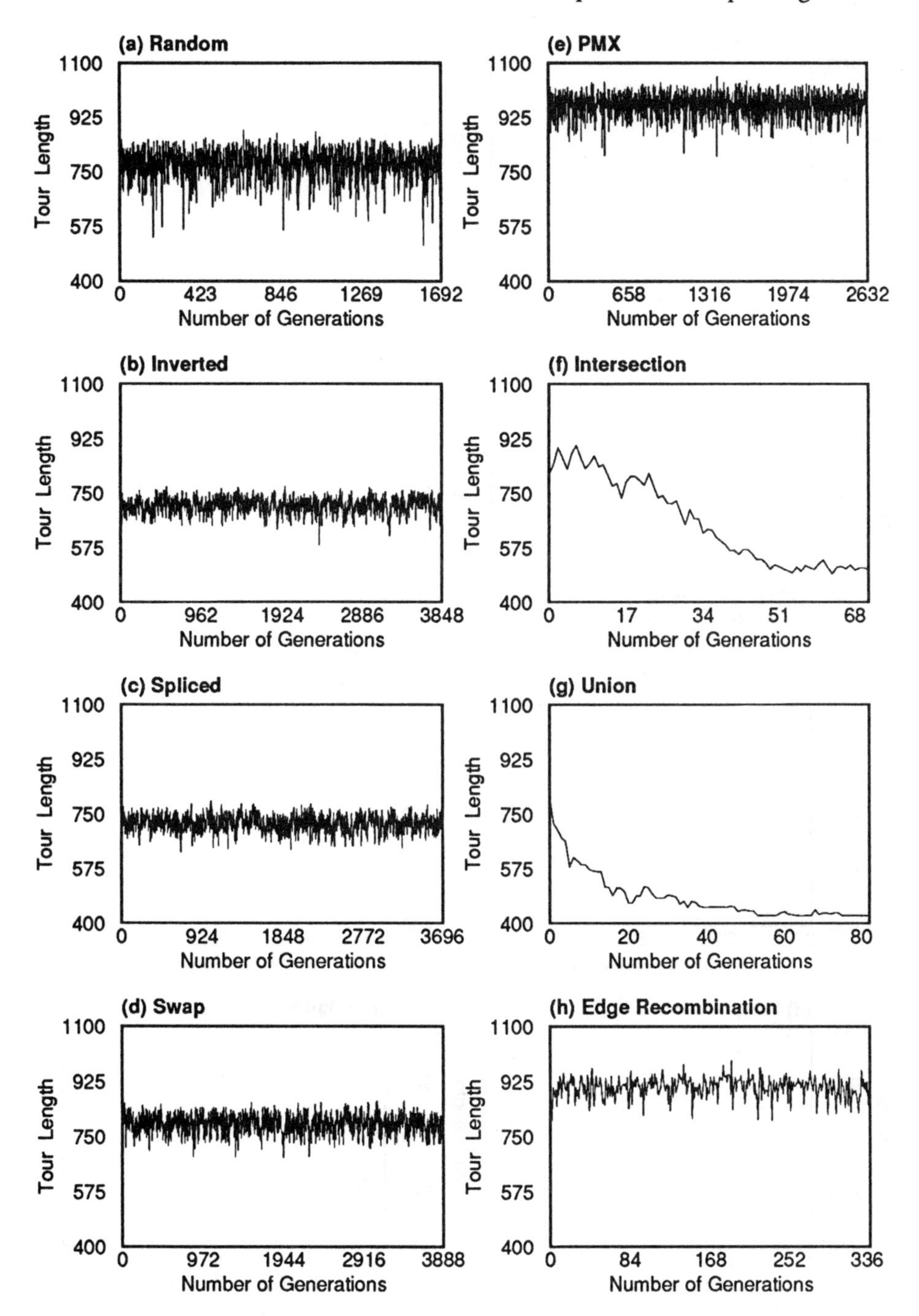

Figure 6: Oliver's 30 City Problem - 45 minute timed run without preserving best solutions of each generation.

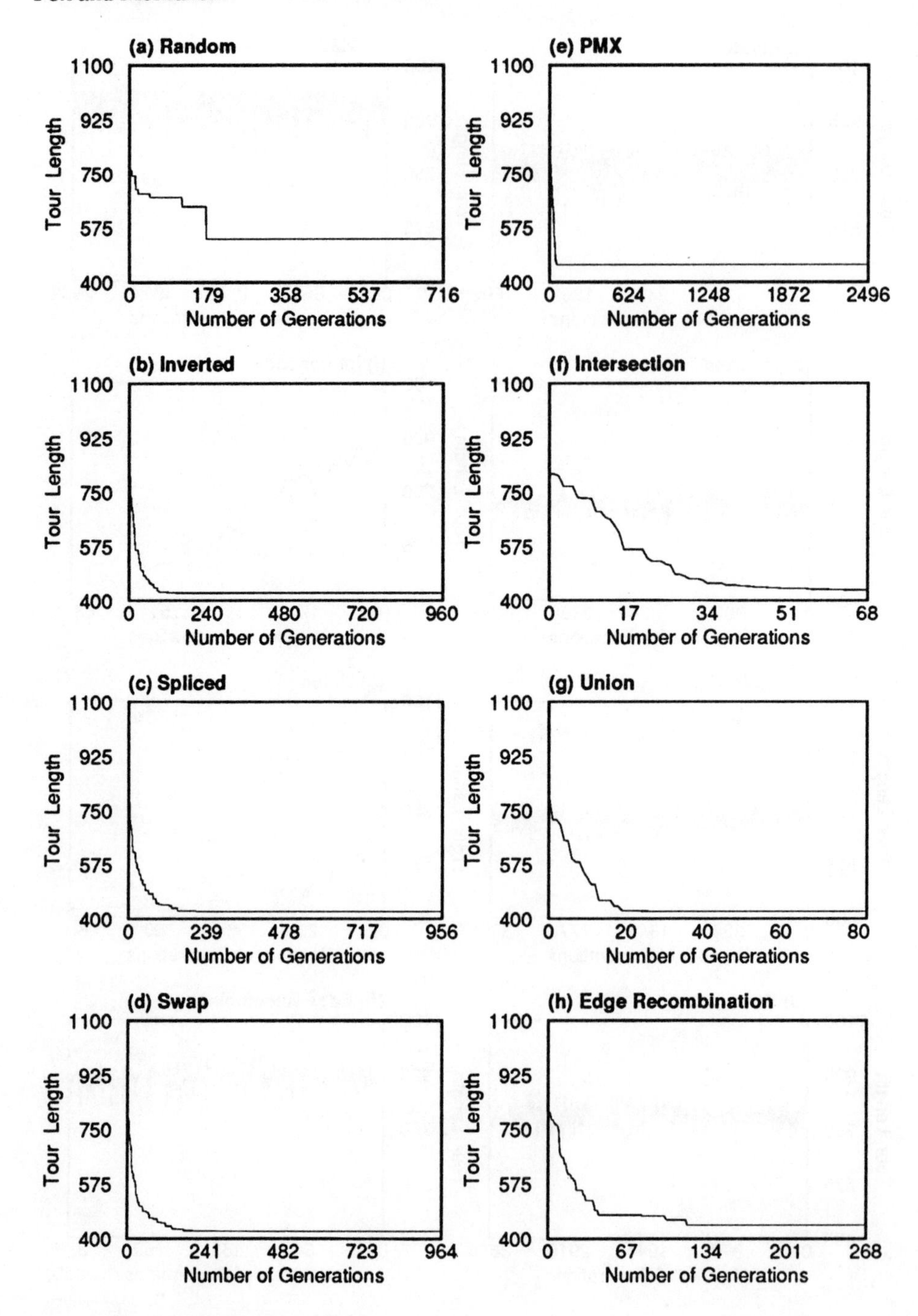

Figure 7: Oliver's 30 City Problem - 45 minute timed run preserving best solutions each of generation.

for any of the operators tested. This could be due to the larger amount of noise inherently provided by sequencing operators as compared to traditional operators.

Although all of the operators performed better when preserving solutions, the Union and Intersection operators managed to make progress even when elitism was not used. One possible explanation is that the Intersection and Union operators are more selective in picking subsequences from the parents for recombination. While the procedure for creating children using ER and PMX are well defined, the subsequences they produce appear arbitrary. Thus, they relay heavier on elitism to preserve good characteristics. In contrast, Intersection and Union are careful in preserving predecessor/successor relationships from each parent and do not rely as much on elitism to preserve these characteristics.

As mentioned earlier, code complexity and actual execution time is important in evaluating a solution. To address these issues, tests were run which used time rather than generation number as the ordinal scale. Some noteworthy characteristics became apparent when timing the operators for 45 minutes. First, all of the unary operators, PMX and ER run significantly faster than Union and Intersection. Intersection and Union complete approximately 80 generations/45 minutes while the other operators complete from eight hundred to three thousand generations. Second, the noise level is higher in the random operator than in the other operators, as could be expected. Third, after hundreds of generations the operators do not appear to get better results.

12 Conclusion

The investigation of reordering operators led to the development of a generic GA testbed. In addition it led into an understanding of the underlying principles of reordering operators and a model in which the basic relationships between elements could be presented. The model can be used to explain the relationships that are preserved by various reordering operators. By looking at the results of using several of the operators under similar conditions, the behavior of the operators become apparent. This insight provides a better understanding of sequencing operators than the procedural explanation which is most often found in the literature.

The relevance of this work to the Job Shop Scheduling Problem is based upon the assumption that an effective means of producing an optimal sequence of cities can be used to produce an optimal sequence of scheduling operations. Although it is apparent that GAs make good progress toward finding solutions, the primary factor that prohibits the use of GAs in real-time scheduling problems is execution time. Currently, the time required to schedule and evaluate a sequence of 1000 scheduling operations is on the order of 30 minutes. One thousand activities is a conservative estimate for the number of activities required by the Onboard Short Term Plan (OSTP) for Space Station Freedom. This implies that a candidate Genetic Algorithm for scheduling in this domain must be effective with as few as 100 individuals, total, over all generations. While improved hardware and software will decrease the amount of time needed to evaluate a solution, the real time savings will come by minimizing the number of solutions to evaluate. This requires that GAs find good solutions in a minimum number of generations with a minimum population size. This can only be accomplished with a thorough understanding of

GAs and recombination operators for sequences.

Acknowledgments

This research has been supported by NASA Contract NAS 9-17885. The GENOA testbed in which the operators were tested is public domain. It is freely available to anyone working on a government project.

References

[Davis 1985] Davis, L., Applying Adaptive Algorithms to Epistatic Domains in *Proceedings of the Ninth International Joint Conference on Artificial Intelligence*, Los Angelos, CA, August 1985 162-164.

[Fox 1989] Fox, B. Mixed Initiative Scheduling, *AAAI - Spring Symposium on AI in Scheduling*, Stanford, CA, March 1989.

[Fox 1990] Fox, B. Non-Chronological Scheduling in *Proceedings AI, Simulation and Planning in High Autonomy Systems*, University of Arizona, March 1990, IEEE Computer Society Press.

[Goldberg 1985] Goldberg, D.E., and Lingle, R. Alleles, Loci, and the Traveling Salesman Problem, in *Proceedings of an International Conferences on Genetic Algorithms and their Applications.* (L. Erlbaum, 1988, original proceedings 1985) 154-159.

[Goldberg 1989] Goldberg D.E. *Genetic Algorithms in Search, Optimization, and Machine Learning.* Reading, MA:Addison-Wesley.

[Oliver 1987] Oliver I.M., Smith D.J., and Holland J.R.C. A Study of Permutation Crossover Operators on the Traveling Salesman Problem, in *Genetic Algorithms and their Applications: Proceedings of the Second International Conference.* (L. Erlbaum, 1987) 224-230.

[Whitley 1989] Whitley D., Starkweather T., and Fuquay D., Scheduling Problems and Traveling Salesmen: The Genetic Edge Recombination Operator.

An Analysis of Multi-Point Crossover

William M. Spears
Naval Research Laboratory
Washington, D.C. 20375 USA
spears@aic.nrl.navy.mil

Kenneth A. De Jong
George Mason University
Fairfax, VA 22030 USA
kdejong@aic.gmu.edu

Abstract

In this paper we present some theoretical results on two forms of multi-point crossover: n-point crossover and uniform crossover. This analysis extends the work from De Jong's thesis, which dealt with disruption of n-point crossover on 2nd order hyperplanes. We present various extensions to this theory, including 1) an analysis of the disruption of n-point crossover on kth order hyperplanes; 2) the computation of tighter bounds on the disruption caused by n-point crossover, by handling cases where parents share critical allele values; and 3) an analysis of the disruption caused by uniform crossover on kth order hyperplanes. The implications of these results on implementation issues and performance are discussed, and several directions for further research are suggested.

Keywords: Genetic algorithm theory, recombination operators

1 Introduction

One of the unique aspects of the work involving genetic algorithms (GAs) is the important role that recombination plays in the design and implementation of robust adaptive systems. In most GAs, individuals are represented by fixed-length strings and recombination is implemented by means of a crossover operator which operates on pairs of individuals (parents) to produce new strings (offspring) by exchanging segments from the parents' strings. Traditionally, the number of crossover points (which determines how many segments are exchanged) has been fixed at a very low constant value of 1 or 2. Support for this decision came from early work of both a theoretical and empirical nature [Holland75, DeJong75].

However, there continue to be indications of an empirical nature that there are situations in which having a higher number of crossover points is beneficial [Syswerda89, Eschelman89]. Perhaps the most surprising result (from a traditional perspective) is the effectiveness on some problems of uniform crossover, an operator which produces on the average $(L / 2)$ crossings on strings of length L [Syswerda89].

The motivation for this paper is to extend the theoretical analysis of the crossover operator to include the multi-point variations and provide a better understanding of when and how to exploit their power. Specifically, this paper will focus on two forms of multi-point crossover: n-point crossover and uniform crossover.

2 Traditional Analysis

Holland provided the initial formal analysis of the behavior of GAs by characterizing how they biased the makeup of new offspring in response to feedback on the fitness of previously generated individuals. By focusing on hyperplane subspaces of L-dimensional spaces (i.e., subspaces characterized by hyperplanes of the form "---a----b---c--"), Holland showed that the expected number of samples (individuals) allocated to a particular kth order hyperplane H_k at time $t + 1$ is given by:

$$m(H_k, t+1) \geq m(H_k, t) * \frac{f(H_k)}{\bar{f}} * (1 - P_m k - P_c P_d(H_k))$$

In this expression, $f(H_k)$ is the average fitness of the current samples allocated to H_k, $\bar{f}$ is the average fitness of the current population, P_m is the probability of using the mutation operator, P_c is the probability of using the crossover operator, and $P_d(H_k)$ is the probability that the crossover operator will be "disruptive" in the sense that the children produced will not be members of the same subspace as their parents.

The usual interpretation of this result is that subspaces with higher than average payoffs will be allocated exponentially more trials over time, while those subspaces with below average payoffs will be allocated exponentially fewer trials. This assumes that there are enough samples to provide reliable estimates of hyperplane fitness, and that the effects of crossover and mutation are not too disruptive. Since mutation is typically run at a very low rate (e.g., $P_m = 0.001$), it is generally ignored as a significant source of disruption. However, crossover is usually applied at a very high rate (e.g., $P_c \geq 0.6$). So, considerable attention has been given to estimating P_d, the probability that a particular application of crossover will be disruptive.

To simplify and clarify the analysis, it is generally assumed that individuals are represented by fixed-length binary strings of length L, and that crossover points can occur with equal probability between any two adjacent bits. For ease of presentation these same assumptions will be made for the remainder of this paper. Generalizing the results to non-binary fixed-length strings is quite straightforward. Relaxing the other assumptions is more difficult.

Under these assumptions, Holland provided a simple and intuitive analysis of the disruption of 1-point crossover: as long as the crossover point does not occur within the defining boundaries of H_k (i.e., in between any of the k fixed defining positions), the children produced from parents in H_k will also reside in H_k [Holland75]. Figure 1 represents this graphically for a 3rd order hyperplane. Note that d_1, d_2, and d_3 represent the 3 defining positions of the 3rd order hyperplane, while P1 and P2 indicate the two parents.

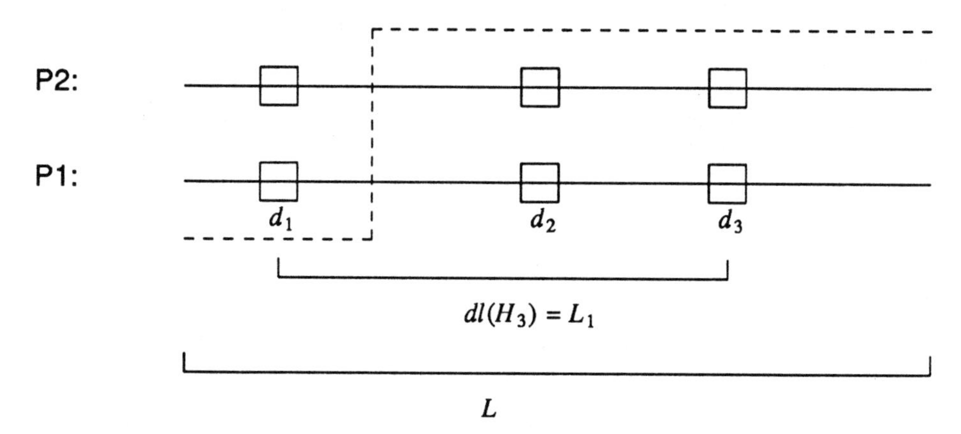

Figure 1: A 3rd Order Hyperplane

If crossover does occur inside the defining boundaries, disruption may or may not result. Disruption will depend on where the crossover point occurs inside the defining boundaries and on the alleles that the parents have in common on the k defining positions. Hence, P_d can be bounded by the probability that the crossover point will fall within the defining boundaries of H_k. Under the assumption of uniformly distributed crossover points, this yields:

$$P_d(H_k) \leq \frac{dl(H_k)}{L-1}$$

where $dl(H_k)$ is the "defining length" of H_k, namely the distance between the first and last of the k fixed defining positions of hyperplane H_k.

This analysis has lead to considerable discussion of the "representational bias" built into 1-point crossover, namely that crossover is much more disruptive to hyperplanes whose defining positions happen to be far apart. It also suggests a plausible role for inversion operators capable of effecting a change of representation in which the defining lengths of key hyperplanes are shortened.

De Jong [DeJong75] extended this analysis to n-point crossover by noting that no disruption can occur if there are an even number of crossover points (including 0) between each of the defining positions of a hyperplane. Hence, we have a bound for the disruption of n-point crossover:

$$P_d(n, H_k) \leq 1 - P_{k,even}(n, H_k)$$

where $P_{k,even}(n, H_k)$ is defined to be the probability that an even number of the n crossover points will fall between each of the k defining positions of hyperplane H_k. De Jong [DeJong75] provided an exact expression for $P_{k,even}$ for the special case of 2nd order hyperplanes (i.e., $k = 2$):

$$P_{2,even}(n, L, L_1) = \sum_{i=0}^{\lfloor \frac{n}{2} \rfloor} \binom{n}{2i} \left[\frac{L_1}{L}\right]^{2i} \left[\frac{L-L_1}{L}\right]^{n-2i}$$

$P_{2,even}(n, L, L_1)$ is the probability that an even number of crossover points will fall within the 2nd order hyperplane defined by L and L_1. Recall that L is the length of the string, while L_1 is the defining length of the hyperplane. The second term of the summation is the probability of placing an even number of crossover points within the 2 defining points. The third term is the probability of placing the remaining crossover points outside the 2 defining points. Finally, the combinatorial term $\begin{bmatrix} n \\ 2i \end{bmatrix}$ represents the number of ways an even number of points can be drawn from the n crossover points.

The family of curves generated by $P_{2,even}$ provide considerable insight into the change in disruptive effects on second order hyperplanes as the number of crossover points is increased. Figure 2 plots the curves for binary strings of length L. Notice how the curves fall into two distinct families depending on whether the number of crossover points is even or odd. Since $P_{2,even}$ guarantees no disruption, we're interested in increasing $P_{2,even}$ whenever possible. By going to an even number of crossover points, we can reduce the representational bias of crossover, but only at the expense of increasing the disruption of the shorter definition length hyperplanes.

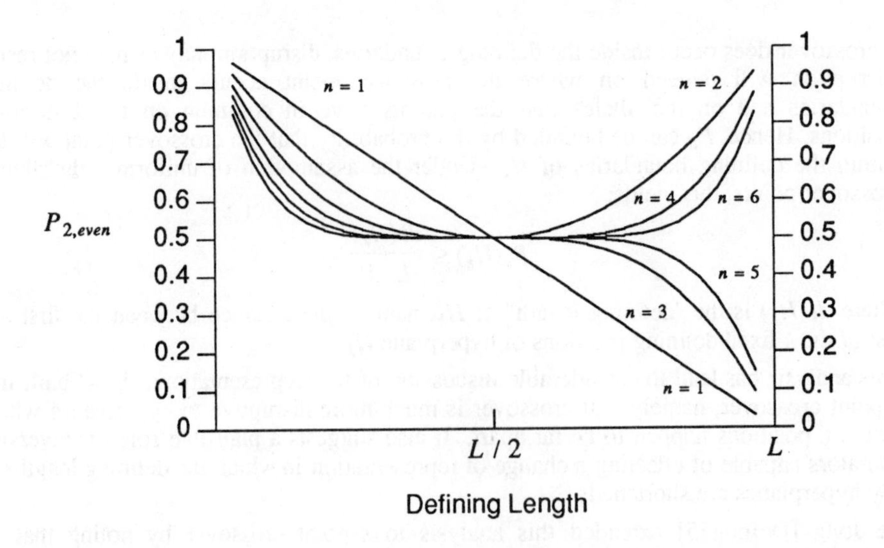

Figure 2. n-point Crossover Disruption on 2nd Order Hyperplanes

If we interpret the area above a particular curve as measure of the cumulative disruption potential of its associated crossover operator, then these curves suggest that 2-point crossover is the best as far as minimizing disruption. These results together with early empirical studies were the basis for using 2-point crossover in many of the implemented systems. Since then, there have been several additional studies focusing on crossover.

Bridges and Goldberg [Bridges85] have extended Holland's analysis of 1-point crossover, deriving tighter bounds on the disruption by taking into account the properties of the second parent and gains in samples in H_k due to disruption elsewhere.

Syswerda [Syswerda89] introduced a "uniform" crossover operator in which P_0 specified the probability that the allele of any position in an offspring was determined by using the allele of the first parent, and $1-P_0$ the probability of using the allele of the second parent. He provided an initial analysis of the disruptive effects of uniform crossover for the case of $P_0 = 0.5$, and compared it with 1 and 2 point crossover. He presented some provocative results suggesting that, in spite of higher disruption properties, uniform crossover can exhibit better recombination behavior, which can improve empirical performance.

Eschelman, Caruana, and Schaffer [Eschelman89] analyze crossover operators in terms of "positional" and "distributional" biases, and present a set of empirical studies suggesting that no n-point, shuffle, or uniform crossover operator is universally better than the others.

These results and other empirical studies motivated us to attempt to clarify the effects of multi-point crossover by extending the current analysis. In this paper we will present the following extensions:

1) An analysis of the disruption of n-point crossover on kth order hyperplanes.

2) The computation of tighter bounds on the disruption caused by n-point crossover, by examining the cases in which parents share common alleles on the hyperplane defining positions.

3) An analysis of the disruption caused by uniform crossover on kth order hyperplanes.

3 Crossover Disruption for Higher Order Hyperplanes

One possible explanation for the conflicting results on the merits of having more crossover points is that De Jong's analysis for the special case of 2nd order hyperplanes simply does not extend to higher order hyperplanes. In this section we attempt to resolve this issue by generalizing De Jong's results to hyperplanes of arbitrary order.

As noted earlier, the disruption probability $P_d(n, H_k)$ of n-point crossover on a kth order hyperplane H_k can be conservatively bounded by $1 - P_{k,even}(n, H_k)$ where $P_{k,even}(n, H_k)$ is the probability that n-point crossover produces only an even number of crossover points between each of the defining positions of H_k.

De Jong's formula for calculating $P_{2,even}$ can be generalized by noting that $P_{k,even}$ can be defined recursively in terms of $P_{k-1,even}$. To see this, consider how $P_{3,even}$ can be calculated in terms of $P_{2,even}$. Figure 3 illustrates the approach graphically.

The probability of n-point crossover generating only an even number of crossover points between *both* d_1—d_2 and d_2—d_3 can be calculated by counting the number of ways an even number of crossover points can fall in between d_1—d_3, and for each of these possibilities requiring an even number to fall in d_1—d_2 (a second order calculation involving L_1 and L_2). More formally, we have:

$$P_{3,even}(n, L, L_1, L_2) =$$

$$\sum_{i=0}^{\left\lfloor \frac{n}{2} \right\rfloor} \binom{n}{2i} \left(\frac{L_1}{L} \right)^{2i} \left(\frac{L-L_1}{L} \right)^{n-2i} P_{2,even}(2i, L_1, L_2)$$

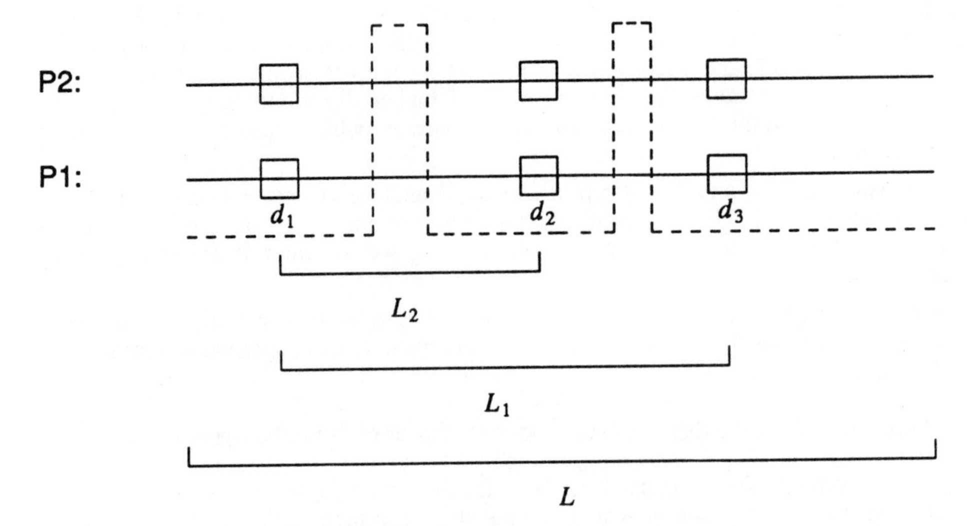

Figure 3. Non-disruptive n-point Crossover

In general, we have:

$$P_{k,even}(n, L, L_1, \ldots, L_{k-1}) =$$

$$\sum_{i=0}^{\left\lfloor \frac{n}{2} \right\rfloor} \binom{n}{2i} \left[\frac{L_1}{L} \right]^{2i} \left[\frac{L - L_1}{L} \right]^{n-2i} P_{k-1,even}(2i, L_1, \ldots, L_{k-1})$$

Figures 4 and 5 illustrate $P_{k,even}$ for hyperplanes of order 3 and 5. Each point on the graph represents an average over all hyperplanes of a particular defining length. Note that, apart from a skewing effect, the curves yield the same interpretation as De Jong's earlier curves for 2nd order hyperplanes: 2 point crossover minimizes disruption. So, extending the analysis thus far does not help in understanding the potential benefits of higher numbers of crossover points (seen in some empirical results).

4 Tighter Estimates on Disruption Probabilities

A second explanation for the conflicting results on the merits of a higher number of crossover points is that the $P_{k,even}$ curves are very weak bounds on P_d. It is possible that P_d itself, if analyzable, would yield different results. In this section we attempt to resolve this issue by providing tighter estimates on P_d.

The primary reason for the weakness of the $P_{k,even}$ bound is that it ignores the fact that many of the cases in which an odd number of crossover points fall between hyperplane defining positions are not disruptive to the sampling process. This occurs whenever the second parent happens to have identical alleles on the hyperplane defining positions which are exchanged by "odd" crossovers. (Note that an "odd" crossover occurs when an odd number of crossover points falls within 2 adjacent defining positions of the hyperplane.) Figure 6 illustrates this in the simple case of 2nd order hyperplanes. Note

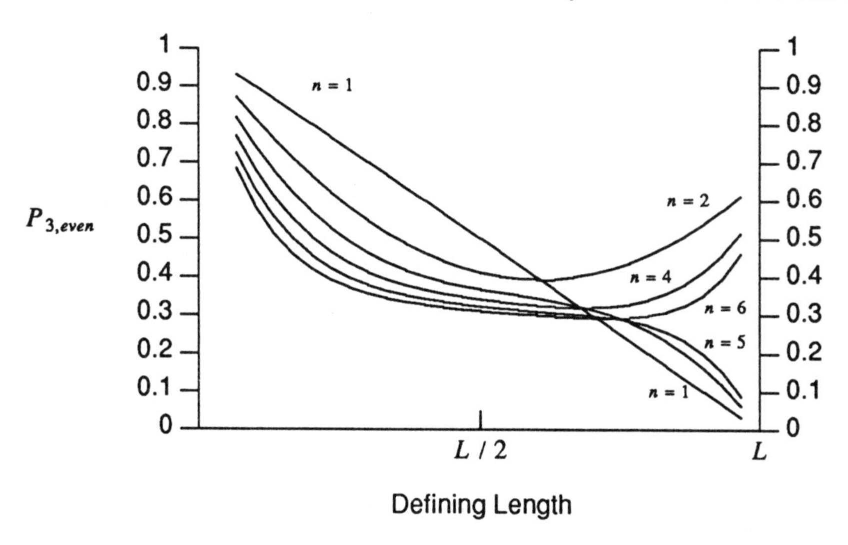

Figure 4. $P_{k,even}$ on 3rd Order Hyperplanes

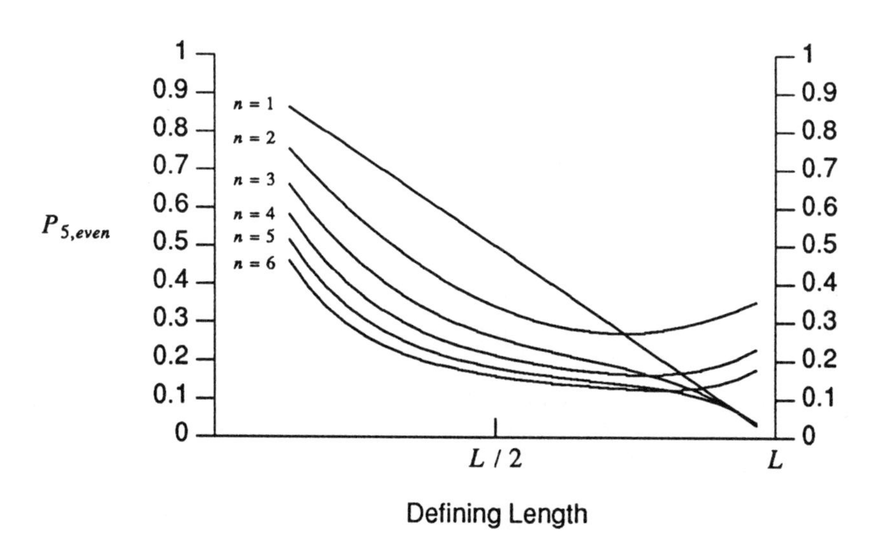

Figure 5. $P_{k,even}$ on 5th Order Hyperplanes

that, in this figure, v_1 and v_2 represent the alleles (i.e., binary values) at those defining positions. Of the 4 possible combinations of matches on the defining positions of H_2, only the first ($-v_1—v_2-, -\bar{v}_1—\bar{v}_2-$) actually results in a disruption.

Deriving an expression for the probability that both parents will share common alleles on the defining positions of a particular hyperplane is difficult in general because of the

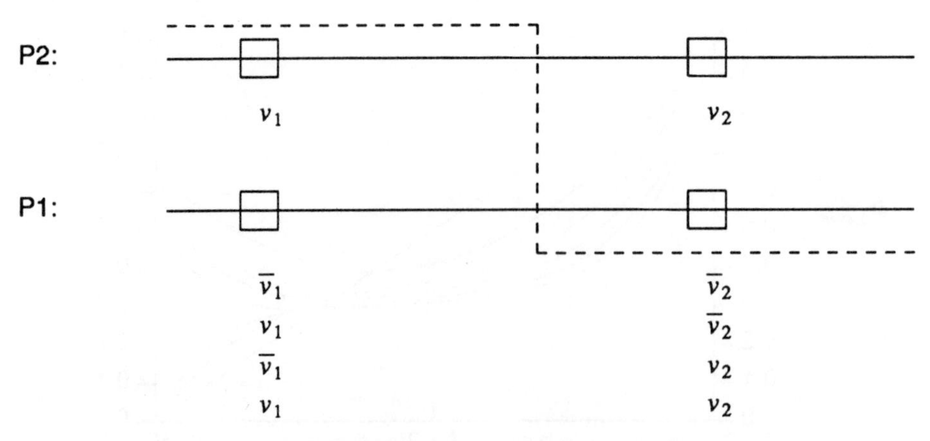

Figure 6. Disruption in "Odd" Crossovers

complexity of the population dynamics. We can, however, get a feeling for the effects of shared alleles on disruption by making the following simplifying assumption: the probability P_{eq} of two parents sharing an allele is constant across all loci.

With this assumption we can generalize $P_{k,even}$ to $P_{k,s}$ (i.e., the probability of *survival*) by including "odd" crossovers which are not disruptive. The generalization is still recursive in form:

$$P_{2,s}(n, L, L_1) = \sum_{i=0}^{n} \binom{n}{i} \left[\frac{L_1}{L} \right]^i \left[\frac{L - L_1}{L} \right]^{n-i} C$$

and

$$P_{k,s}(n, L, L_1, \ldots, L_{k-1}) = \sum_{i=0}^{n} \binom{n}{i} \left[\frac{L_1}{L} \right]^i \left[\frac{L - L_1}{L} \right]^{n-i} P_{k-1,s}(i, L_1, \ldots, L_{k-1})$$

Notice that we are now summing over all crossover distributions (both even and odd), but have added a "correction" factor C at the "bottom" of the recursion to sort out the desired cases. C must be defined, then, for each path through the recursion. If each n is even at every level in that path, then there are an even number of crossover points between each of the defining positions. In this case, we define C to be 1, ensuring that all the even cases are counted as before. Suppose, however, that n is odd at some level in a path. Then there must be two adjacent defining positions that contain an odd number of crossover points. If C were defined to be 0 when this situation occurred, we would have exactly the same formulation as $P_{2,even}$ and $P_{k,even}$. However, we want to include those cases where the alleles of the parents on the hyperplane defining positions match in such a way that an "odd" crossover will not be disruptive. At the point where the recursion "bottoms out", a particular distribution of crossover points is completely specified. This, in turn enables one to identify how many of the given hyperplane's defining positions are being exchanged by this particular "odd" crossover. If both parents match on these positions, no disruption occurs.

As we saw in Figure 6, this will be the case for 2nd order hyperplanes if the parents match on either the first or the second or both defining positions. Hence, setting $C = P_{eq} + P_{eq} - (P_{eq})^2$ specifies the proportion of non-disruptive "odd" crossovers. If we assume that $P_{eq} = 0.5$ for example, then $C = 0.75$. This indicates that 75% of the "odd" crossovers are non-disruptive, which agrees with the prior discussion for Figure 6.

This same observation is true for kth order hyperplanes. If an "odd" crossover results in m of the k defining positions being exchanged, no disruption will occur if: 1) the parents match on all m positions being exchanged, or 2) if they match on all $k - m$ positions not being exchanged, or 3) they match on all k defining positions. Hence, the general form of the correction is:

$$C = P_{eq}{}^{m} + P_{eq}{}^{k-m} - P_{eq}{}^{k}$$

Figure 7 illustrates this for one particular "odd" crossover on 4th order hyperplanes.

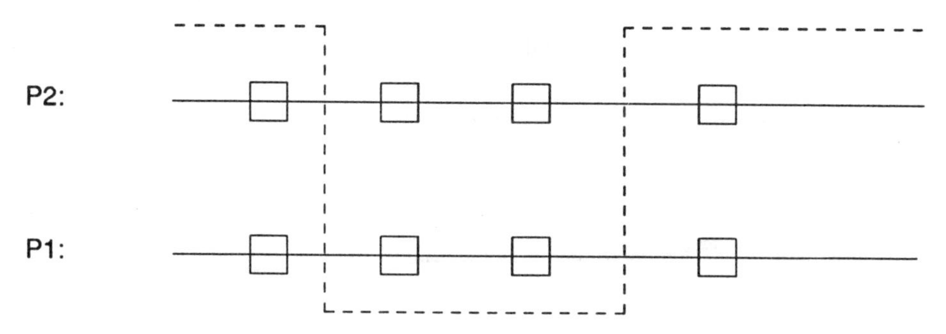

Figure 7. Non-disruptive "Odd" Crossover on 4th Order Hyperplanes

In this case,

$$C = P_{eq}{}^{2} + P_{eq}{}^{2} - P_{eq}{}^{4}$$

If $P_{eq} = 0.5$, then $C = (7 / 16)$ reflects the proportion of cases in which this particular crossover will not be disruptive.

Figures 8 and 9 show the effects of counting the non-disruptive "odd" crossovers. Figure 8 assumes a value of $P_{eq} = 0.5$, which is likely to hold in the early generations when matches are least likely. Figure 9 assumes a value of $P_{eq} = 0.75$ to get a feeling of the effect as the population becomes more homogeneous. Note that in both cases, the amount of expected disruption has been significantly reduced and the relative difference in disruption among different numbers of crossover points is reduced as well. At the same time, note that the curves for the various number of crossover points have held their relative position with respect to one another.

These results help explain the fact that in some empirical studies little or no difference in effect is seen by varying the number of crossover points between, say, 1 and 16. It does not appear to explain why in some situations more crossover points and, in particular, uniform crossover seems to perform significantly better.

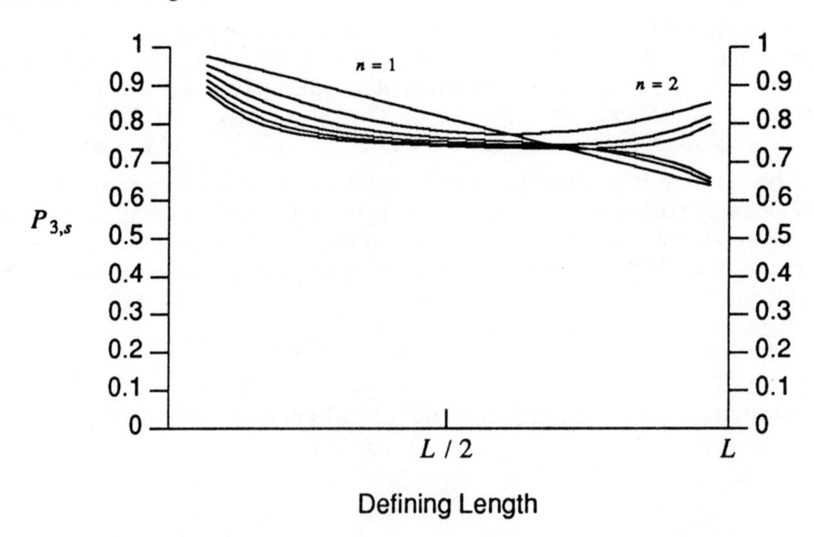

Figure 8. $P_{k,s}$ on 3rd Order Hyperplanes with P_{eq} = 0.5

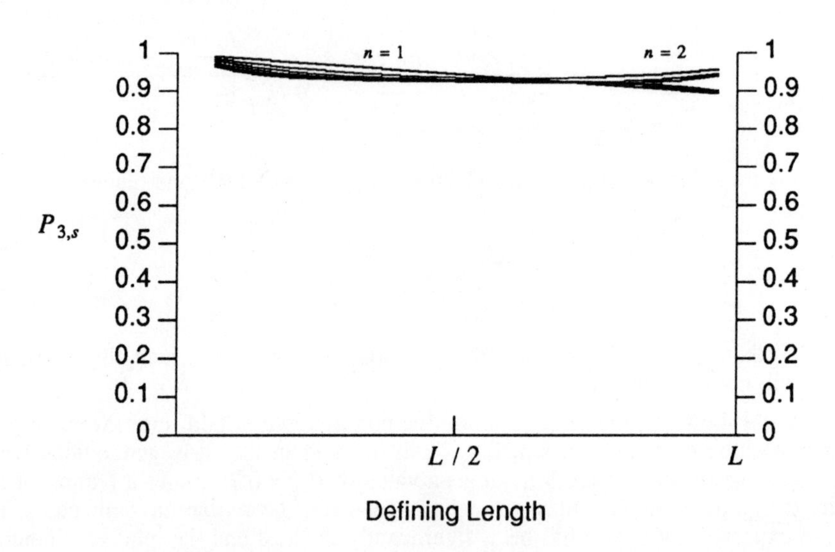

Figure 9. $P_{k,s}$ on 3rd Order Hyperplanes with P_{eq} = 0.75

5 Analyzing Uniform Crossover

Syswerda [Syswerda89] defined a family of "uniform" crossover operators which is a variant of a notion that has been informally experimented with in the past: to produce offspring by randomly selecting at each loci the allele of one of the parents. By defining P_0 to be the probability of using the first parent's allele, offspring can be produced by flipping a P_0 biased coin at each position. (Other informal studies viewed the process as a random walk and defined P_0 as the probability of switching over to the other parent. The two views are equivalent if and only if $P_0 = 0.5$.)

A good way of relating uniform crossover to the more traditional n-point crossover is to think of uniform crossover as generating a mask of 0s and 1s, indicating which parent's allele is to be used at each position. As we scan the mask from left to right, a switch from 0 to 1 or from 1 to 0 represents a crossover point. For example, the mask 0011100 defines a 2-point crossover operation. If $P_0 = 0.5$, all masks are equally likely. If we examine the n-point crossover operations defined by this set of masks, we see immediately that they are binomially distributed around $((L-1)/2)$. For example, the set of all 4-bit masks defines:

2 0-point crosses
6 1-point crosses
6 2-point crosses
2 3-point crosses

If $P_0 \neq 0.5$, the masks are no longer uniformly distributed, but contain on the average longer runs of 0s or 1s. From the point of view of n-point crossover, the effect is to skew the binomial distribution toward 0.

We are now in a position to analyze the disruption properties of uniform crossover in the same manner as the analysis of n-point crossover in the preceding sections. We note that the notion of an even number of crossover points between the defining positions of hyperplane H_k corresponds to masks which have either all 0s or all 1s on the defining positions of H_k. Hence, the corresponding conservative bound on the disruption of uniform crossover is given by:

$$P_d(H_k) \leq 1 - P_{k,even}(H_k)$$

where

$$P_{k,even}(H_k) = (P_0)^k + (1-P_0)^k$$

If $P_0 = 0.5$ for example, then

$$P_{k,even}(H_k) = (\frac{1}{2})^{k-1}$$

for all hyperplanes of order k. Notice that, unlike the traditional n-point crossover, there is no representational bias with uniform crossover in the sense that all hyperplanes of order k are equally disrupted regardless of how long or short their defining lengths are.

As before, we can get a tighter estimate of P_d if we include non-disruptive "odd" crossovers. For uniform crossover this corresponds to those masks which are not either all 0s or all 1s on the hyperplane defining positions, but are non-disruptive because the parents share common alleles on those particular positions. More formally, we have

$$P_{k,s}(H_k) = P_{k,even}(H_k) + \sum_{i=1}^{k-1} \binom{k}{i} (P_0)^i \ (1-P_0)^{k-i} \ (P_{eq}{}^i + P_{eq}{}^{k-i} - P_{eq}{}^k)$$

where P_{eq} is the probability of matching alleles, as before. Note that the last term in the expression is identical to the correction C defined earlier for the n-point crossover analysis. If the above is rewritten more concisely, $P_{k,s}$ can be expressed in a form similar to that derived for the n-point analysis:

$$P_{k,s}(H_k) = \sum_{i=0}^{k} \begin{bmatrix} k \\ i \end{bmatrix} (P_0)^i \ (1-P_0)^{k-i} \ ({P_{eq}}^i + {P_{eq}}^{k-i} - {P_{eq}}^k)$$

Figure 10 illustrates the relationship between uniform crossover and *n*-point crossover for 3rd order hyperplanes. Note that, as expected, uniform crossover does not minimize disruption but, at the cost of higher disruption, removes any representational bias. This helps to explain why uniform crossover can yield performance improvements in some cases. Consider situations in which the critical low order hyperplanes happen to be widely separated in a particular representation. Uniform crossover significantly reduces the disruption pressure on these critical hyperplanes at the expense of more disruption on the adjacent (but non-critical) low order hyperplanes. However, in the reverse situations in which the representation happens to place critical positions close together, 1 and 2 point crossover is more effective.

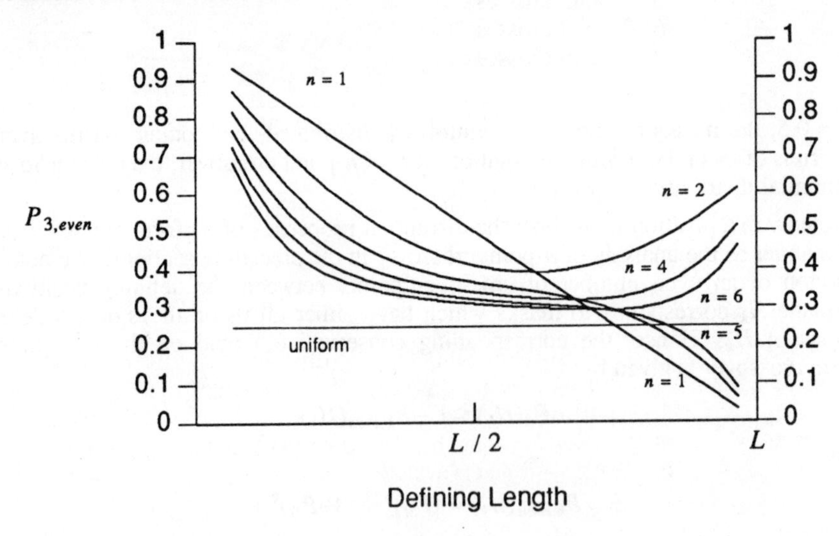

Figure 10. Disruption of Uniform Crossover

6 Is Disruption Always Bad?

So far, the analysis of crossover has focused on its potential for sampling disruption with the implication that disruption is bad. Sampling disruption is important for understanding the effects of crossover when populations are diverse (typically early in the evolutionary process). However, when a population becomes quite homogeneous, another factor becomes important: whether the offspring produced by crossover will be different than their parents in some way (thus generating a new sample) or just clones. This property of crossover has been dubbed "crossover productivity" and is easy to measure. Figure 11 illustrates how significantly the "productivity" of 2-point crossover can drop off as evolution proceeds. The horizontal axis indicates the number of generations the GA has run (i.e., we use a generational GA). The vertical axis indicates the number of crossovers, at each generation, that produced offspring different from their parents. Since $P_c = .6$, and the population size is 100, the maximum productivity is 60. The problem examined, HC11, is a boolean satisfiability problem explained in

[Spears90]. The problem has 55 binary variables, and has one unique solution with a fitness of 1.0.†

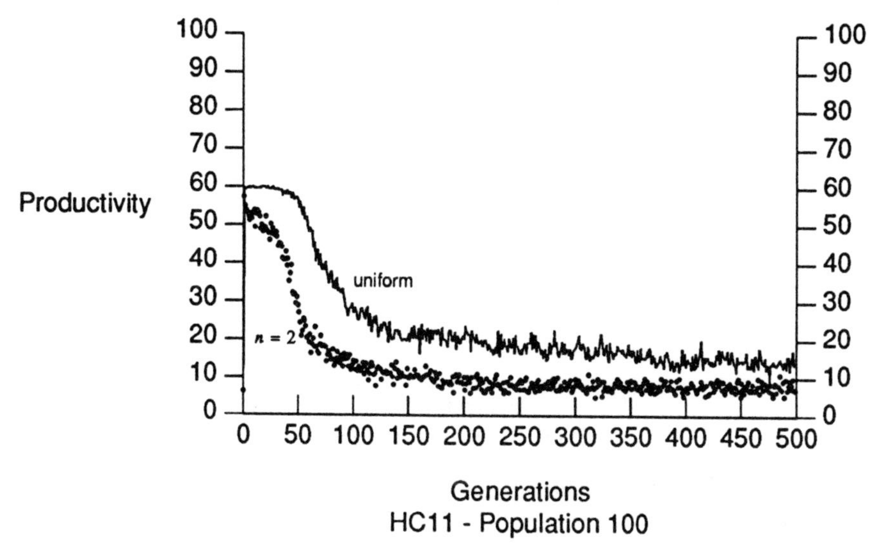

Figure 11. Productivity of 2-point & Uniform Crossover

If we try to formally compute the probability that the offspring will be different than their parents, the computation is precisely the same as the previous disruption computations. To see this, consider two parents whose alleles differ on only 4 loci. In order for crossover to produce new offspring, some but not all of those alleles must be exchanged. The probability of this occurring is just $P_d(H_4)$. In other words, those operators that are more disruptive are also more likely to create new individuals from parents with nearly identical genetic material.

This observation helps explain some of the other experimental results in which higher crossover rates performed better. Figure 12 is an example of one such result. Again, the horizontal axis represents generations. The vertical axis represents the best individual seen. Notice that 2-point crossover converges more quickly, but to a lower plateau than uniform crossover which converges more slowly to a better solution.

This suggests two additional directions for research. First, note that it may be possible to have the best of both worlds by modifying 2-point crossover to be less likely to produce clones. This can be achieved in a brute force way by repeated calls to crossover until non-clones are produced, or in more sophisticated ways such as Booker's reduced surrogate approach [Booker85]. Figure 13 illustrates the effect of the brute force technique on one particular example. Notice that this change has little effect during the early generations when children are most likely to be different anyway. However, the increased "productivity" in the later stages slows the early convergence seen before.

† All experimental results are averaged over at least 10 independent runs.

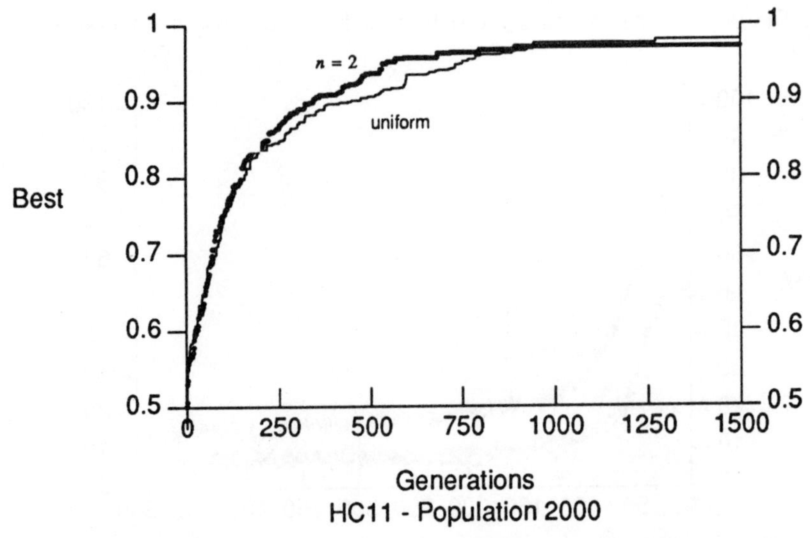

Figure 12. Productivity-related Performance of 2-point & Uniform Crossover

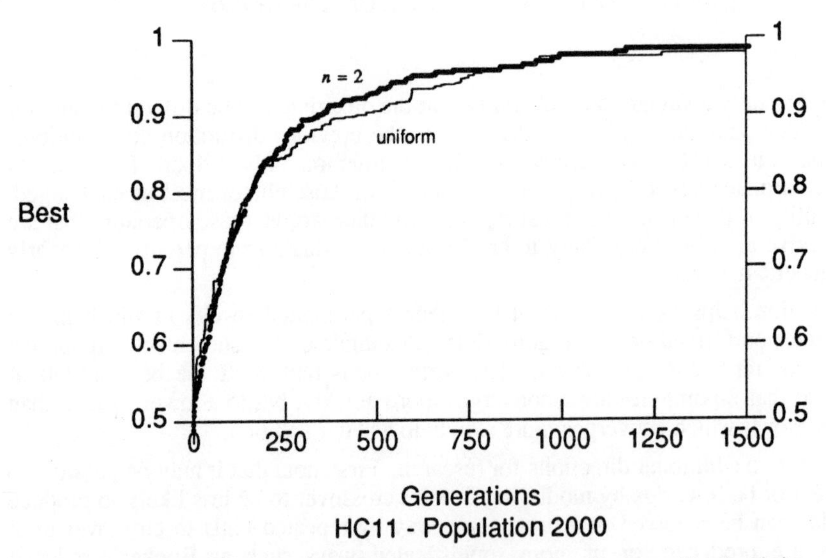

Figure 13. 2-point Crossover Augmented to Increase Productivity

The second direction for future research is the obvious interaction of multi-point crossover and population size. Smaller population sizes tend to converge faster to levels of homogeneity which reduce crossover productivity. With larger population sizes the effects appear to be much less dramatic. This suggests a way to understand the role of multi-point crossover. With small populations, more disruptive crossover operators such as uniform or *n*-point ($n >> 2$) may yield better results because they help overcome the limited information capacity of smaller populations and the tendency for more homogeneity. However, with larger populations, less disruptive crossover operators (2-point) are more likely to work better, as suggested by Holland's original analysis.

7 Conclusions and Further Work

The extensions to the analysis of *n*-point and uniform crossover presented in this paper provide additional insight into the role and effective use of these operators. At the same time, this analysis has suggested some directions for further research. The authors are currently involved in extending the results presented here to include the interacting effects of population size and crossover productivity. The view we are taking is that there is very little likelihood of finding globally correct answers to questions such as the choice of population size and crossover operators. Our goal is to understand these interactions well enough so that GAs can be designed to be self-selecting with respect to such decisions.

Acknowledgements

We would like to thank Diana Gordon for her help in probability theory, and Alan Schultz for many helpful comments and discussions.

References

Booker, Lashon B. (1987). Improving Search in Genetic Algorithms, *Genetic Algorithms and Simulated Annealing*, Morgan Kaufmann Publishing.

Bridges, C. & Goldberg, D. (1985). An Analysis of Reproduction and Crossover in a Binary-Coded Genetic Algorithm, *Proc. 3rd Int'l Conference on Genetic Algorithms*, Lawrence Erlbaum Publishing.

De Jong, Kenneth A. (1975). *An Analysis of the Behavior of a Class of Genetic Adaptive Systems*, Doctoral thesis, Dept. Computer and Communication Sciences, University of Michigan, Ann Arbor.

Eschelman, L., Caruana, R. & Schaffer, D. (1989). Biases in the Crossover Landscape, *Proc. 3rd Int'l Conference on Genetic Algorithms*, Morgan Kaufman Publishing.

Goldberg, David E. (1989). *Genetic Algorithms in Search, Optimization & Machine Learning*, Addison-Wesley Publishing Company, Inc.

Holland, John H. (1975). *Adaptation in Natural and Artificial Systems*, The University of Michigan Press.

Spears, William M. (1990). *Using Neural Networks and Genetic Algorithms as Heuristics for NP-Complete Problems*, Masters Thesis, Department of Computer Science, George Mason University, Fairfax, Virginia, 1990.

Syswerda, Gilbert. (1989). Uniform Crossover in Genetic Algorithms, *Proc. 3rd Int'l Conference on Genetic Algorithms*, Morgan Kaufman Publishing.

Evolution in Time and Space - The Parallel Genetic Algorithm

Heinz Mühlenbein
GMD Schloss Birlinghoven
D-5205 Sankt Augustin1

Abstract

The parallel genetic algorithm (PGA) uses two major modifications compared to the genetic algorithm. Firstly, selection for mating is distributed. Individuals live in a 2-D world. Selection of a mate is done by each individual independently in its neighborhood. Secondly, each individual may improve its fitness during its lifetime by e.g. local hill-climbing. The PGA is totally asynchronous, running with maximal efficiency on MIMD parallel computers. The search strategy of the PGA is based on a small number of active and intelligent individuals, whereas a GA uses a large population of passive individuals. We will investigate the PGA with deceptive problems and the traveling salesman problem. We outline why and when the PGA is succesful. Abstractly, a PGA is a parallel search with information exchange between the individuals. If we represent the optimization problem as a fitness landscape in a certain configuration space, we see, that a PGA tries to jump from two local minima to a third, still better local minima, by using the crossover operator. This jump is (probabilistically) successful, if the fitness landscape has a certain correlation. We show the correlation for the traveling salesman problem by a configuration space analysis. The PGA explores implicitly the above correlation.

Keywords: parallel optimization, parallel genetic algorithm, deceptive problems, traveling salesman

1 Introduction

Random search methods based on evolutionary principles have been proposed in the 60's. They did not have a major influence on mainstream optimization. We believe that this will change. The unique power of evolutionary algorithms shows up with parallel computers. An example is our parallel genetic algorithm PGA introduced in 1987 [MGSK87]. It runs especially efficiently on parallel computers. Moreover our research indicates that parallel searches with information exchange between the searches are often better than a single search. Thus the PGA is a truly parallel algorithm which combines the hardware speed of parallel processors and the software speed of intelligent parallel searching.

We have successfully applied the PGA to a number of problems, including function optimization [MSB91] and combinatorial optimization. In this paper we try to explain the search strategy of the PGA in discrete optimization problems. Numerical results for the traveling salesman problem TSP, the quadratic assignment problem and the graph partitioning problem can be found in [GS89], [Müh89] and [vLM91].

In all these applications we have used the same algorithm with only slight modifications. We have taken the largest published problem instances known to us. The PGA found solutions, which are comparable or even better than any other solution found by other heuristics. This is proof by experimental result. We hope that future researchers will start where we have left off. The evaluation of heuristics for difficult combinatorial optimization problems has to be done with large problems. The results on small toy problems cannot be extrapolated to large problems.

The most difficult part of a random search method is to explain why and when it will work. We will do this analysis for the PGA with the deceptive problems and the TSP.

Throughout the paper we will use biological terms. We believe, that this is a source of inspiration and helps to understand the PGA intuitively. We do not claim that modelling natural evolution will automatically lead to a good optimization method.

The outline of the paper is as follows. In section 2 the problem of a parallel search with linkage is introduced. Evolutionary algorithms and genetic algorithms are reviewed in section 3. The importance of a spatial population structure on evolution is discussed in section 4. The next four sections show the influence of hill-climbing and crossing-over on the PGA. Analytical and numerical results are presented for deceptive problems proposed by Goldberg [Gol89b]. The more difficult TSP is discussed in sections 9 and 10. First the influence of the hill-climbing strategy is shown, then crossing-over is discussed by a configuration space analysis.

2 Parallel search and optimization

In this paper we consider the following problem:

OPT 1 *P: Given a function $F : X \mapsto R$, where X is some metric space. Let S be a subspace of X. We seek a point x in S which optimizes F on S or at least yields an acceptable approximation of the suprenum of F on S.*

Many optimization methods have been proposed for the solution of this problem. We will investigate parallel optimization methods. A parallel optimization method of parallelism N is characterized by N different search trajectories, which are performed in parallel. It can be described as follows

$$x_i^{t+1} = G_i(x_1^t, ..x_N^t, F(x_1^t), ..F(x_N^t))i = 1, ..., N \qquad (1)$$

The mapping $G = (G_1, ...G_N)$ describes the linkage or information exchange between the parallel searches. If the N searches are independent of each other we just have

$$x_i^{t+1} = G(x_i^t, F(x_i^t)) \qquad (2)$$

A parallel search method which combines the information from two searches can be described as follows

$$x_i^{t+1} = G_i(x_{i-1}^t, x_i^t, F(x_{i-1}^t), F(x_i^t))i = 1, ..., N \qquad (3)$$

The basic questions of parallel search methods can now be stated

- Are N parallel searches of time complexity t as efficient as a single search of time complexity $N * t$?
- Are N linked searches more efficient than N independent searches?
- How should the linkage be done?

In order to understand these questions intuitively, we leave the abstract mathematical description and turn to a natural search metaphor. The advantage of using a metaphor is that it leads to a qualitative understanding of the problem and the algorithm.

In this paper we use the following simple metaphor: The search is done by N active individuals, x_i describes the position of individual i and $F(x_i)$ its current value, representing the height in an unknown landscape. How should the population of individuals search the unknown landscape?

Many different models are possible. The individuals could be geographers who want to find the highest mountain. There is fog all over the place. The geographers can communicate with each other by broadcasting their information. How should they best do it? Many different strategies can be thought of. In a search algorithm we would then try to mimic their behavior.

In this paper, we will investigate search algorithms which mimic evolutionary adaptation found in nature. Each individual is identified with an animal, which searches for food and produces offspring. In evolutionary algorithms, $F(x_i)$ is called the fitness of individual i, x_i^{t+1} is an offspring of x_i^t, and G is called the selection schedule.

3 Evolutionary algorithms and genetic algorithms

A survey of search strategies based on evolution has been done in [MGSK88]. We recall only the most important ones. A generic evolutionary algorithm can be described as follows

Evolutionary algorithm

STEP1: Create an initial population of size $M = N * O$

STEP2: Compute the Fitness $F(x_i) i = 1, \ldots, M$

STEP3: Select N individuals according to some selection schedule

STEP4: Create O offspring of each of the N individuals by small variation

STEP5: If not finished, return to STEP2

Variants of this algorithm have been invented by many researchers, see for example [Rec73], [Sch81]. An evolutionary algorithm is a random search which uses selection and variation. The linkage of the parallel searches is only implicit by the selection schedule. Searches with bad results so far are abandoned and new searches are started in the neighborhood of more promising searches.

In biological terms, evolutionary algorithms model natural evolution by asexual reproduction with mutation and selection. Search algorithms which model sexual reproduction are called genetic algorithms. They were invented by Holland [Hol75]. Recent surveys can be found in [Gol89a] and [Sch89].

Genetic Algorithm

STEP0: Define a genetic representation of the problem

STEP1: Create an initial population $P(0) = x_1^0, ..x_N^0$

STEP2: Compute the average fitness $\overline{F} = \sum_i^N F(x_i)/N$. Assign each individual the normalized fitness value $F(x_i^t)/\overline{F}$

STEP3: Assign each x_i a probability $p(x_j, t)$ proportional to its normalized fitness. Using this distribution, select N vectors from $P(t)$. This gives the set $S(t)$

STEP4: Pair all of the vectors in $S(t)$ at random forming $N/2$ pairs. Apply crossover with probability p_{cross} to each pair and other genetic operators such as mutation, forming a new population $P(t+1)$

STEP5: Set $t = t + 1$, return to STEP2

In the simplest case the genetic representation is just a bitstring of length n, the "chromosome". The positions of the strings are called "locus" of the chromosome. The variable at a locus is called "gene", its value "allele". The set of chromosomes is called the "genotype"" which defines a "phenotype" (the individual) with a certain fitness. We will later show with examples why and when crossover guides the search.

A genetic algorithm is a parallel random search with centralized control. The centralized part is the selection schedule. The selection needs the average fitness of all individuals. The result is a highly synchronized algorithm, which is difficult to implement efficiently on parallel computers.
In our parallel genetic algorithm, we use a distributed selection scheme. This is achieved as follows. Each individual does the selection by itself. It looks for a partner in its neighborhood only. The set of neigborhoods defines a spatial population structure.

Our second major change can now easily be understood. Each individual is active and not acted on. It may improve its fitness during its lifetime by performing a local search.
A generic parallel algorithm can be described as follows

Parallel genetic algorithm

STEP0: Define a genetic representation of the problem

STEP1: Create an initial population and its population structure

STEP2: Each individual does local hill-climbing

STEP3: Each individual selects a partner for mating in its neighborhood

STEP4: An offspring is created with genetic crossover of the parents

STEP5: The offspring does local hill-climbing. It replaces the parent, if it is better than some criterion (acceptance)

STEP6: If not finished, return to STEP3.

It has to be noted, that each individual may use a different local hill-climbing method. This feature will be important for problems, where the efficiency of a particular hill climbing method depends on the problem instance.

In the terminology of section 2, we can describe the PGA as a parallel search with a linkage of two searches. The linkage is done probabilistically constrained by the neighborhood. The information exchange within the whole population is a diffusion process because the neighborhoods of the individuals overlap.

In a parallel genetic algorithm, all decisions are made by the individuals themselves. Therfore the PGA is a totally distributed algorithm without any central control.

There have been several other attempts to implement a parallel genetic algorithm. Most of the algorithms run k identical standard genetic algorithms in parallel, one run per processor. They differ in the linkage of the runs. Tanese [Tan89] introduces two *migration* parameters: the *migration interval*, the number of generations between each migration, and the *migrationrate*, the percentage of individuals selected for migration. The subpopulations are configured as a binary n-cube. A similar approach is done by Pettey et al. [PS89]. In the implementation of Cohoon et al. [CHMR87] it is assumed that each subpopulation is connected to each other. The algorithm from Manderick et al. [MS89] has been derived from our PGA. In this algorithm the individuals of the population are placed on a planar grid and selection and crossing-over are restricted to small neighborhoods on that grid.

All but Manderick's algorithm use subpopulations that are densely connected. We will show in the next section why restricted connections like a ring are much better for the genetic algorithm. All the above parallel algorithms do not use hill-climbing, which is one of the most important parts of our PGA.

An extension of the PGA, where subpopulations are used instead of single individuals, has been described in [MSB91]. This algorithm out-performs the standard GA by far in the case of function optimization. It is also a better search method than most of the standard mathematical methods.

We will explain in the next sections the three important parts of the search strategy of the PGA - the spatial population structure, hill-climbing and crossing-over.

4 Diversification by a spatial population structure

Genetic algorithms suffer from the problem of premature convergence. In order to solve this problem, many genetic algorithms enforce diversification explicitly, violating the biological metaphor. A popular method is to accept an offspring only if it is more than a certain factor different from all the members of the population.

The PGA tries to introduce diversification more naturally by a spatial population structure. Fitness and mating is restricted to neighborhoods called *demes*. This name has been introduced in population genetics, where it is well known that a population with a spatial structure has more variety than a panmictic population. The importance of this fact on evolution has been however highly controversially discussed.

Wright [Wri32] has argued that the best way to avoid being hung up on a low fitness peak is to have the population broken up into many nearly isolated subpopulations. Wright's theory has two phases. In the first phase of the evolution, the allele frequencies drift to some extent in each subgroup. One subgroup might by chance drift into a set of gene frequencies that correspond to a higher peak. Then the second phase sets in. This subgroup now has a higher fitness than other subgroups and will tend to displace them until eventually the whole population has the new, favorable gene combination. Then the whole process starts again.

Fisher [Fis58], in contrast, argued that no such theory is needed. In a highly multidimensional fitness surface, the peaks are not very high and are connected by fairly high ridges, always shifting because of environmental changes. According to Fisher, the analogy is closer to waves and troughs in an ocean than in a static landscape. Alleles are selected because of their average effects, and a population is unlikely to ever be in such a situation that it can never be improved by direct selection based on additive variance.

The difference between these two views is not purely mathematical, but physiological. Does going from one favored combination of alleles to another often necessitate passing through genotypes that are of lower fitness? Fisher argued that evolution typically proceeded in a succession of small steps, leading eventually to large differences by the accumulation of small ones. According to this view, the most effective population is a large panmictic one in which statistical fluctuations are slight and each allele can be fairly tested in combination with many others alleles. According to Wright's view, a more favorable structure is a large population broken up into subgroups, with migration sufficiently restricted (less than one migrant per generation) and size sufficiently small to permit appreciable local differentiation.

Three different mathematical models for spatially structured populations have been proposed

- the island model
- the stepping stone model

- the isolating by distance model

In the island model, the population is pictured as subdivided into a series of randomly distributed demes among which migration is random.

The stepping-stone model deals with discrete demes, separated into distinct subpopulations. Migration takes place between neighboring demes only.

The isolation by distance model treats the case of continuous distribution where effective demes are isolated by virtue of finite home ranges (neighborhoods) of their members. For mathematical convenience it is asssumed that the position of a parent at the time it gives birth relative to that of its offspring when the latter reproduces is normally distributed.

Felsenstein [Fel75] has shown that in many cases the above models lead to unrealistic clumping of individuals and concluded, that they are biologically irrelevant. There have been many attempts to investigate spatial population structures by computer simulations, but they did not have a major influence.

The issue raised by Wright and Fisher is still not settled. Because of the difficulty of solving the problem by mathematical analysis, population genetics has unfortunately lost interest in the problem. A good survey of the different population models can be found in Felsenstein [Fel76].

The PGA implements Wright's model. The two phases of Wright's theory can actually be observed. But the second phase performs differently. The biggest changes of the population occur at the time after migration between the subpopulations. Recombinations between immigrants and native individuals will occasionally lead to higher peaks which were not found by any of the subpopulations during isolation.

The creative forces of evolution take place at migration and few generations afterwards. Wright's argument that better peaks are found just by chance in small subpopulations does not capture the essential facts.

The above described phases can be easily shown in the application function optimization. We take as example searching the global minimum of a two dimensional function. The function is called Shekel's foxholes or DeJong's function F5. It has 25 local minima. The values of the minima are in the bottom row 1,2,3,4,5, in the second row 6,7,.. and so on. The value at the plateau is 500. In this example we have a fitness landscape which violates Fisher's assumption. To get from one valley to another one the population has to climb to the plateau, there are no other connections (because it is a minimization problem the valleys take over the role of the peaks in maximization).

We try to solve this problem with four subpopulations of 10 individuals each. Migration takes place after 10 generations. In figure 1 the progress of two subpopulations is shown. The number denotes the generation number. After generation ten subpopulation one is concentrated in four valleys, subpopulation two in two valleys. Subpopulation one has the y value of the optimum, subpopulation 2 the x value. After migration and recombination, subpopulation one expands into five valleys, subpopulation two into four valleys. Subpopulation one has now the genetic material to discover the lowest valley by recombination. At generation 12 subpopulation 1 has discovered the second lowest valley and it finds the global minimum at genera-

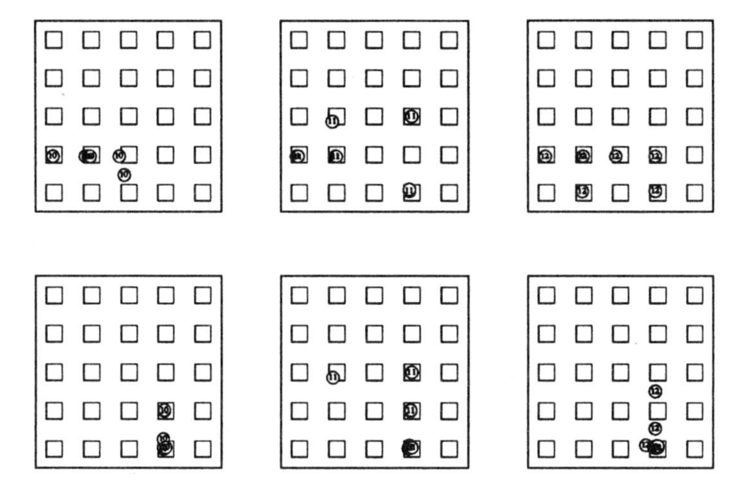

Figure 1: Evolution of two subpopulations

tion 15. Subpopulation one is now concentrated in the two lowest valleys. A detailed discussion about the PGA in function optimization can be found in [MSB91].

We summarize the most important conclusion of this section.

Subpopulations isolated for a certain time keep the diversity of the population high. After migration new promising areas can be discovered by crossing-over.

In most of our applications we have used a population structure called ladder. Our ladder is circular, both ends are closed. Why not use a more obvious structure like a torus or a still more densely connected structure? The answer is simple. In a ladder the variety of the population remains higher than in torus. In a torus of size n*n it takes $O(n)$ steps for an item of information to propagate to all places, in a ladder of the same size it takes $O(n^2)$ steps.

We now turn to the analysis of crossing-over.

5 The schema theorem revisited

Many researchers in the genetic algorithm community refer to the schema theorem to explain the search strategy of genetic algorithms. For this reason it is also called the "Fundamental Theorem of Genetic Algorithms" [Gol89a]. The usual interpretation of the fundamental theorem claims that a genetic algorithm with binary representation makes an optimal allocation of the sample points given the information at hand. We believe that it is not worthwhile to discuss all the diffuse interpretations in detail. Instead, we will use a "deceptive" problem to show in this specific case what is wrong with these interpretations of the schema theorem. We will use the same deceptive problem in the next section to explain the search strategy of the PGA in simple terms.

The theory of deceptive problems has been developed by Goldberg et al. [Gol89b]. We take as example the three bit function, which is defined in the following table.

bit	value	bit	value
111	30	100	14
101	0	010	22
110	0	001	26
011	0	000	28

The function is called order-3 deceptive because we have

$$f_s(0**) > f_s(1**); f_s(*0*) > f_s(*1*); f_s**0 > f_s(**1)$$

$$f_s(00*) > f_s(11*), f_s(10*), f_s(01*)$$

$$f_s(0*0) > f_s(1*1), f_s(0*1), f_s(1*0)$$

$$f_s(*00) > f_s(*11), f_s(*10), f_s(*01)$$

and

$$\mathbf{f(000) < f(111)}$$

f_s denotes the fitness of a schema. It is defined as the average of the fitness of all points belonging to that schema, for example

$$f_s(0**) = 1/4(f(011) + f(010) + f(001) + f(000))$$

$$f_s(0**) = 19 > f_s(1**) = 11$$

The above three-bit function itself is too small to demonstrate any search strategy. Therefore Goldberg has proposed more difficult functions, which are constructed from the above three-bit function. The function **E10** consists of ten three-bit functions, where the three bits are grouped together. In the "ugly" function **U10** the three-bit subfunctions reside far apart at locations i,$i+10$,$i+20$ for $i = 1$ to 10. In both functions the fitness is defined as the sum of the subfunctions.

Functions E10 and U10 are defined on 2^{30} points. The schema $(1*..*)$ is defined on 2^{29} points. The exact fitness of the schema is given by the average of the 2^{29} function values. In a real genetic algorithm we have a much smaller population size, say 2^{12} points. Let us assume that half of the points will be members of our schema. The small number of 2^{11} points will be used by the genetic algorithm to estimate the fitness of the above schema. Holland [Hol89] correctly uses the notion of $f_s(t)$ to show that the fitness of the schema s depends on the population at generation t.

This observation leads to our main objection to the mainstream interpretation of the schema theorem.

The estimate of the fitness of a schema is equal to the exact fitness in very simple applications only. Therefore interpretations using the exact fitness cannot be applied in connection with the schema theorem. But if the estimated fitness is used in the interpretation, then the schema theorem is almost a tautology, only describing proportional selection.

This objection is similar in spirit to that of Grefenstette et al.[GB89].

Without any additional comment, we just state our second major objection against using the schema theorem to explain the search strategy of genetic algorithms. The schema theorem estimates only the disruption, i.e. the possibility of losing good substrings. The question of why the genetic algorithm builds better and better substrings by crossing-over is ignored.

The search strategy of a genetic algorithm and the PGA especially can be explained much more easily. The introduction of schemata only creates artificial confusion.

The search strategy is mainly driven by the crossover operator. It can best be described as a *scatter search* [Glo77], where new points are sampled between probabilistically chosen parent points. Because of the selection operator, the points get closer and closer. The closer the parent points are to each other, the smaller is the sampling area for new points. Furthermore, substrings which are the same in the two parent points will be transfered to the offspring. Selection and crossing-over together lead to an automatic concentration of the individuals into a promising area.

We will investigate the main components of a PGA - crossing-over and hill-climbing - with the deceptive problems E10,U10,E20 and U20. First we will make an analysis of very simple search methods. Later we will compare the results with the parallel genetic algorithm.

6 Statistical analysis of the deceptive problems

One of the simplest random search methods is *multistart hill-climbing*. An initial configuration is randomly generated and then a hill-climbing algorithm is applied. A more clever strategy is *iterated hill-climbing*. Here a new configuration for hill-climbing is generated by a mutation of the given configuration. A new configuration is accepted if it is better than the old one. This strategy can be seen as a simple evolutionary algorithm in the space of all local optima. Multistart hill-climbing has been also investigated by Ackley [Ack87]. Unfortunately he called it iterated hill-climbing. But our notion is in accordance to the notion in mathematical optimization.

We will analyse these two strategies with two hill-climbing strategies - *next ascent nahc* and *steepest ascent sahc*. In next ascent hill-climbing, the first bit of a sequence of bits is flipped if it gives an improvement. In steepest ascent hill-climbing, the bit giving the best improvement is flipped. In both cases, the iteration is stopped if no improvement can be obtained. The best-ascent hill-climbing strategy tests more configurations than the next ascent strategy. But in our ugly deceptive problem only one of the three two bit configurations leads to the optimal substring (111) with next ascent, whereas all three two bit configurations lead to the optimal string with steepest ascent. The region of attraction is larger for steepest ascent than for next ascent.

The statistical analysis of multistart hill-climbing is very simple. We define

$$prob(success) = \frac{hill.conf}{total.conf}$$

Hill.conf denotes the number of configurations leading to the global optimum if the hill-climbing strategy is applied. For an ugly deceptive problem with $n = 3m$ we get

$$hill.conf(nahc) = 2^{-m}$$
$$hill.conf(sahc) = 2^{-2m}$$

The number of tested configurations is $3m$ for nahc and $3m(m+1)$ for sahc in the worst case. Therefore we get the following result.

Theorem: *Multistart with steepest ascent hill-climbing finds the optimum with probability* 2^{-m}. *The amount of computation to find the optimum with probability almost one is* $3m(m+1) * 2^m$. *Multistart with next ascent hill-climbing needs* $3m * 2^{2m}$ *steps.*

A good hill-climbing strategy pays off the larger the problem is. We will later show that the same observation is true for the PGA.

The statistical analysis of stochastic iterated hill-climbing is more complex. In order to show the basic idea of the analysis, we start with a simple non-deceptive problem. We make the assumption that the optimum can be reached by a sequence of one bit flips. Let p be the probability of mutating one bit of the n-bit string. Let k be the current number of bits wrong. The probability of getting nearer to the optimum is the product of the probability of not flipping one of the right $n-k$ bits and the probability of flipping at least one of the k wrong bits. Ignoring higher order terms we obtain approximately

$$prob_n(k, better) = (1-p)^{n-k} * (1-(1-p)^k) \tag{4}$$

The same analysis can also be done for the deceptive problems. We will first analyse steepest ascent hill-climbing.

We start our analysis with the observation, that after hill-climbing the string will only consists of the substrings of the two local minima (000) and (111). Two bits at least have to be flipped together to jump from (000) to (111) and vice versa. One bit flips in a substring will not lead to any changes. The probability of at least two bits flipped in three trials is $x = 3p^2(1-p) + p^3$. The two bit flips have to occur within a substring.

In the general case, let x be the probability of jumping into the attraction region of the local minima (111) from (000), given the hill-climbing method. Let y be the probability of jumping into the attraction area of (000) from (111). Ignoring higher order terms we approximately obtain for $l = 1, 2, ...$ substrings wrong

$$prob_m(l, better) = (1-y)^{(m-l)}(1-(1-x)^l) \tag{5}$$

For next ascent hill-climbing we have $x = p^2$ and $y = 2p - p^2$. The above probability is independent of the position of the substring in the global string. In iterative hill-climbing there is no difference between a deceptive problem and an ugly deceptive problem.

Strategy	p	ET	sd	ET(6)	Conf.	sd	Nor.Conf
it-nahc-U10	0.1	1250	663	1300	37480	20010	4875
it-sahc-U10	0.1	98	54	104	11144	5989	1203
it-sahc-U10	0.18	47	24	55	7932	3995	835
it-nahc-U20	0.04	7504	3365	7703	450340	201900	51754
it-sahc-U20	0.04	684	347	681	136325	68542	14259
it-sahc-U20	0.12	144	71	152	64374	31294	6567
it-sahc-U20	0.3	953	474	1052	525267	286433	53103

Table 1: Iterated hill-climbing (128 runs each)

$ET_m(l) = 1/prob_m(l)$ is the expected number of trials needed to get at least one additional subfunction correct. In *iterated hill-climbing* we use the simple search strategy of accepting a new configuration only if it is better than the old configuration. Then the total number of trials ET(m) to reach the optimum can be computed as

$$ET(m) = ET_m(1) + ET_m(2) + .. + ET_m(initial) \tag{6}$$

Initial denotes the number of suboptimal subfunctions after the first hill-climbing. The worst case is $initial = m$. ET(m) depends on the mutation rate p. In order to minimize ET(m) all the values $ET_m(l)$ should be minimized. The optimal mutation rate for a given l can easily be computed from (5). However, this procedure is unfair, because the number of wrong substrings has to be known. The largest number of trials is always needed for the last step $ET_m(1)$. Therefore a good estimate of the mutation rate is the optimal mutation rate for $ET_m(1)$. This can easily be computed. We get for steepest ascent hill-climbing

$$x = \frac{1}{m}$$

For the mutation rate p we obtain approximately

$$p = \sqrt{\frac{1}{3m}} \tag{7}$$

In table 1 numerical experiments with different mutation rates are summarized for $m = 10$ and $m = 20$.

Conf denotes the total number of configurations evaluated, Nor.Conf are normalized configuration evaluations. These are derived as follows: During hill-climbing we need only a small number of table look-ups to evaluate a configuration, whereas m table look-ups are normally needed. In Nor.Conf we count m hill-climbing configuration evaluations as one normal configuration.

ET denotes the average number of trials to reach the optimum. It agrees quite well with the number computed by formula (6). The results confirm our statistical analysis. A mutation rate too high or too low gives an increase of the number of trials. Formula (7) gives a good estimate for the optimal mutation rate.

Furthermore, a good hill-climbing strategy pays off. At equal mutation rates, steepest ascent hill-climbing needs only one fourth of the configuration evaluations of next ascent hill-climbing.

Our analysis can easily be extended to higher order deceptive problems like Whitley's order-4 deceptive problem [WS90]. We will not do this here, but turn to the PGA.

7 The PGA and the deceptive problems

We will explain the search strategy of the PGA and its major components by solving the easy deceptive problems E10 and E20 and the "ugly" deceptive problems U10 and U20. The following features will be investigated:

- no hill-climbing
- next ascent hill-climbing
- steepest ascent hill-climbing
- 2-point cyclic crossing-over
- uniform crossing-over

Our 2-point cyclic crossing-over is implemented as follows. First the starting point of the crossing-over section is randomly drawn. Then the crossing-over interval is drawn (between 20% and 60% of the length of the string). In uniform crossing-over, each bit of the parent strings is chosen with probability 0.5. Table 2 shows the results of our numerical experiments. The mutation rates are $p = 0.1$ for $m = 10$ and $p = 0.04$ for $m = 20$.

The table can be used for many individual comparisons. We summarize the major results:

- A PGA with hill-climbing performs much better than without(wohc) hill-climbing, especially if normalized configurations are considered
- The performance of 2-point crossing-over depends on the coding. The dependency is not so strong with steepest ascent hill-climbing.
- The performance of uniform crossing-over is independent of the coding.
- A good hill-climbing strategy pays off

These results are in agreement with our results on difficult combinatorial optimization problems [Müh89] and function optimization [MSB91].

We also make a short comparison with other simulations of deceptive problems. In our PGA simulations we used a very small population size (16 or 32) in order to reduce the amount of computation. Goldberg [Gol89b] investigated the deceptive problems E10 and U10 with a standard GA of population size 2000. He reported 40000 function evaluations for the simple GA and 40600 for his messy genetic algorithm mGA, both for the function E10. The simple GA could not solve the ugly function U10, because it was run with a mutation rate of 0.

Strategy	Pop	Gen.	sd	Conf.	sd	Nor.Conf
2pc-wohc-E10	32	667	62	21398	1970	
2pc-wohc-U10	32	9979	3528	319379	112918	
2pc-nahc-E10	16	83	18	40500	8600	5245
2pc-nahc-U10	16	346	53	167400	25900	21600
2pc-sahc-E10	16	11	2	27273	4086	2890
2pc-sahc-U10	16	11	3	36066	8431	3771
2pc-nahc-E20	32	168	45	325632	85900	34990
2pc-sahc-E20	32	23	4	198678	24093	20558
2pc-sahc-U20	32	31	5	450138	66690	45010
uc–wohc-E10	32	2957	739	94669	23649	
uc–wohc-U10	32	2788	567	89260	18135	
uc–nahc-E10	16	148	27	71150	13200	7420
uc–nahc-U10	16	151	37	73152	17700	7480
uc–sahc-E10	16	7	4	22830	9804	2380
uc–sahc-E20	32	437	41	842112	86200	90490
uc–sahc-E20	32	22	5	272292	32976	27850

Table 2: PGA results (10 runs each)

Whitley [WS90] made investigations with his GENITOR II, also with a population size of 2000. He did not report the number of tested configurations, so no detailed comparisons can be made to our work. Moreover the non-distributed GENITOR was run with a mutation rate of 0. In this case it was not able to solve the "ugly" deceptive problem U10. In the distributed run an adaptive mutation rate was used. The algorithm now solved the problem. We conjecture that the main reason for this success is the mutation, not the population structure.

Our statistical analysis has shown the importance of mutation for the deceptive problems. It is not useful to compare GA runs without mutation to runs with mutation! *Mutation is a very important component of a genetic algorithm. It is a common mistake to believe that mutation is not important because the mutation rate is so small.*

8 Iterated hill-climbing vs. the PGA

The three analysed search strategies can ce described as follows. Multistart hill-climbing only generates new initial configurations for hill-climbing. There is no linkage between the searches. Iterated hill-climbing generates a new configuration in the neighborhood of the old one. If by chance an improvement is obtained, the new configuration is accepted. Iterated hill-climbing can be considered as a more restricted form of multistart hill-climbing. In the PGA new configurations are mainly generated by crossing-over. In this search strategy new configurations are built out of the components of the population of existing configurations. The recombination of the components is done randomly. How do these search strategies compare?

With the deceptive problems, multistart hill-climbing performs much worse than the other two. But iterated hill-climbing needs less computation than our PGA if a good mutaion rate is used. Why using a parallel genetic algorithm at all? The answer is parallelism.

Let us discuss the results of the experiments it-nahc-U10 and it-sahc-U20 with the PGA experiments uc-nahc-U10 and uc-sahc-U20. In both cases the PGA needs about twice the amount of computation of iterated hill-climbing. But it runs on $16(U10)$ and $32(U20)$ processors in parallel. The optimum is found after a small number of generations. The speedup of the PGA compared to iterated hill-climbing is about half the number of processors. This is a reasonable speedup for a parallel search method. But we would like to mention that in the problem domain optimization of difficult functions we could report instances of a superlinear speedup [MSB91]. This does not happen here because our deceptive problems are not difficult enough. In such problems the PGA simply does the same evaluations but in parallel. At the termination of the PGA most of the strings of the population have just one or two subfunctions incorrect.

Furthermore the comparison is a little unfair. Iterated hill-climbing strongly depends on the mutation rate. The PGA is very robust with respect to the mutation rate. We always use as mutation rate just the inverse of the length of the chromosome. In contrast the computation of a reasonable mutation rate for iterated hill-climbing needs information about the problem.

Goldberg [GDK90] suggested two extensions of the deceptive problems. One extension is to vary the size of the subfunctions, the other to vary the scale. A varying scale does not affect iterated hill-climbing because of the hard selection used. A more difficult problem for iterated hill-climbing would be deceptive problems which consists of subfunctions with different sizes, maybe nine order-3 deceptive subfunctions and one order-6 deceptive function. In this case the optimal mutation rate cannot be computed.

Our analysis has shown that deceptive problems can be nicely used to explain the search strategies of the different algorithms. But the original deceptive problems are not difficult enough to serve as good benchmarks for real optimization problems. The reason is that the local minima are too regularly spread in the fitness landscape. A mutation of just two bits leads from each of the local maxima to a better one. In our opinion it is not very useful to make the deceptive functions in small steps more and more complex and then to develop algorithms which solve these problems. A better method is to investigate the search strategies with more classical difficult benchmark optimization problems. We will do this in the next sections with the traveling salesman problem.

9 The traveling salesman problem

The famous traveling salesman problem (TSP) can be easily stated.

OPT 2 (TSP) *Given are n cities. The task of the salesman is to visit all cities once so that the overall tourlength is minimal.*

This problem has been investigated in [MGSK88] ,[GS89] and [GS91] with the PGA. The genetic representation is straightforward. The gene at locus i of the chromosome codes the edge (or link) which leaves city i. With this coding, the genes are not independent from each other. Each edge may appear on the chromosome only once, otherwise the chromosome would code an invalid tour. A simple crossing-over will also give an invalid tour. This is the reason why this simple genetic representation has not been used in genetic algorithms. The researchers tried to find a more tricky representation in order to apply a simple crossover operator.

We take the opposite approach. We use a simple representation, but an intelligent crossover operator. The crossover operator for the TSP is straightforward. It inserts part of chromosome A into the corresponding location at chromosome B, so that the resulting chromosome is the most similar to A and B. A genetic repair operator then creates a valid tour.

We call our crossover operator MPX, the maximal preservative crossover operator. It preserves subtours contained in the two parents. The pseudocode is given below.

PROC crossover (receiver, donor, offspring)

> Choose position $0 <= i < nodes$ and length $b_{low} <= k <= b_{up}$ randomly.
> Extract the string of edges from position i to position $j = (i + k)$ MOD *nodes* from the mate (donor). This is the crossover string.
> Copy the crossover string to the offspring.
> Add successively further edges until the offspring represents a valid tour. This is done in the following way:
>
> IF an edge from the receiver parent starting at the last city in the offspring is possible (does not violate a valid tour)
> THEN add this edge from the receiver
> ELSE IF an edge from the donor starting at the last city in the offspring is possible
> THEN add this edge from the donor
> ELSE add that city from the receiver which comes next in the string, this adds a new edge, which we will mark as an implicit mutation.

We want to recall, that in the PGA the crossover is not done in the space of all TSP configurations, but in the space of all local minima. Our local search is a simple version of the famous 2-opt heuristic developed by Lin-Kernighan [Lin65]. Two edges are exchanged randomly. If the resulting tour is shorter, the exchange is accepted. Then the next two edges are exchanged until no exchange of two edges gives a better tour.

10 Performance evaluation for the TSP

Two different performance measures at least are used for comparing heuristics. The first measure is to compare the best solution for very large problems which each of the heuristics was able to get. The second measure is to compare the quality of the solutions after a fixed amount of time. A detailed study of the performance of the PGA based on the first measure has been done in [GS91]. In this study the

Instance	t	SA	Mult2Opt	MultLk	Gen2Opt	GenLK
EUR100	60	2.59	3.23	0.0	1.15	0.0
GRO442	4100	2.60	9.29	0.27	3.02	0.19
GRO532	8600	2.77	8.34	0.37	2.99	0.17
GRO666	17000	2.19	8.67	1.18	3.45	0.36

Table 3: Relative deviation from optimal tour length (in %)

importance of the population structure ladder on the quality of the solution was shown.

The importance of the local search in the PGA has been investigated by Ulder et al. [UPvL+91]. They asked the question: Given a fixed amount of time, is it better to use a simple, but fast local search for a large number of generations or to use a sophisticated local search for a small number of generations. In their experimental setup, the genetic algorithm had to converge within a certain time limit. Convergence is achieved when all tours in the current population have the same length or the length of the best tour did not improve within five successive generations. This condition forced the population sizes of the genetic algorithm to be very small. The experiment was done on a sequential computer.

Ulder et al. compared the standard 2-opt with the more complicated Lin-Kernighan [LK73] neighborhood. In the latter case they used a pair of improvement operators, the dynamical k-swap and the additional 4-swap as described in [LK73]. The results are shown in table 3.

The problem instances starting with GRO can be found in [GH88]. GRO532 is the Padberg-Rinaldi problem mentioned above. The reference point is given by simulated annealing SA according to the cooling schedule of Aarts et al. [AK89]. The results demonstrate the advantage of the MPX crossing-over. The genetic versions Gen2Opt and GenLK of 2-Opt and Lin-Kernighan, respectively, perform clearly better than their multistart companions, despite the penalty of the convergence constraint. Moreover, GenLK is superior to the other algorithms. This shows the advantage of using a sophisticated local search method in the PGA.

The absolute quality of the above solutions is worse than the solutions we obtained with the PGA [GS91]. This can be explained by the experimental setup, which required a convergence within the time limit given by simulated annealing.

In combining these results on local search and our results on population structure in the PGA, it seems obvious that a good PGA implementation should use at least two different local search methods. Most of the population should use a simple, but fast local search. But one or more processors should perform a very good local search like the Lin-Kernighan heuristic. Each time a new best-so-far solution has been found, one of the processors with the sophisticated local search method should be initialized with that solution.

We have not yet done this implementation for the TSP, because we believe that highly sophisticated implementations of an *iterated Lin-Kernigham algorithm* will yield semi-optimal solutions faster than even our PGA. The reason is that the Lin-

Kernigham heuristic is very good for Euclidean TSP-problems. The situation is different with other combinatorial problems like e.g. the graph partitioning problem, where the Lin-Kernigham heuristic performs much worse.

In the next section we investigate by a configuration space analysis why Gen2Opt is better than Mult2Opt.

11 Configuration space analysis of the TSP

The analysis of the TSP is more difficult than that of the deceptive problems because the genetic coding is more complex. In the TSP we have $n-1$ alleles at each loci instead of two. Furthermore, the alleles are not independent. In our configuration space analysis the following questions will be investigated

- How many different edges take part in 2-opt local minima?
- How big is the overlap between two 2-opt local minima
- Which edges are most likely contained in 2-opt local minima?

It is impossible to answer these questions by analytical methods. The configuration space analysis can only be done by computation. We have chosen TSP problems of order 40,60,80 and 100 as problem instances. The cities have been placed randomly. It is outside the scope of this paper to discuss whether TSP problems with random drawn cities are representative for the class of all TSP problems. We have to mention that our conjectures are not valid for degenerate TSP instances like cities arranged on a grid or on a circle. On a circle, there exists only one 2-opt tour. On a grid, the 2-opt tours have a special structure.

A similar configuration space analysis has been done by Kirkpatrick [KT85]. He investigated a different TSP problem, where all the distances are drawn randomly. This is a non geometric TSP problem.

Our analysis is based on 800 different 2-opt tours, which have been generated from different initial tours. Table 4 gives the number of different edges and the average distance between two tours. The difference is defined as the Hamming distance i.e the number of different edges. AHD denotes the average distance, HD10 the average distance of the best 10 tours, HD50 the average distance of the best 50 tours.

n	no edges	AHD	HD10	HD50
40	140	13.2	7.1	9.0
60	234	20.0	11.4	13.9
80	340	28.0	17.5	21.4
100	417	38.0	24.8	30.2

Table 4: Different edges and average distance in 2-opt solutions

The data of the table suggest a conjecture.

Conjecture 1 *The number of edges of 2-opt local minima of random TSP problems is bound approximately by* $4.3 * n$ *The average distance between two 2-opt tours is approximately* $1/3 * n$.

We are well aware, that the above conjectures can only be proved by exhaustive search. But we have so much evidence, that we firmly belive that the conjecture is correct. Kirkpatrick [KT85] conjectured a bound of $3 * n$ for his TSP problem and an average distance of $1/3 * n$. In addition, our table shows that the better the tours are, the more similar they are.

The bound on the number of edges implies the following. If we combine all 2-opt tours into a single supergraph, then on the average $2 * 4.3 = 8.6$ edges are only connected to a given city. This result is interesting in itself and deserves further study.

Our main question is, why the MPX crossover is so successful on TSP problems. The answer is surprisingly simple. It is based on the following observation.

Conjecture 2 *The edges of the best 2-opt tours of random TSP problems are most likely contained in other 2-opt tours also.*

In order to explain this conjecture we define the *frequency of occurrence* of the edges of a tour. It is defined as follows. For each of the given 2–opt tours we compute how often its edges are in other 2-opt tours also. We sum these numbers and divide the sum by the product of the total number of tours and the number of cities. The frequency of occurrence is $1/n$, if the edges of all tours are different , and 1, if all tours are equal. The following table shows the result

n	aver	max	min	r
40	0.62	0.69	0.53	-0.78
60	0.67	0.73	0.56	-0.78
80	0.64	0.70	0.56	-0.85
100	0.62	0.67	0.55	-0.79

Table 5: Frequency and correlation r of pathlength to frequency

The average frequency of occurence is approximately $2/3n$. This result can be interpreted as each edge being contained on the average in 2/3 of the other 2-opt tours. The correlation coefficient r tests the hypothesis of a correlation between the pathlength and the frequency. The computed values indicate that tours with small pathlength have a higher frequency of occurence in the pool.

This fact is also shown in the next table. Here the pool of edges is restricted to the best 10 2-opt tours, the best 100 2-opt tours etc.

N	aver100	aver 80	aver60	aver40
10	0.78	0.80	0.83	0.84
100	0.69	0.72	0.74	0.74
200	0.67	0.70	0.72	0.70
400	0.65	0.68	0.70	0.67
800	0.62	0.64	0.67	0.62

Table 6: Frequency of occurrence for the best N tours

The table shows that the better the solutions, the more similar they are. This observation does not imply, that crossing-over of just two strings is the best way to do the linkage. We have shown in [Müh89] that a voting recombination of seven parents works very well in the quadratic assignment problem. A comparison of voting recombination of several TSP tours vs. crossing-over of two tours is a topic of future research.

We summarize the result of this analysis in two statements.

The combined power of selection and crossover *Better 2-opt tours are more similar. By combining two good tours, the probability of obtaining a still better tour is higher than by combining two arbitrary 2-opt tours.*

The ideal PGA *Combine the genetic material of two individuals to produce better individuals. By selection of the better individuals decrease the genetic variety of the population, so that the chance of producing a better individual remains high. But take care that the genetic material of the best individual remains in the gene pool*

Unfortunately, the PGA does not know the best individual beforehand. If selection eliminates some of the genetic material of the best individual, it will converge to suboptimal solutions.

12 Conclusion

This paper has described the *clean* parallel genetic algorithm PGA, clean in the sense that it relies on a single strategy only. The clean algorithm has been successfully applied in a number of difficult optimization problems. In the paper we have shown the importance of a population structure, a good hill-climbing strategy and a problem dependent crossover operator. We have mentioned some extensions which will make the PGA still more effective. The most important ones are

- some individuals do a different local search
- the population structure may change during the run

A computational optimal PGA would start with very simple and fast local search methods. The search space is broadly explored mainly by crossing-over. In the middle of the search, the PGA should also use individuals with very good search methods. At the end of the algorithm, similar individuals should die out.

It is very easy to extend the PGA to real life optimization problems with conflicting goals. Here a group of species can be used for solving problems in cooperation or in competition.

The PGA is also a contribution to evolutionary theory. It has shown the importance of a spatial population structure for the evolution of a species in a rough fitness landscape. The success of the PGA suggest exploring other problem solving metaphors also. Why not use the market economy as a parallel search metaphor? A comparison of problem solving by a market framework and by biological evolution on the same set of artificial problems would give further insight into economy and biology.

References

[Ack87] D. Ackley. *A Connectionist Machine for Genetic Hillclimbing.* Kluver Academic Publisher, Boston, 1987.

[AK89] E. Aarts and J. Korts. *Simulated Annealing and Boltzmann Machines.* John Wiley, New York, 1989.

[CHMR87] J.P. Cohoon, S.U. Hedge, W.N. Martin, and D. Richards. Punctuated equilibria: A parallel genetic algorithm. In J.J. Grefenstette, editor, *Proceedings of the Second International Conference on Genetic Algorithms*, pages 148–154. Morgan-Kaufman, 1987.

[Fel75] J. Felsenstein. A pain in the torus: Some difficulties with models of isolation by distance. *Amer. Natur.*, 109:359–368, 1975.

[Fel76] J. Felsenstein. The theoretical population genetics of variable selection and migration. *Ann. Rev. Genet.*, 10:253–280, 1976.

[Fis58] R. A. Fisher. *The Genetical Theory of Natural Selection.* Dover, New York, 1958.

[GB89] J.J. Grefenstette and J.E. Baker. How genetic algorithms work: A critical look at implicit parallelism. In H. Schaffer, editor, *3rd Int. Conf. on Genetic Algorithms*, pages 20–27. Morgan-Kaufmann, 1989.

[GDK90] D.E. Goldberg, K. Deb, and B. Korb. Messy genetic algorithms revisited: Studies in mixed size and scale. *Complex Systems*, 4:415–444, 1990.

[GH88] M. Grötschel and O. Holland. Solution of large-scale symmetric traveling salesman problems. Technical report, Institut f. Ökonometrie und Operations Research, University of Bonn, Report No. 88506-OR, 1988.

[Glo77] F. Glover. Heuristics for integer programming using surrogate constraints. *Decision Sciences*, 8:156–166, 1977.

[Gol89a] D.E. Goldberg. *Genetic Algorithms in Search, Optimization and Machine Learning.* Addison-Wesley, 1989.

[Gol89b] D.E. Goldberg. Messy genetic algorithms: Motivation, analysis, and first results. *Complex Systems*, 3:493–530, 1989.

[GS89] M. Gorges-Schleuter. Asparagos: An asynchronous parallel genetic optimization strategy. In H. Schaffer, editor, *3rd Int. Conf. on Genetic Algorithms*, pages 422–427. Morgan-Kaufmann, 1989.

[GS91] M. Gorges-Schleuter. *Genetic Algorithms and Population Structures - A Massively Parallel Algorithm.* PhD thesis, University of Dortmund, 1991.

[Hol75] J.H. Holland. *Adaptation in Natural and Artificial Systems.* Univ. of Michigan Press, Ann Arbor, 1975.

[Hol89] J.H. Holland. Searching nonlinear functions for high values. *Appl. Math. and Comp.*, 32:255–274, 1989.

[KT85] S. Kirkpatrick and G. Toulouse. Configuration space analysis of travelling salesman problems. *J.Physique*, 46:1277–1292, 1985.

[Lin65] S. Lin. Computer solutions of the traveling salesman problem. *Bell. Syst. Techn. Journ.*, 44:2245–2269, 1965.

[LK73] S. Lin and B. W. Kernighan. An efficient heuristic for the traveling salesman problem. *Operations Research*, 21:298–516, 1973.

[MGSK87] H. Mühlenbein, M. Gorges-Schleuter, and O. Krämer. New solutions to the mapping problem of parallel systems - the evolution approach. *Parallel Computing*, 6:269–279, 1987.

[MGSK88] H. Mühlenbein, M. Gorges-Schleuter, and O. Krämer. Evolution algorithms in combinatorial optimization. *Parallel Computing*, 7:65–88, 1988.

[MS89] B. Manderick and P. Spiessens. Fine-grained parallel genetic algorithm. In H. Schaffer, editor, *3rd Int. Conf. on Genetic Algorithms*, pages 428–433. Morgan-Kaufmann, 1989.

[MSB91] H. Mühlenbein, M. Schomisch, and J. Born. The parallel genetic algorithm as function optimizer. *Parallel Computing*, 16, 1991.

[Müh89] H. Mühlenbein. Parallel genetic algorithm, population dynamics and combinatorial optimization. In H. Schaffer, editor, *3rd Int. Conf. on Genetic Algorithms*, pages 416–421, San Mateo, 1989. Morgan Kaufmann.

[PS89] C. Peterson and B. Söderberg. A new method for mapping optimization problems onto neural networks. *Int. J. Neural Syst.*, 1:995–1019, 1989.

[Rec73] I. Rechenberg. *Evolutionsstrategie - Optimierung technischer Systeme nach Prinzipien der biologischen Information.* Fromman Verlag, Freiburg, 1973.

[Sch81] H.-P. Schwefel. *Numerical Optimization of Computer Models.* Wiley, Chichester, 1981.

[Sch89] J.D. Schaffer, editor. *Genetic Algorithms and Their Application*, San Mateo, 1989. Morgan Kaufmann.

[Tan89] R. Tanese. Distributed genetic algorithm. In H. Schaffer, editor, *3rd Int. Conf. on Genetic Algorithms*, pages 434–440. Morgan-Kaufmann, 1989.

[UPvL+91] N.L.J. Ulder, E. Pesch, P.J.M. van Laarhoven, H.-J. Bandelt, and E.H.L. Aarts. Improving tsp exchange heuristics by population genetics. In R. Maenner and H.-P. Schwefel, editors, *Parallel Problem Solving from Nature.* Springer-Verlag, 1991.

[vLM91] G. von Laszewski and H. Mühlenbein. A parallel genetic algorithm for the graph partitioning problem. In R. Maenner and H.-P. Schwefel, editors, *Parallel Problem Solving from Nature.* Springer-Verlag, 1991.

[Wri32] S. Wright. The roles of mutation, inbreeding, crossbreeding and selection in evolution. In *Proc. 6th Int. Congr. on Genetics*, pages 356–366, 1932.

[WS90] D. Whitley and T. Starkweather. Genitor ii: a distributed genetic algorithm. *J. Expt. Theor. Artif. Intell.*, 2:189–214, 1990.

Author Index

Key Word Index